ADVANCED RESEARCH METHODS FOR THE SOCIAL AND BEHAVIORAL SCIENCES

Written by an interdisciplinary team of global experts covering diverse research methods – including research design, research tools, and statistical techniques – this volume focuses on advanced research methods for anyone working in the social and behavioral sciences. The information needed to perform research in the laboratory, the field, or online is mapped out to provide specific applications and tools in applying each method. The issues surrounding reliability, validity, and obtaining consent are explained alongside detailed descriptions of the impact of preknowledge on participant behavior, the ways researchers unintentionally influence participants, and tips for administering suspicion probes and debriefings. The book then lays out biophysiological measures, eyetracking methods and technologies, the construction of questionnaires, and reaction time methodologies without assuming too much prior knowledge. The basics of Bayesian analysis, item response analysis, social network analysis, and meta-analysis are also summarized as the editors combine innovative methods and statistics to showcase how to perform quality research.

JOHN E. EDLUND is Associate Professor of Psychology at the Rochester Institute of Technology, USA, and serves as the Research Director of Psi Chi: The International Honor Society in Psychology. He has won numerous awards related to teaching and is passionate about the dissemination of psychological knowledge to the world.

AUSTIN LEE NICHOLS is the Director of Research for the Connection Lab in San Francisco, USA. Prior to this, he worked in various faculty positions around the world in both psychology and management. He has won awards for his teaching, research, and service from various international organizations.

ADVANCED RESEARCH METHODS FOR THE SOCIAL AND BEHAVIORAL SCIENCES

Edited by

JOHN E. EDLUND
Rochester Institute of Technology

AUSTIN LEE NICHOLS
Connection Lab

CAMBRIDGE
UNIVERSITY PRESS

University Printing House, Cambridge CB2 8BS, United Kingdom

One Liberty Plaza, 20th Floor, New York, NY 10006, USA

477 Williamstown Road, Port Melbourne, VIC 3207, Australia

314–321, 3rd Floor, Plot 3, Splendor Forum, Jasola District Centre,
New Delhi – 110025, India

79 Anson Road, #06–04/06, Singapore 079906

Cambridge University Press is part of the University of Cambridge.

It furthers the University's mission by disseminating knowledge in the pursuit of
education, learning, and research at the highest international levels of excellence.

www.cambridge.org
Information on this title: www.cambridge.org/9781108425933
DOI: 10.1017/9781108349383

First published 2019

Printed in the United Kingdom by TJ International Ltd, Padstow Cornwall

A catalogue record for this publication is available from the British Library.

Library of Congress Cataloging-in-Publication Data
Names: Edlund, John E., 1978– editor. | Nichols, Austin Lee, 1984– editor.
Title: Advanced research methods for the social and behavioral sciences /
    edited by John E. Edlund, Rochester Institute of Technology, Austin Lee Nichols,
    Connection Lab.
Description: Cambridge, United Kingdom ; New York, NY : Cambridge University Press,
    2019. | Includes index.
Identifiers: LCCN 2018045905 | ISBN 9781108425933 (hardback) | ISBN 9781108441919 (pbk.)
Subjects: LCSH: Social sciences–Methodology. | Psychology–Research–Methodology.
Classification: LCC H61 .A39325 2019 | DDC 001.4/2–dc23
LC record available at https://lccn.loc.gov/2018045905

ISBN 978-1-108-42593-3 Hardback
ISBN 978-1-108-44191-9 Paperback

Dedication

I dedicate this book to my awesome wife and children. Without you, all that I have and do would not be possible.

John E. Edlund

First, I would like to dedicate this book to the person who provided the greatest support for me through the editorial process, my wife. I would also like to thank the countless people who provided input or feedback along the way as well as my coeditor, John Edlund, for his persistence and patience throughout the process.

Austin Lee Nichols

Contents

Contents

List of Figures

List of Tables

Contributors

Rachel Adams Goertel – Roberts Wesleyan College

Ginette C. Blackhart – East Tennessee State University

Mary G. Carey – University of Rochester Medical Center

Heather J. Carmack – University of Alabama

Esther Choi – Public Health Management Corporation

Travis D. Clark – Valley City State University

Karen L. Dugosh – Public Health Management Corporation

John E. Edlund – Rochester Institute of Technology

Sinikka Elliott – University of British Columbia

Susan E. Embretson – Georgia Institute of Technology

Timothy M. Errington – Center for Open Science

Ignacio Ferrero – University of Navarra

David S. Festinger – Philadelphia College of Osteopathic Medicine, Public Health Management Corporation

James P. Goertel – State University of New York at Brockport

Rosanna E. Guadagno – Alliant International University

Clifford E. Hauenstein – Georgia Institute of Technology

Jeremy D. Heider – Southeast Missouri State University

Milica Miočević – Utrecht University

Austin Lee Nichols – Connection Lab

Elena C. Papanastasiou – University of Nicosia

Jeff B. Pelz – Rochester Institute of Technology

Michael C. Philipp – Massey University

Javier Pinto – Universidad de Los Andes

Eric Quintane – Universidad de Los Andes

Jason T. Reed – Empirisoft Corporation

Sebastian Leon Schorch – Universidad de Los Andes

Chloe Sierka – Public Health Management Corporation

John J. Skowronski – Northern Illinois University

Courtney K. Soderberg – Center for Open Science

Heather M. Stassen – Cazenovia College

David B. Strohmetz – University of West Florida

Rens van de Schoot – Utrecht University, North-West University

Fons J. R. van de Vijver – Tilburg University, University of Queensland, North-West University

Eric J. Vanman – University of Queensland

Michael F. Wagner – Northern Illinois University

Gregory D. Webster – University of Florida

1 A Brief Orientation to Research Methods and Statistics for the Social and Behavioral Sciences

John E. Edlund

Austin Lee Nichols

Science is a struggle for truth against methodological, psychological, and sociological obstacles.

(Fanelli & Ioannidis, 2013)

When teaching about research methods in our classes, the first thing we discuss is the nature of science itself and what makes the scientific method the best way to learn about the world. In 1959 Karl Popper wrote the seminal treatise on the nature of the scientific method. Central to Popper's treatise is the falsifiability of a theory. Popper argues that we start with observations of the world; from there he suggests that scientists start building theories about the world. We then collect additional observations that can test the veracity of the theory. If the theory holds up, we have greater trust in the theory. If the theory fails, we should abandon it in light of a better theory that can account for the additional observations. The falsifiability of a theory is a critical component of science (and what distinguishes science from pseudoscience).

Ultimately, the scientific method (as initially suggested by Popper) is the best way we can learn about the world. As such, his approach underpins all of the methods used in the social and behavioral sciences. Before we get into each of these topics in detail, we thought it best to overview some concepts that are important on their own, but especially necessary to understand before reading the chapters that follow. To do so, we first discuss what good science is. Next, we review some of the basic background material that we expect readers to be familiar with throughout the book. From there, we turn to the nature of data. Finally, we orient the reader to the basic statistics that are common in the social and behavioral sciences.

1.1 What Is Good Science?

Good science is not only characterized by being falsifiable – good science should be universal, communal, disinterested, and skeptical (Merton, 1973). Merton suggested that good science would share all of these traits and would advance the scientific endeavor. The first trait science should have is universality. This norm suggests that science should exist independently of the scientists doing the research. As such, anyone investigating a particular problem should be able to obtain the same results. Of course, science has not always lived up to this norm (e.g., Eagly & Carli, 1981).

Merton's next trait of good science is communality. Communality refers to the idea that scientific evidence does not belong to the individual scientists – rather it belongs to everyone. As such, secrecy and failing to share your material and/or data violates this norm. This suggests that when scientists investigate a particular finding, they should share those results regardless of what they are. Many governments have embraced communality as part of their grant-awarding protocol and require recipients to share their data in a publicly accessible forum (European Research Council, 2018; National Institutes of Health, 2003).

The third trait of good science is disinterestedness. This suggests that the scientific enterprise should act for the benefit of the science rather than for the personal enrichment of the scientists. That is, science should work for the betterment of others rather than to protect one's own interest. One logical consequence of this (which is challenging to implement in practice) is that reviewers of work critical of their own research should set aside their views and potential impacts on themselves. Additionally, this suggests that research should not be incentivized for obtaining certain results. However, in practice, this has proven to be a challenge (e.g., Oreskes & Conway, 2010; Redding, 2001; Stevens et al., 2017).

The final trait that Merton suggests is a component of good science is organized skepticism. This suggests that any scientific claim should be subjected to scrutiny prior to being accepted. This norm was taken further by Carl Sagan, who suggested that "Extraordinary claims require extraordinary evidence" (Sagan, 1980). As such, any scientific report should face critical review in light of the accumulated scientific wisdom. Sadly, this norm has also failed in many instances (e.g., power posing: Cesario, Jonas, & Carney, 2017).

Using Popper's (1959) falsifiability criterion and Merton's (1973) norms, science can teach us about our world. Although incorrect findings do occur and get published, the self-correcting nature of science allows us to learn more and to benefit from that increased knowledge. In moving beyond the philosophy of science, we need to start defining key terms, practices, and approaches that are used in the social sciences.

1.2 Research Methods

This next section will lay out a basic primer on terminology. Certainly, many of these terms will be familiar to you, but our goal is to bring everyone up to speed for the subsequent chapters.

1.2.1 Quantitative and Qualitative Research

One of the first key distinctions to make relates to the basic approach to scientific research. Roughly speaking, research can be broken down into two classes: quantitative and qualitative. Qualitative research is typically more in depth – these studies will explore the totality of the situation; the who, what, where, when, why, and how of a particular question. One common example of qualitative research is a case study. In a case study, the particular target is fully explored and contextualized. Other examples of qualitative research include literature reviews, ethnographic studies, and analyses of historical documents. Qualitative methods may also make use of interviews (ranging from unstructured participant reports, to fully structured and guided interviews) and detailed coding of events, and prescribed action plans. Innumerable fields use qualitative methods and most commonly employ this method when first starting an investigation of a certain research question. Qualitative research is then often used to generate theories and hypotheses about the world.

Quantitative research, the focus of this book, typically will not look at a given participant or phenomenon in as much depth as qualitative research does. Quantitative research is typically driven by theory (in many cases, by suggestions from qualitative research). Importantly, quantitative research will make mathematical predictions about the relationship between variables. Quantitative research may also explore the who, what, where, when, why, and how (and will generally make predictions whereas qualitative research generally does not).

In many cases, researchers will make use of both qualitative and quantitative methods to explore a question. For example, Zimbardo used both methods in the seminal Stanford prison study (Haney, Banks, & Zimbardo, 1973). In this volume, we will focus on quantitative methods; certainly, many concepts overlap and readers will end up better prepared for learning about qualitative methods after reading this volume. However, our primary focus is on quantitative methods and concepts.

1.2.2 Theories and Hypotheses

Simply put, a scientific theory is an idea about how the world works. We have already discussed that a scientific theory must be falsifiable. This is an immutable requirement of theory. Theories about the world can be driven by qualitative or quantitative observations. Importantly, a theory will allow for the generation of testable hypotheses. A hypothesis is a specific prediction of how two variables will relate (or fail to relate).

These fairly simple definitions hide a great deal of complexity. A "good" theory is much more than simply falsifiable. Good theories and hypotheses share a number of traits when compared to poor theories and hypotheses. Simply put, a good theory will allow for unique predictions about how the world is, and demonstrate that it can predict relationships better than other theories can. Mackonis (2013) specified four traits that characterize good theory:

Simplicity. A good theory will be parsimonious – convoluted theories are generally lacking and would benefit from refinement or the development of a better theory. This is sometimes termed Occam's Razor.

Breadth. Good theories will generally be larger in scope than rival theories (in terms of situations in which the theory can make predictions). Certainly, there are few theories which are truly expansive (perhaps the generalized theory of evolution being the largest in scope). However, better theories will be more expansive than more limited theories (all things being equal).

Depth. Good theories will be less variant across levels of analysis. For instance, a theory that specifies testable mechanisms for how the theory works will be preferred to theories that fail to specify mechanisms (and instead have the action occurring in a "black box").

Coherence. Good theories will also conform to generally accepted laws of the universe (although these laws are subject to change themselves based on the development of new theories about the universe).

To illustrate these concepts (and the concepts that follow), we would like to first offer a theory and some hypotheses. Our theory is that readers of this volume are great people who will know more about research methods in the social sciences as a result of reading this volume. This theory is lacking in scope (only applying to a limited number of people), but it is a parsimonious theory that offers concrete predictions that no other theory does. It is also certainly a theory that could be falsified (if you aren't great!). Given the specificity of this theory, two hypotheses naturally flow from the theory: a) compared to some other group, readers of this volume are "great" (we will discuss defining this term in the next section) and b) readers will know more about research methods after reading this volume than they did before.

1.2.3 Operationalization and Constructs

Good theories will suggest ways in which two phenomena are related to one another. These phenomena are often referred to as constructs. We might specify that two concepts are related to one another in some way (at this moment, in the abstract). This is termed conceptualization. These concepts are typically non-observable in their essence, although we can measure aspects of a particular construct

with a good operationalization. Operationalization is the specific definition of a construct for a particular study. Operationalization can vary between researchers – reasonable people can design very different ways of operationalizing the same variable. You might ask, isn't this a bad thing? No! In fact, this will allow us to have more confidence in the results we obtain (we will return to this concept later in this chapter as well as in Wagner & Skowronski, Chapter 2).

To continue our earlier example, the constructs are defined in the hypothesis. We are exploring the "greatness" of the reader along with the reader's knowledge about research methods. These are our theoretical constructs. We must first have a personal understanding, or conceptualization, of what each of these constructs is. What exactly is "greatness"? As authors, we went back and forth in terms of discussing this concept as we wrote in this chapter (i.e., is the concept even definable?). To us, greatness is an idea that we think people will agree on that ties in aspects of likeability, talent, and skill. Next, we need to decide how to operationalize these variables. Obviously, given our conceptualization, there are many ways we could operationalize greatness. One way we could operationalize this construct would be to measure friendliness (likeability). Certainly, friendliness has been measured (Reisman, 1983) and would capture an aspect of greatness. Perhaps we might want to operationalize the variable as intelligence (given our interest in how a textbook might change levels of greatness). There are innumerable intelligence measures available (e.g., Flanagan, Genshaft, & Harrison, 1997). We could also design our own greatness measure for this study (see Stassen & Carmack, Chapter 12 for a detailed exposition on the construction of a scale). Any of these operationalizations might be acceptable for testing whether using this text contributes to greatness. We are also interested in testing whether the reader knows more about research methods as a result of reading this text. In this case, the conceptualization of content knowledge is likely understood and shared by many. Similarly, many would suggest an operationalization of a quiz or test that would assess factual knowledge of research methods. Of course, there are hundreds (or thousands) of possible questions that could be assessed; some may be excellent and some may not. For the purposes of this chapter, let us say that we decided to operationalize greatness with a new measure that we created using items representative of likeability, talent, and skill and that we chose last year's research methods final exam to operationalize content knowledge.

1.2.4 Variables

Once we have operationalized our constructs, we need to start using two key terms associated with our research (which will become key when we turn to statistics). The **independent variable (IV)** is the variable that we believe causes another variable when we do inferential statistics. IVs can be experimentally manipulated (two or more conditions are manipulated and assigned to the participants), quasi-experimental (two or more non-randomly assigned conditions; biological sex is a commonly reported quasi-experimental variable), or simply be measured variables that are hypothesized to affect another variable. In our example, we could experimentally compare students who learn research methods due to this volume compared to students who learn research methods via another text (and

as long as students were randomly assigned to textbooks, we could say that any difference was caused by the different text). As such, each book (this volume or the other volume) would represent different levels of the IV. The **dependent variable(s) (DV)** is the variable that the IV is proposed to act upon. This is the variable that many researchers would expect to show a difference as a result of the IV. In our example, the DV would be content knowledge (i.e., the test/quiz score).

1.2.5 Reliability and Validity

Reliability and validity are key concepts in the social sciences. In fact, they are so key that the next chapter in this book is dedicated to exploring these issues more fully than can be covered here. For simplicity's sake, we will say that **reliability** is the consistency of measurement. Are you getting the same (or very similar) value across multiple modes and administrations? **Validity** is the meaning of your measurement. Validity asks whether you are getting values that correspond to your underlying construct and theoretical variables (see Wagner & Skowronski, Chapter 2 for a detailed exposition on these issues).

1.3 Data

Before we can cover the kinds of statistics encountered in the social sciences, we need to define the kinds of data generated in these studies. The first kind of data is **nominal** data (sometimes called categorical data). Nominal data consists of discrete groups that possess no rankings. Biological sex is one commonly used example of nominal data. There is no intrinsic ordering to sex; you could just as easily say females come before males or vice versa. Additionally, when you assign quantitative values to sex, any number could reasonably be used (females could = 1, males could = −1, or any other mutually exclusive categorizations). Another example of nominal data is the reading of this volume (readers of the volume = 1, and non-readers of this volume = 0).

The next kind of data that we can work with is called **ordinal** data. Ordinal data possesses all of the traits of nominal, but ordinal data adds a specific ordering to the categories. Military rank is a classic example of ordinal data. For instance, the lowest category of Naval Officer rank is Ensign, followed by Lieutenant Junior Grade, Lieutenant, etc. The categories are discrete and ordered, but, beyond that, simply knowing the rank will not provide any additional information.

The third kind of data encountered in the social and behavioral sciences is **interval** data. Interval data possesses all of the traits of nominal data, but adds in equal intervals between the data points. A common example of interval data is temperature in Fahrenheit. For instance, the difference between 22°F and 23°F is the same as the difference between 85°F and 86°F. Importantly, zero simply represents a point on the continuum with no specific characteristics (e.g., in an interval scale you can have negative values, such as −5°F). This is the first kind of data that can be referred to as continuous.

The final kind of data that we can encounter is **ratio** data. Ratio data possesses all of the traits of interval data but incorporates an absolute zero value. A common example of ratio data is reaction time to a stimulus. In this setting, time cannot be negative and a person's reaction time to a particular stimulus might be measured at 252 milliseconds.

1.4 Statistics

The final section of this volume explores new statistical trends in the social and behavioral sciences. Chapter 18 explores Bayesian statistics, Chapter 19 explores item response theory, Chapter 20 explores social network analysis, and Chapter 21 introduces meta-analysis. However, before we introduce you to these advanced forms of statistics, we want to ensure that you have a grounding in other common statistical procedures that are often encountered in social and behavioral science research. In this volume, we will focus on linear relationships (and statistics) as these are the most common statistics used in these fields.

Before discussing these statistics, we need to briefly discuss the common assumptions of the data. Many of the statistics discussed are based on the **normal distribution**. This is the idea that, for a particular variable with a large enough sample drawn from a population, the values of that variable will concentrate around the mean and disperse from the mean in a proportional manner. This distribution is sometimes referred to as a bell curve (see Figure 1.1). Importantly, only interval and ratio data can potentially obtain a bell curve. As such, unless otherwise noted, the statistics we discuss in this chapter are only appropriate for these kinds of data as DVs (although it is worth explicitly noting that nominal and ordinal data can be used as IVs in these analyses).

There are a number of commonly reported descriptive statistics in the social and behavioral sciences. The following list is certainly not exhaustive, but it represents the most commonly used descriptive statistics. Additionally, for brevity's sake, we will not discuss the intricacies of the calculation of these statistics

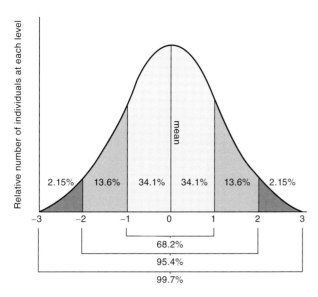

FIGURE 1.1 The normal curve

1.4.1 Measures of Central Tendency

One of the kinds of information that we will want to know about our data is where the center of that data is. There are three common measures of central tendency in the social and behavioral sciences. The **mean** is the mathematical average of all of the values. However, it is important to know that the mean can potentially be impacted by a few outlier variables (we will discuss this when we discuss standard deviation), and as such, should be carefully inspected. The **median** is the central number in a distribution (the middle number). The median is often less impacted by outliers. The **mode** is the most common value in a distribution. The mode is not be impacted by outlier values. Of particular note, in the typical normal distribution (when the sample is large enough), the mean, median, and mode will be the same value (approximately).

(for further details, we encourage you to explore our suggested readings); we will instead conceptually discuss these statistics.

1.4.2 Measures of Distribution

The second kind of information that we typically seek in our data are measures of the distribution of values (around the central value). The simplest value commonly reported is the range. The **range** is simply the distance from the smallest value to the largest value. One of the appeals of the range is that it is easy to calculate and understand. However, this value does not tell us how the values between the extremes are distributed and can also be affected by outliers. The most common measure of distribution reported in the social sciences is standard deviation. **Standard deviation** is the measure of how much divergence there is, on average, in values across the distribution. Mathematically, standard deviation is the square root of its variance (how far a value is from the mean). As such, it is a measure of how far away from the mean the "typical" value is in the distribution. Finally, skew and kurtosis describe how far from normality (i.e., the bell curve pictured in Figure 1.1) the observed data is. Specifically, **skew** is a measure of how asymmetrical the

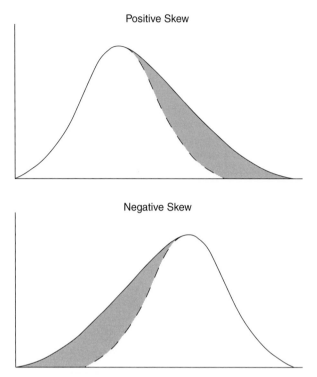

Positive Skew

Negative Skew

FIGURE 1.2 Positive and negative skew

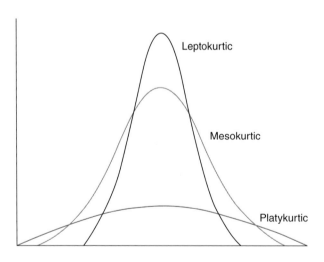

Kurtosis

Leptokurtic

Mesokurtic

Platykurtic

FIGURE 1.3 Types of kurtosis

distribution is around the mean (see Figure 1.2). Skew values range from positive numbers that describe a distribution skewed to the right (i.e., the right tail is longer than the left tail) to negative values that describe a distribution skewed to the left (i.e., the left tail is longer than the right tail). **Kurtosis** is a measure of the central peak of the data (see Figure 1.3). Higher kurtosis values indicate a higher central peak in the data (termed "Leptokurtic") whereas lower values indicate a relatively flat peak (termed "Platykurtic"). A normal kurtoic curve is termed "Mesokurtic."

1.5 Statistical Analyses

As we move on to discussing correlation and inferential statistics, we need to briefly discuss statistical significance and the assumptions of analyses. In null hypothesis testing, we *actually* test the likelihood that the observed distribution would occur if the null hypothesis (that there is no difference between groups)

was true. In the social and behavioral sciences, we typically set our level of confidence for a type 1 error (i.e., that there is an effect when, in reality, there isn't – α) as $p < .05$. Of course, there is nothing magical about that likelihood as it is simply an arbitrarily chosen (but accepted) value (Rosnow & Rosenthal, 1989). The other kind of error that we are concerned about is a type 2 error (saying there is no difference between the groups when, in reality, there is – β). Researchers are concerned with obtaining sufficient power to detect effects to avoid the possibility of type 2 errors. See Figure 1.4 for a humorous illustration of the two types of errors.

FIGURE 1.4 Type 1 and type 2 errors

However, simply reporting a *p*-value along with a statistical test variable is no longer considered sufficient. Many journals now require the reporting of confidence intervals and effect sizes. **Effect sizes** are measures of how "meaningful" the difference is. There are numerous measures of effect size that can be reported ranging from Cohen's *d*, to Cohen's *g*, to a partial eta squared, to others. **Confidence intervals (CIs)** are a projection of where the true value is likely to occur (typically with a 95 % chance). This is often phrased, in practice, as 95 % confidence, but we cannot interpret this to mean that there is 95 % probability that the interval contains the true value. Instead, what we can simply state is that there is a 95 % chance that the interval contains the "true" value. That is, if repeated samples were taken and the 95 % confidence interval was computed for each sample, 95 % of the intervals would contain the population value. CIs can be calculated to reflect on differences between the means, point values, or other effects sizes.

All statistical tests rest on certain assumptions. Generally speaking, the analyses we will discuss in this chapter will assume interval or ratio data that is normally distributed (low levels of skewness and kurtosis). As such, an inspection of skewness and kurtosis (and a transformation of data, if necessary) should be a part of all analyses (see LaMothe & Bobek, 2018 for a recent discussion of these issues). The good news is that the majority of analyses used by social and behavioral scientists are robust against violations of normality (i.e., even if you run your analyses on data that somewhat violates these assumptions, your chances of a type 1 error will not increase; Schmider, Ziegler, Danay, Beyer, & Bühner, 2010).

1.5.1 Moderation and Mediation

Two terms that describe how your IVs may influence your DV(s) are moderation and mediation. **Moderation** occurs when the level of an IV influences the observed DV. Moderators simply specify under what kinds of circumstances would a particular effect be likely (or unlikely) to occur. An example of moderation might be how great people read this volume. You have two groups (roughly dividing people in the world into great and non-great) and the great people are statistically more likely to read this book when assigned in a class setting. That's moderation!

Mediation, on the other hand, explains how a particular variable leads to an observed change in the DV. That is, a mediating variable is the mechanism through which a particular change can take place. For instance, let's say we measure research methods knowledge in students prior to taking a methods course and after (via a test). Reading this volume may be the mechanism by which students learn and do better (on a similar but different test). If this volume fully explains that relationship, reading this volume would be said to fully mediate the relationship between the changes from pre-test to post-test. However, we like to think that your instructor is valuable as well and is contributing to your learning. In that case, the reading of the volume would be said to partially mediate the relationship between scores on the pre-test and post-test (and that your instructor's contributions would also partially mediate, or explain, that relationship).

1.5.2 Correlation

In its simplest terms, correlation is a measure of the degree of linear association between two (or more) variables. Correlations range from -1 to $+1$, where the closer to an absolute value of 1, the stronger the association is. As such, a correlation of $-.85$ is stronger than one of .40. A positive correlation coefficient indicates a relationship where, as one variable increases in size, the other variable increases in size as well. An example of a positive correlation would be that between the amount of studying a student does for an exam and the number of correct answers on an exam (as amount of studying goes up, we would expect the number of correct answers on an exam to go up). This would be depicted as $r = .65$. A negative correlation coefficient suggests that, as one variable increases in size, the correlated variable will decrease. An example of a negative correlation would be the relationship between amount of studying for an exam and the number of incorrect answers on an exam (as the amount of studying goes up, we would expect the number of incorrect answers on an exam to go down). This would be depicted as $r = -.70$.

The most common type of correlation in the social sciences is a Pearson correlation (depicted as r). A correlation coefficient of .5 (or higher) would be considered a large correlation, a .3 would be considered a medium-sized correlation, and a correlation of .1 would be considered small (Cohen, 1988). Importantly, Pearson correlations can only be analyzed with interval or ratio data; to run correlations-based analyses on ordinal data, you would likely use a Spearman Rho analysis (which generates results similar to a Pearson correlation, in many instances).

1.5.3 T-Test

Perhaps the most commonly used statistic (historically) in the social and behavioral sciences is the t-test. A t-test evaluates the difference between two groups on a continuous dependent variable. Perhaps the most common analysis is the **independent samples t-test**, where the data is drawn from two discrete groups. Following from our earlier example, we could conduct an independent samples t-test on two groups of students, one that had not previously read this volume and one sample that had read this volume. In this case, our DV could be test scores. Here, we found a difference between the students who had read a different volume ($M = 90.56$, $SD = 7.98$) and students who had not read the volume ($M = 75.14$, $SD = 12.65$), $t (98) = 7.29$, $p < .001$, $CI_{\text{mean difference}}$ 11.22–19.61).

Another common type of t-test is the **paired samples t-test**. In the paired samples t-test, you compare two samples that are meaningfully linked (paired) together. For instance, another way of testing the effectiveness of reading this volume would be to do a pre-test and a post-test on the same students. One advantage to this approach is an increase in statistical power (by eliminating person variance), but disadvantages exist as well (such as subject attrition). In this case, our DV would still be test scores. Here, we found a difference from before students had read the volume ($M = 52.92$, $SD = 8.50$) to scores after students had read the volume ($M = 90.56$, $SD = 7.98$), $t (49) = 29.2$, $p < .001$, $CI_{\text{mean difference}}$ (35.05–40.22).

1.5.4 ANOVA

One of the limitations of a t-test is that you can only do analyses between two discrete groups. Although one could certainly do a series of sequential t-tests on more than two groups, that approach would increase the odds of a type 1 error. An alternative approach (which is also very commonly used in the social and behavioral sciences) that does not have the limitation of looking at only two groups at one time is an **analysis of variance (ANOVA)**. Similar to a t-test, an ANOVA tests the difference between group means.

A common type of ANOVA is the one-way ANOVA. This type of ANOVA is the most conceptually similar to the t-test, with the simple addition of more than two groups. When a one-way ANOVA is performed on two groups, the statistical conclusion reached will be the same as a t-test. When analyzing three or more groups, if a significant effect is found, a researcher can only conclude that there is a difference in at least two of the groups; to determine which groups are different, you will need to run follow-up tests (called post-hoc tests).

Another common type of ANOVA is the repeated measures ANOVA. In the repeated measures ANOVA, one of the IVs relates to multiple administrations of a similar instrument (commonly as a pre- and post-test). An advantage to the repeated measures ANOVA is that it can allow for testing of effects over time and increased statistical power (resulting from the ability to partial out individual-level variance). Conceptually, this is very similar to the paired samples t-test.

Another common type of ANOVA is the factorial ANOVA. In the factorial ANOVA, you can introduce two (or more) manipulations. One of the distinct advantages to this analysis approach is that you can test for interactions. An interaction between two variables occurs when the level of one variable has a meaningful impact on the level of the second variable (see Moderation above). For instance, in our previous example, we were looking at students who had (or hadn't) read this volume and how they did on a test. We could also look at previous methods experience (none or some). We could then test the independent effects of reading the volume, the effect of previous methods experience, and the potential interaction between the two variables. In this case, we found that this volume was most useful for students with previous methods experience (M = 95.64, SD = 3.03), similar levels of performance were seen by students who had no previous experience but did read the book (M = 85.48, SD = 8.19) and students who had experience but had not read the book (M = 83.76, SD = 8.69), and the lowest levels of performance were seen in students who have no previous experience and did not read the book (M = 66.52, SD = 9.82), $F_{\text{interaction}}(1, 96)$ = 5.048, p < .05, η^2_{partial} = 0.05. We also found that there was a significant main effect of book (M_{ourbook} = 90.56, SD = 1.11, $M_{\text{otherbook}}$ = 75.14, SD = 1.11), F (1, 96) = 95.74, p < .001, η^2_{partial} = 0.50 and a significant main effect of previous experience ($M_{\text{previousexperience}}$ = 89.70, SD = 1.11, $M_{\text{noexperience}}$ = 76.00, SD = 1.11), F (1, 96) = 75.57, p < .001, η^2_{partial} = 0.44. Had this data been analyzed only using t-tests, the important conclusion (derived from the interaction), that both experience and reading the volume lead to the highest levels of performance, would have been missed. It is also worth noting that factorial ANOVA can be combined with a repeated measures ANOVA for analysis purposes (obtaining the combined advantages of both approaches).

1.5.5 Regression

Ultimately, both t-tests and ANOVA represent specialized (and generally easier to understand) versions of regression. Regression analyses attempt to explain the relationship between any set of IVs on a DV. Importantly, regression analyses allow you to move beyond having an IV that is required to be sorted into discrete groups (as required by t-tests and ANOVAs). As such, you can look at variables that are naturally continuous (perhaps length of time spent studying for an exam).

There are a number of kinds of regression that can be used to evaluate your data. The most common type of regression is a linear regression (which is the underpinning of t-test and ANOVA). Beyond reporting the test statistics, we commonly report the mean at $+1$ and -1 SD on the DV (to help the reader in interpreting the data). So for instance, in our previous example, we could regress the number of hours spent studying for a test on the observed test score. We observed a significant effect where students who studied more had higher tests scores ($m_{+1SD} = 89.06$) than students who studied less ($m_{-1SD} = 76.40$), $b = 1.48$, $SE_b = .22$, $p < .001$, $R^2 = .31$.

It is worth noting that linear regression has the same assumptions that ANOVA and t-test have (i.e., continuous data that is normally distributed). However, not all forms of regression have that requirement. For instance, logistic regression allows you to analyze nominal data as the dependent variable (so long as there are only two levels, such as correct or incorrect). There are also other forms of regression that can be used to analyze your data (although these are distinctly less common in the social and behavioral sciences). Perhaps the best book to learn regression is Cohen, Cohen, West, and Aiken (2003).

1.5.6 Factor Analysis

At times, you will want to make composite variables of your responses. For instance, we have been discussing using a test as a DV throughout this chapter. Tests, however, by their very nature are composite variables. As a researcher, you will want to know how well the individual components work and whether they are associated in the way you think they are. One common analysis will be a reliability analysis. The most commonly used analysis in the social sciences is a Cronbach's alpha. As Chapter 2 will note, you must have reliability in your measures to be able to do any analyses. Wagner and Skowronski will cover the relative pros and cons of this approach in depth.

However, the basic reliability of the individual components will not tell you whether your items are actually associated with one another in measuring an underlying construct. To determine whether a particular set of questions are associated with one another, you can do a statistical technique called **factor analysis**. In factor analysis, you attempt to reduce the number of component items into a cluster, called a factor. For instance, in our example, you could easily imagine a single factor representing all of the test questions (this could be titled "Knowledge of Research Methods"). Conversely, a set of four factors could emerge that corresponds to the major sections of this volume ("Performing Good Research," "Understanding Issues Present Throughout the Research Process," etc.). Ultimately, the results of your factor analysis will inform how you analyze your data.

There are two major approaches to factor analysis. The first is called **exploratory factor analysis (EFA)**. In an EFA, you do not have a priori factor structures in mind; you simply let the mathematical associations emerge as they appear (for this, you will have a number of options in terms of the mathematics involved; in the social and behavioral sciences, a maximum likelihood analysis is the most common). In an EFA, you have several options to determine how many factors are present in the data. The first is the Eigenvalue test, where any factor with an Eigenvalue greater than one is considered an important factor. The second is called the Skree test, where you look at plotted Eigenvalues and you determine the number of factors by looking for the elbow in the Skree plot (and the factors above the elbow are considered important factors). An additional consideration is whether you will want to look at a rotated or unrotated set of factors. In many cases, a rotated solution is easier to interpret. A distinct advantage of an EFA is that you do not need a priori views of how your individual items might relate to one another.

The second major approach to factor analysis is called a **confirmatory factor analysis (CFA)**. With a CFA, you have an a priori factor structure that you are looking to confirm. The a priori factor structure could emerge from theory or previous studies. In a CFA, you will specifically test how well your proposed model fits the data. With CFA, there are a number of commonly used fit metrics, such as the GFI, CFI, RMSEA (or any other number of fit measures). An advantage of the CFA approach is that it allows you to see how well your model fits relative to an absolute standard. Regardless of which approach you take (EFA or CFA) on a particular dataset, you should never do both on the same data!

1.5.7 Chi-Square

The final commonly used statistical technique that we will briefly review in this chapter is chi-square. One of the distinct advantages of a chi-square analysis is that many of the assumptions required for regression-based analyses are not requirements when using chi-square. For instance, you can do a chi-square analysis on two sets of nominal data. As with all of the other techniques reviewed in this chapter, there are multiple permutations available. The most commonly used technique in the social and behavioral sciences is the Pearson chi-square. In the Pearson chi-square, you test to see if your observed frequencies are different from the expected distribution if the null hypothesis were true. In our example, we might find that the proportion of students who used this volume passed the test at a higher rate (19/20) than the proportion of students who did not use this volume (5/20), $\chi^2(1, N = 40) = 20.41, p < .001$.

1.6 Conclusion

This chapter has presented an abbreviated overview on the philosophy of science, key terminology for doing science in the social and behavioral sciences, and a very brief overview of common statistical terms and analyses. There is a lot more to come in the chapters that follow. In the first part of this volume, "Performing Good Research," we provide you with some considerations associated with doing good research. You need to pay attention to issues surrounding reliability and validity (Wagner & Skowronski, Chapter 2). You also need to decide if you want to do your research in the laboratory (Nichols & Edlund, Chapter 3), in the field (Elliott, Chapter 4), and/or online (Guadagno, Chapter 5). In the next part, "Understanding Issues Present Throughout the Research Process," you will read about issues surrounding the obtaining of consent (Festinger, Dugosh, Choi, & Sierka, Chapter 6), the influence preknowledge can have on behavior in studies (Papanastasiou, Chapter 7), the ways experimenters can unintentionally influence their participants (Strohmetz, Chapter 8),

and how to use suspicion probes and debriefings (Blackhart & Clark, Chapter 9). Part Three covers the "The Social and Behavioral Scientist's Toolkit," detailing common biophysiological measures (Vanman & Philipp, Chapter 10), eyetracking methods and technologies (Pelz, Chapter 11), the construction of questionnaires (Stassen & Carmack, Chapter 12), and reaction time methodologies (Heider & Reed, Chapter 13). The fourth part, "Emerging Issues in Social and Behavioral Science Research," covers replication in the social sciences (Soderberg & Errington, Chapter 14), research ethics (Ferrero & Pinto, Chapter 15), interdisciplinary research (Adams Goertel, Goertel, & Carey, Chapter 16), and cross-cultural research (Van de Vivjer, Chapter 17). The final part of this volume, "New Statistical Trends in the Social and Behavioral Sciences," covers the basics of Bayesian analysis (Miočević & Van de Schoot, Chapter 18), item response analysis (Hauenstein & Embretson, Chapter 19), social network analysis (Schorch & Quintane, Chapter 20), and concludes with an overview of meta-analysis (Webster, Chapter 21).

KEY TAKEAWAYS

- To do science, you must first understand what science is and what characterizes good science and theory.
- Good conceptualizations and operationalizations are key to good science.
- You should be familiar with the basics of social and behavioral science methods,

such as the kinds of data, IV, DV, reliability and validity.
- You should also be familiar with common statistical analyses, such as descriptive statistics, t-tests, ANOVA, regression, factor analysis, and chi-square.

SUGGESTED READINGS

Cohen, J., Cohen, P., West, S. G., & Aiken, L. S. (2003). *Applied Multiple Regression/ Correlation Analysis for the Behavioral Sciences* (3rd ed.). Mahwah, NJ: Lawrence Erlbaum Associates.

Saldana J., & Omasta, M. (2018). *Qualitative Research: Analyzing Life*. Thousand Oaks, CA: Sage.

Tabachnick, B. G., and Fidell, L. S. (2013). *Using Multivariate Statistics* (6th ed.). Boston: Pearson.

REFERENCES

Cesario, J., Jonas, K. J., & Carney, D. R. (2017). CRSP special issue on power poses: What was the point and what did we learn? *Current Research in Social Psychology, 2*(1), 1–5.

Cohen, J. (1988). *Statistical Power Analysis for the Behavioral Sciences* (2nd ed.). Hillsdale, NJ: Lawrence Erlbaum Associates.

Cohen, J., Cohen, P., West, S. G., & Aiken, L. S. (2003). *Applied Multiple Regression/ Correlation Analysis for the Behavioral Sciences* (3rd ed.). Mahwah, NJ: Lawrence Erlbaum Associates.

Eagly, A. H., & Carli, L. L. (1981). Sex of researchers and sex-typed communications as determinants of sex differences in influenceability: A meta-analysis of social influence studies. *Psychological Bulletin, 90*(1), 1–20. http://dx.doi.org/10.1037/ 0033-2909.90.1.1

European Research Council. (February, 2018). Open Access. Retrieved from https:// erc.europa.eu/funding-and-grants/managing-project/open-access

Fanelli, D., & Ioannidis, J. P. A. (2013). US studies may overestimate effect sizes in softer research. *Proceedings of the National Academy of Sciences of the United States of America, 110*(37), 15031–15036.

Flanagan, P., Genshaft, J. L., & Harrison, P. L. (Eds.) (1997). *Contemporary Intellectual Assessment: Theories, Tests, and Issues.* New York: Guilford Press.

Haney, C., Banks, C., & Zimbardo, P. (1973). Interpersonal dynamics in a simulated prison. *International Journal of Criminology & Penology, 1*(1), 69–97.

LaMothe, E. G., & Bobek, D. (2018). A modern guide to preliminary data analysis and data cleansing in behavioural accounting research. In T. Libby & L. Thorne (Eds.), *The Routledge Companion to Behavioural Accounting Research* (pp. 327–348). New York: Routledge/Taylor & Francis Group.

Mackonis, A. (2013). Inference to the best explanation, coherence, and other explanatory virtues. *Synthese, 190*, 975–995.

Merton, R. K. (1973). *The Sociology of Science: Theoretical and Empirical Investigations.* Chicago, IL: University of Chicago Press.

National Institutes of Health. (2003, October). NIH Sharing Policies and Related Guidance on NIH-Funded Research Resources. Retrieved from http://grants.nih.gov/grants/policy/data_sharing/data_sharing_guidance.htm

Oreskes, N., & Conway, E. M. (2010). *Merchants of Doubt: How a Handful of Scientists Obscured the Truth on Issues from Tobacco Smoke to Global Warming.* London: Bloomsbury Press.

Popper, K. R. (1959). *The Logic of Scientific Discovery.* Oxford: Basic Books.

Redding, R. E. (2001). Sociopolitical diversity in psychology: The case for pluralism. *American Psychologist, 56*(3), 205–215.

Reisman, J. M. (1983). SACRAL: Toward the meaning and measurement of friendliness. *Journal of Personality Assessment, 47*(4), 405–413.

Rosnow, R. L., & Rosenthal, R. (1989). Statistical procedures and the justification of knowledge in psychological science. *American Psychologist, 44*(10), 1276–1284.

Sagan, C. (1980). *Cosmos: A Personal Voyage.* TV Series.

Schmider, E., Ziegler, M., Danay, E., Beyer, L., & Bühner, M. (2010). Is it really robust? Reinvestigating the robustness of ANOVA against violations of the normal distribution assumption. *Methodology: European Journal of Research Methods for the Behavioral and Social Sciences*, 6(4), 147–151.

Stevens, S. T., Jussim, L., Anglin, S. M., Contrada, R., Welch, C. A., Labrecque, J. S., Motyl, M., ... & Campbell, W. K. (2017). Political exclusion and discrimination in social psychology: Lived experiences and solutions. In: J. T. Crawford & L. Jussim (Eds.), *The Politics of Social Psychology* (pp. 215–249). New York: Routledge.

Part One: Performing Good Research

2 Reliability and Validity of Measurement in the Social and Behavioral Sciences

Michael F. Wagner

John J. Skowronski

As scientists, one of our tasks is to measure stuff. Chemists measure atomic weight; physicists measure the speed of light. Developing good measures is crucial to science. Science was dramatically advanced by the electron microscope, the Hubble telescope, and functional magnetic resonance imaging (fMRI). Good measures are *crucial* to good science.

Two principles are fundamental to good measurement. The first is that a measure must be **valid**: It must measure what it claims to measure (e.g., Kelly, 1927). The second fundamental principle is that a measure must be **reliable**: The measure ought to produce about the same reading each time it is used (Nunnally, 1978). Some argue that these are linked: A measure needs to exhibit reliability before it can be considered to be valid. For example, a scale with a loose spring that produces wildly different readings each time it weighs the same object would probably not produce a valid weight for the object (although a valid weight might be produced once in a while). However, note that even perfect reliability does not ensure validity: A scale that is always 50 pounds off may be reliable, but it does not produce a valid measure of an object's weight.

Stated more formally, at least some degree of reliability is a necessary condition, but not a sufficient condition, for the validity of a measure.

Establishing that measures are reliable and valid is important to the extent to which one trusts research results. For example, many are attracted to implicit measures of memories, beliefs, and attitudes. Some believe that implicit measures assess knowledge that is not easily available via responses to self-report scales (implicit knowledge, e.g., Dienes, Scott, & Wan, 2011). Moreover, because the purpose of implicit measures is hard for respondents to ascertain, their use may minimize the influence of factors that may lower validity, such as response biases introduced by either social desirability concerns or efforts at self-presentation (see also Strohmetz, Chapter 8). These validity concerns help to explain why implicit measurement has proliferated in areas such as assessment of an individual's self-concept, their stereotypes, and their prejudices.

However, because implicit measures often measure a construct indirectly, some worry that they might not purely assess the construct of interest. Indeed, there is vigorous debate about the reliability and validity of many implicit measures (e.g., Blanton, Klick, Mitchell, Jaccard, Mellers, & Tetlock, 2009; Fazio & Olson, 2003; Krause, Back, Egloff, & Schmukle, 2011). For example, an early version of the Implicit Association Test attempted to assess the extent to which an individual might be prejudiced against a given group. However, that version of the test was criticized because its measures may have been sensitive to both an individual's own level of dislike for a given group, as well as to an individual's knowledge of how much a society dislikes a given group (see Olson & Fazio, 2004). One implication of this complication is that a person might potentially be labeled by the test as "personally prejudiced" when their high score may only reflect the knowledge that the group in question is societally disfavored. It is worth noting that recent versions of the stereotype-IAT have been developed that may avoid this issue.

This measure ambiguity issue is especially important for social and behavioral scientists because, in their research, what is being measured is often unclear. This is less of a problem in many non-social and behavioral science studies. For example, when assessing object length, there is clarity about the object being measured and consensus about the meaning of the measure. Indeed, a meter was originally defined as 10^{-7} part of one half of a meridian (Delambré, 1799; National Institute of Standards and Technology [NIST], 2017), but today is defined as the length of the path traveled by light in a vacuum during the time interval of 3.33564095^{-7} seconds (NIST, 2017). Thus, when measuring object lengths, there are standard, consensually acceptable methods available to make such measurements.

In contrast, social and behavioral scientists often measure **hypothetical constructs**. These are concepts or processes that are hypothesized, but that (at least at the moment) cannot be directly viewed or assessed. For example, a researcher might guess that a person is prejudiced. However, prejudice is not directly viewable, but is merely hypothesized to exist. One can get information about an individual's prejudice level only by looking at the influences that prejudice supposedly exerts on other things (e.g., behaviors). Moreover, when assessing hypothetical constructs, social and behavioral scientists are obliged to take extra steps to show that their measures are actually measuring the constructs. This is not a problem when measuring wooden board lengths – there is little doubt about what the measures are assessing and there is consensual agreement about appropriate measures. This is not the case when measuring hypothetical constructs, such as prejudice. With such constructs, social and behavioral scientists need to convince both themselves and others that a measure does actually assess their hypothetical constructs of interest. The latter portion of this chapter will provide a brief primer on how such **validity** evidence is obtained.

However, gathering and providing validity evidence is only one task faced by researchers. A researcher also needs to convince themselves and others that their measures are reliable. The term **reliability** refers to the extent to which a measure yields the same results on repeated trials. This is crucial: the trustworthiness of a study's results partly depends on trust in the measurement tool. Trust in the tool is affected by the extent to which the tool produces repeatable measurements. The next portion of this chapter will discuss how social and behavioral scientists obtain and present evidence of a measure's reliability.

We caution that our presentations of reliability and validity are both brief and elementary. Those who want to know more about construct measurements (e.g., how to interpret them and various theories underlying such measurements) should consult additional sources (e.g., Furr & Bacharach, 2013; Price, 2017).

2.1 The Pursuit of Evidence for Reliability

A program of research should provide evidence about the reliability of its measurement procedures. There are many approaches to establishing measure reliability (see Watson & Petrie, 2010). *We advocate using as many as possible.* Doing so can provide an especially complete, and relatively unbiased, picture of a measuring procedure's reliability.

2.1.1 The Test-Retest Method

One approach to reliability is the **test-retest method**. To illustrate this method, imagine that we are developing a measurement procedure in which we employ 100 different types and brands of length-measuring devices (e.g., tape measures, folding rulers, measuring sticks, laser measures). We intend to measure wooden board lengths by using each device once for each board, then combining the results of each of the 100 devices. Establishing the test-retest reliability of this measuring procedure is as easy as "one, two, three": 1) do each set of 100 measurements once, then 2) do the whole set of 100 measurements again, then 3) compare the results of

steps 1 and 2. For example, in a simple study, one might acquire 100 wooden boards of different lengths, measure each board twice using the 100-measure procedure, and compare the results produced by each pair of measurement procedures used on each board. The measurement procedure would be deemed reliable if it produced approximately the same result on each of the two measures taken from all 100 boards.

One straightforward examination method looks at the average discrepancy score – how much the difference between the boards deviates from 0 – between each pair of measures for each of the 100 wooden boards. A difference score of 0 for a measurement pair indicates perfect reliability (some call this **agreement**; see Bland & Altman, 1986). However, error often creeps into measurement, so there will likely be difference scores that are non-zero. One hopes that errors are small in magnitude, and that the distribution of errors is "fair" (e.g., some a bit above 0, and others a bit below).

The extent to which the measure is reliable can be statistically assessed by using a t-test. For each wooden board, subtract measure 2 (calculated from the second set of 100 different measures) from measure 1 (calculated from the first set of 100 measures) and calculate the absolute value of that difference (eliminating negative difference scores). This produces a **discrepancy score** for each board, which can be entered into a t-test that assesses whether the discrepancy score is significantly greater than 0. If a significant discrepancy score is obtained, then one would conclude that the measurement procedure is not reliable.

However, how good is "good enough"? That depends on researcher needs. For some applications, a level of reliability/agreement that is "close enough" may suffice. Other precision-dependent applications may require high agreement/reliability. The degree of agreement/reliability demanded in the t-test is altered by controlling a priori the p-value used in the test (.05, .01, .001, etc.). The smaller the a priori p-value, the greater the level of agreement/reliability that is demanded from the measurement procedure.

This discrepancy approach can be usefully extended. For example, one can build Bland–Altman plots (see Bland & Altman, 1986). To do so, one constructs a graph that plots the difference scores on the vertical (y) axis against the average of the two measures on the horizontal (x) axis. Examination of such plots allows one to see if the magnitude of the discrepancy scores systematically changes with changes in the magnitude of the object/thing/construct/variable being measured. In our wooden board example, this might plausibly occur if there are bigger discrepancy scores when the boards are long as opposed to short.

While the discrepancy approach makes intuitive sense as an approach to reliability, social and behavioral scientists rarely use it. One reason may be that social and behavioral scientists are not especially comfortable with variations in the a priori determinations of how much agreement there should be for a scale to be called "reliable" enough. However, a bigger problem is that measuring hypothetical constructs is not exactly like measuring boards. To illustrate this lack of equivalence, instead of measuring boards, let's imagine measuring prejudice against Klingons (a fictional race from the fictional Star Trek Universe, e.g., see www.startrek.com/database_

article/klingons [2018]). We can develop a measurement procedure that is similar to the one used to measure boards. For example, we might build a 100-item self-report scale. Each item on the scale (e.g., "I hate Klingons") asks a respondent to provide a response intended to reflect the degree to which they harbor anti-Klingon prejudice. In this way, each self-report item corresponds to each of our 100 different board-measuring devices. Moreover, as in our board example, we can assess prejudice level in 100 different respondents. To take this parallel further, just as for each board, we can average each participant's responses to our 100 Klingon-prejudice items to get an "average" measure of prejudice for each respondent. Finally, just as with our board-measuring procedure, we can try to establish measure reliability by asking people to respond to our measure twice. Conceptually, if people provide exactly the same (or very similar) responses to the Klingon-prejudice scale on each of the two occasions, then scale reliability will be demonstrated.

However, some agreement might not be "real." For example, memories from the time 1 measurement procedure might influence results produced by the time 2 measurement procedure. For example, imagine that a participant thinks: "I remember this item. I made a strongly disagree response last time. I don't want to look wishy-washy, so I'll give the same response again." Such thinking may artificially decrease discrepancy scores (it could also increase discrepancy scores if respondents want to look different at time 2).

How do researchers respond to this potential problem? They can vary the methods used to assess reliability (this can be termed **triangulation**) and look for similarities (this can be termed **convergence**) in results provided by the different methods.

2.1.2 Use of Parallel Forms to Assess Reliability

For example, researchers might not use the entire 100-item measure twice. Instead, they might randomly select fifty items and give those at time 1, and then have people respond to the other fifty items at time 2. The item sets used at time 1 and at time 2 can differ across respondents. For example, for test 1, a researcher can intentionally construct different sets of fifty items for each respondent, or the researcher can let a computer program randomly select a set of fifty test 1 items for each respondent (all respondents would then get the fifty items not selected for them at test 1 for test 2). The results provided by each of these two versions can then be compared. If the measurement procedure is reliable, people who score high in anti-Klingon prejudice measured at time 1 ought to score high in measured prejudice at time 2. This approach can be termed assessing reliability via the use of **parallel forms**.

However, note that something important has changed here: The researcher cannot compute discrepancy scores because the items are not exactly the same for each half of the test. Nonetheless, all is not lost. Social and behavioral scientists might try to assess reliability by correlating (e.g., via the Pearson r statistic) responses to the two score sets. If the scale is reliable, then individuals with a high time 1 score should also have a high time 2 score. However, there is a potential problem with this approach: Pearson's r will yield high correlations if the two sets of measures evince the same ordering on measure 1 and

measure 2, *even if the actual values of the measures are different.*

Let us illustrate this. For simplicity, let's only consider three wooden boards. Imagine that on trial 1, the 100-measure procedure measures the three wooden boards at 5 inches, 6 inches, and 7 inches. On trial 2, imagine that the 100-measure procedure measures the same three boards at 9 inches, 10 inches, and 11 inches. A Pearson correlation between these two sets of values yields a value of 1. This value might mislead a researcher into concluding that the measurement procedure is highly reliable. It clearly is not. The board scores, while in the *same order* on the two tests, do not indicate a high level of **agreement** in the exact values of the scores.

Alternative statistical procedures have been developed to apply correlational statistical methods to better assess test-retest reliability. For example, Lin (1989) proposed and developed a method known as the **concordance correlation coefficient**. Other statistical procedures that can be applied are **intraclass correlation (ICC)** measures (for an overview, see Koo & Li, 2016). While the mathematics that are applied in these correlation-derived reliability statistics vary, they are united in their attempt to take into account both measure ordering and agreement. Hence, in our view, these methods are better suited to assess test-retest reliability than the simple Pearson correlation. That's not to say that the Pearson correlation is useless – but researchers can probably do better.

Because you are astute readers, you are probably asking "Why couldn't you have just used these advanced correlation procedures instead of the discrepancy procedure for the test-retest approach?" Well, you are right – in the test-retest procedure, one could get an estimate of reliability by using one of the advanced correlation techniques (concordance correlation, intraclass correlation). Indeed, this latter approach has the advantage of providing an easy to understand index of reliability. The correlation statistic ranges from -1 to 1, and the higher the value, the stronger the reliability estimate. Because of this ease of interpretation, many researchers prefer these sophisticated correlation techniques over the discrepancy score approach. However, because we like triangulation and convergence, we advocate using as many of the different approaches as one can manage to apply.

While the parallel forms approach eliminates some potential biases, it retains others. These include biases that might be caused by changes in a respondent between time 1 and time 2. This may be a problem of special concern for social and behavioral scientists. Indeed, research shows that the levels of hypothetical constructs, such as personality traits and attitudes, can meaningfully fluctuate across time (see Hamaker, Nesselroade, & Molenaar, 2007). Sometimes, such changes are easily understandable. For example, a Klingon might have saved earth between time 1 and time 2, so people's prejudice levels toward Klingons might shift dramatically in the time between the tests. This would produce high discrepancy scores, but those scores would not reflect low test reliability (note that a similar artificially low reliability score could occur with wooden boards if someone sawed pieces off the boards between time 1 to time 2). Another concern is that people's understanding of the test might change from time 1 to time 2. For example,

some participants might not realize that the test measures anti-Klingon prejudice until after they finished taking it the first time. This realization might change their responses to the test when it is taken the second time around. Such a change in understanding might again serve to raise discrepancy scores, artificially lowering the measured agreement/reliability of the test.

In fact, it is easy to imagine a host of variables contributing to differences in response patterns obtained for the first test administration and the second test administration. What if the first test administrator was a human and the second administrator was a Klingon? What if the first test was taken in a quiet room and the second taken when the researchers down the hall were playing a loud crying baby audio file? What if the first test was taken at the start of a semester and the second test taken just before finals? All of these variables might serve to increase discrepancy scores calculated from two administrations of the same test given at two different times to the same respondents.

2.1.3 Assessing Reliability via Internal Consistency

Another approach to reliability can combat these potential problems. This **internal consistency** approach uses the subscale generation idea employed in the parallel forms approach but eliminates the time lag between the two administrations of the test. For example, imagine that we take the 100-item anti-Klingon scale and give it all at once. However, also imagine that, before the test is administered, we split the scale into two fifty-item subsidiary scales (using the same methods as described for the development of parallel forms). Results from the two sub-scales can be compared, and reliability can be assessed via the advanced correlational methods that we have mentioned (e.g., intra-class correlations, concordance). Most import-antly, these results would not be contaminated by any changes introduced during the passage of time or by changes in the testing situation.

However, when using this approach, social and behavioral scientists sometimes do not use these advanced correlational methods. Instead, they often calculate **Cronbach's alpha** (see Cortina, 1993; Feldt, Woodruff, & Salih, 1987; Sijitsma, 2009). The advantage of this statistic's approach to reliability is that it eliminates the need for researchers to manu-ally construct subsets of items to use for reli-ability calculations. Instead, Cronbach's alpha automatically approximates the reliability esti-mate produced by using Pearson's r to **correl-ate all possible** split-half forms with all possible complements of these forms. In our Klingon prejudice scale case, that would be the average of all the correlations calculated from all possible form A/form B pairs. That's a *lot* of possible form combinations, and that's a good thing. Obviously, this approach bypasses any bias that a researcher might inadvertently introduce when manually con-structing form A/form B pairs. Moreover, interpretation is easy – the higher the Cron-bach's alpha, the higher the perceived reliabil-ity of the scale. However, the disadvantage of this approach is that one is forced to use Pear-son's r as an index of reliability, which, in our opinion, is not optimal. Nonetheless, the Pear-son's approach to reliability is better than none at all.

Convention among many social and behav-ioral scientists is that an alpha value of .7 to .9

is often deemed to be adequate. A Cronbach's alpha of less than .5 is viewed by most researchers as inadequate. However, in our view, these conventions are overly restrictive. Why? In our view, the value that is acceptable in the context of any given research project may vary. For example, the value of alpha is known to vary by scale length – short scales tend to yield smaller Cronbach's values than long scales. Thus, while administering the longest scale available may maximize reliabil-ity, but doing so may sometimes not be pos-sible. For example, in a time-constrained context, a researcher may have to use an abbreviated version of a longer psychometric-ally validated scale (see Nichols & Webster, 2013). One likely consequence of this shortened scale is a reduced Cronbach's alpha. If the researcher is pursuing a fre-quently replicated result, that reliability reduction might be judged as an acceptable price to pay for a shortened procedure, even if Cronbach's alpha estimate falls below .5.

An additional context for consideration of acceptable alpha values is the strength of the "signal" one is searching for with the scale relative to the amount of "noise" that can affect scale responses. Sometimes, the signal that one is searching for may be very weak because the measure at issue may be affected by many things. This can be easily understood in the context of our wooden board measure-ment example. Let's assume that a researcher is trying to measure the length of wooden boards while on a tiny boat that is being tossed about in a hurricane. The pitching and yawing introduced by the storm is equiva-lent to adding "noise" to the attempt to measure actual board length (the "signal"). In this noise-filled measurement context, the reliability of the 100-measure board measurement procedure is bound to go down relative to a comparatively "noiseless" environment (e.g., measures made in a stable and quiet lab on dry land). Thus, when measures are influenced by a great deal of noise, it may be the case that a relatively modest reliability estimate might be the best that can be expected. Clearly, this modest estimate is not optimal. However, when considerable noise influences measurement procedures, modest reliability may be the best that can be achieved.

This same idea often applies to the measures used by social and behavioral scientists. For example, responses to the anti-Klingon prejudice measure could be affected by such things as elements of a respondent's self-concept, concerns about how the respondent looks to other people, concerns about society's values, and response tendencies introduced by respondent personality characteristics, such as agreeableness and neuroticism. Thus, it might be hard to measure the anti-Klingon prejudice influence (the signal) on item responses in the context of all of these other influences (the noise). Moreover, unlike the board example, in which noise can be quieted by moving the measurement procedure to dry land, when making psychological measurements, there is often only so much that one can do to quiet the noise. Hence, modest reliability statistic values might be the norm when measuring the multi-influenced hypothetical constructs pursued by many social and behavioral scientists.

2.1.4 Reliability in Measures Made by Different Researchers

There is another twist to the story. In all of our previous examples, the measurement tool has been used by the same person. This is not always the case, as measurement tools will often be used by different people. One other reliability concern is the extent to which multiple users agree in the measures that they produce using a given measurement procedure.

For example, each of two researchers could administer the 100-item Klingon-prejudice scale to two different sets of 100 respondents each. An intraclass correlation procedure (again, a better choice than Pearson's *r*) could explore whether the anti-Klingon prejudice scores significantly varied by researcher. A high ICC coefficient (approaching 1) indicates little researcher impact on the measures produced.

These between-researcher effects might be expected to be small when procedurally standardized self-report scales are used. In these cases, the opportunity for researchers to impact the results is typically (though not always) relatively small. However, one might sometimes be very worried about the extent to which a researcher might contribute to the measures (for more on this, see Strohmetz, Chapter 8). For example, imagine a case in which cultural anthropologists record samples of behaviors in various cultural contexts and try to extract important cultural norms and standards from the recorded behaviors. To aid their efforts, such researchers will often use some kind of guide, such as a behavior codebook. Careful researchers typically assign two or more individuals (coders) to do the coding and typically report evidence suggesting that the coders could reliably use the codes – that is, the coders both coded their observations in the same way. Low agreement may suggest that the coders' own idiosyncratic norms and standards might have influenced their coding.

One way to tally the extent to which coders agree is to calculate the **joint probability of agreement**. However, this statistic can be a bit misleading, for it does not reflect how often coders might agree, simply by chance. For example, if there were only two coding categories, an agreement rate of 50% reflects chance-level agreement. In our view, **Cohen's kappa** (for two coders) and **Fleiss' kappa** (for more than two coders) are preferred agreement statistics because they adjust for these chance common codings. **Krippendorff's alpha** has similar advantages, and can be used when the coding scheme contains multiple categories that are naturally ordered (e.g., small to large or weak to strong).

2.1.5 Why Report Evidence about Reliability?

When reporting results of research, it can be important to report measure reliability to readers (typically in a research report's methods section or results section). Doing so might spur research. On learning that a given measurement procedure is not optimally reliable, people might start thinking about how the measurement procedure might be improved. This may not seem very sexy to many who want to know about constructs such as prejudice and who don't give a hoot about how to measure such things. However, remember that significant and important contributions to science come from those who find better ways to measure stuff.

The second reason for reporting reliability information (typically in a format such as α = .74) is that it provides a context that readers can use to think about research results. For

example, assume that a) you conducted an experiment designed to reduce anti-Klingon prejudice; b) your statistical analysis did not support the efficacy of your manipulation; but c) the reliability of the measure was low. It may be that your manipulation reduced anti-Klingon prejudice, but your measure was so unreliable that it failed to detect the effect of the manipulation. This may spur a researcher to develop a more reliable method of measuring anti-Klingon prejudice. By lowering measurement "noise" and increasing reliability, in a subsequent experiment, this new measurement method may respond to the effects of your manipulation.

2.2 The Pursuit of Evidence for Construct Validity

When measuring concrete stuff, no one typically doubts that the measurement procedure is measuring the concrete object – the measure has high **face validity**. The presence of the concrete object is perceivable during measurement, so it seems obvious that the data from the measurement procedure is measuring the object. Moreover, for many physical measures (length, time) there are consensually accepted tools for measurement. For example, when measuring the lag between stimulus presentation and response, no one typically questions whether a standard time measurement device (e.g., a stopwatch) is a valid measure. However, an especially careful researcher might do a preliminary validity study in which they compare the measures provided by their measuring device against an accepted standard. For example, prior to conducting a study in which a watch is used to time stamp entries into a log-book, a researcher might compare their watch readings to the readings provided by the atomic clock of the National Institute of Standards and Technology and adjust (or calibrate) their watch based on the results of the comparison.

In comparison, providing evidence for the validity of measures assessing hypothetical constructs requires an especially high degree of effort and attention from researchers. The logic is straightforward. If a measure "behaves" as the construct suggests that it "should," then a researcher can conclude that a measure is a valid measure of the construct. What does the statement about a measure "behaving as it should" mean (an idea that some call **criterion validity**)? As noted by Borsboom, Mellenbergh, and Van Heerden (2004), and by Kane (2001), to pursue criterion validity, one needs to have a well-specified a priori theory about when and how and why and in what circumstances a hypothetical construct will influence other variables.

2.2.1 Construct Validity and Convergence with Other Measures

With the theory in place, one can turn to data collection. Strong inferences can be made about the hypothetical construct when the data "fits" the expected action of the hypothetical construct. For example, we might believe that our Klingon prejudice measure is actually measuring anti-Klingon prejudice if people who the scale says have high anti-Klingon prejudice display other Klingon prejudice relevant tendencies, such as: a) scoring high on other measures thought to measure anti-Klingon prejudice (**convergent validity**); b) predicting other measures thought to be dependent on hatred of Klingons, such as

refusing to sit next to Klingons, refusing to marry them, and telling anti-Klingon jokes (**predictive validity**); and c) being in certain categories or groups, such as being a member of the anti-Klingon society (predicting group memberships in expected ways can also be called **concurrent validity**).

2.2.2 Construct Validity and Divergence with Other Measures

The measure of interest should *not* be linked to, or influence, stuff that ought to be unrelated to anti-Klingon prejudice. This is called **discriminant validity**. For example, one might think that there is no reason for anti-Klingon prejudice to be linked to the personality trait of neuroticism. If this is the case, then studies should show that there is no correlation between scores on the anti-prejudice scale and scores on a scale that measures neuroticism. Moreover, manipulations designed to alter an individual's level of neuroticism should not alter a person's scores on the Klingon prejudice scale. When a line of research involving many studies provides evidence of discriminant validity, combined with studies that provide strong evidence of criterion validity, confidence in the construct validity of the measure can become quite high.

2.2.3 From the Trees to the Forest: Using the Various Measures of Validity

These ideas reflect Cronbach and Meehl's (1955) conception of validity as relying on a **nomological network**. Evidence for construct validity is garnered when studies show that observed relationships are consistent with theorized relationships among respective constructs. One technique that is often employed to explore these networks is the **multitrait-multimethod matrix** (MTMM; Campbell & Fiske, 1959). A MTMM matrix is simply a table that presents correlations among various measures. However, the table is constructed using a particular organizational scheme. The organizational scheme is to enter the data into a correlation matrix table using both the trait assessed and the method used to assess the trait, grouping all the entries by trait and method (e.g., trait 1-method 1, trait 1-method 2, trait 1-method 3, trait 2-method 1, etc.). This same scheme is used for both axes of the table. Such a table allows researchers to easily see whether a measure is strongly correlated with other measures which it should correlate with (convergent validity), and whether it only weakly correlates with (or does not correlate with) measures with which it ought not to be linked (discriminative validity). One other advantage is that this table allows a researcher to see if the method of measurement matters to the evidence that is collected. If this is the case, then different traits that are assessed using the same method will be strongly correlated. Similarly, you can also get a sense of discriminant validity by looking at correlations (hopefully low) among constructs that are supposedly unrelated.

An example of such a matrix appears in Table 2.1, which depicts data from measures of the characteristics (administration, feedback, consideration) of managers from the perspective of supervisors, the managers themselves, and employees (in this example,

Table 2.1 Example of a multitrait-multimethod correlation matrix from Mount

	Supervisor			Self			Subordinate		
	A	F	C	A	F	C	A	F	C
Supervisor									
A	1.00								
F	.35	1.00							
C	.10	.38	1.00						
Self									
A	**.56**	.17	.04	1.00					
F	.20	**.26**	.18	.33	1.00				
C	−.01	−.03	**.35**	.10	.16	1.00			
Subordinate									
A	**.32**	.17	.20	**.27**	.26	−.02	1.00		
F	−.03	**.07**	.28	.01	**.17**	.14	.26	1.00	
C	−.10	.14	**.49**	.00	.05	**.40**	.17	.52	1.00

From Mount (1984; reproduced from davidkenny.net/cm/mtmm.htm). A = Administration, F = Feedback, C = Consideration.

these are the different "methods"). The correlations that appear in boldface in the table are thought to reflect convergent validity – these are measures of the same manager trait but from different sources. Visual inspection suggests that, though the convergent validity correlations are positive, they are also small in magnitude. Such results suggest that the measures are being influenced by variables other than the manager characteristics of interest (e.g., the validity evidence is weak). Though visual inspection is sometimes sufficient, sophisticated statistical analyses (factor analyses) can be brought to bear on such data. However, full explication of the analysis of MTMM matrices is beyond the scope of this chapter. For an entrée to this area, see Kenny and Kashy (1992).

2.2.4 Considerations in the Documentation and Use of Validity

In practice, it can be difficult to make a strong case for the construct validity of a measure. This is because many measures are not "pure" assessment devices but are "contaminated" by unwanted influences. For example, two substantial worries for many self-report instruments are that people may want to present themselves in the best possible light to other people (**social desirability bias**), and people may want to respond in ways that fit how they think about themselves (**self-presentation bias**). Other measurement difficulties emerge because constructs may often "covary" – this means they are often present in a given measure at the same time. For example, people who exhibit high levels of depression also tend to exhibit high levels of anxiety, and depressive disorders and anxiety disorders tend to covary (e.g., Lamers et al.,

2011; also see Caspi et al., 2014). Thus, when measuring depression, it can be hard to eliminate anxiety as the construct that influences responses.

Doing so may be facilitated by the use of data-exploration techniques, such as factor analysis and latent variable modeling. For example, confirmatory factor analyses (CFAs) have been applied to MTMM by specifying that the observed MTMM is explained by a measurement model consisting of two constructs: construct variance and method variance. From this measurement model, one can estimate construct variance, method variance, and covariance between observed residuals of prediction (which would indicate a relationship between the trait measured and the method used). For example, there might be residual correlations between variables in the MTMM if there were a self-selection effect in different methods, whereby participants who score higher on the trait of interest also tend to self-select into a particular study method. In such a case there would presumably be some exogenous, theoretical reason to explain residual correlations or, at the construct level, relationships between the trait and the construct.

However, even these kinds of analyses might not suffice to establish a measure's construct validity. As noted by Borsboom et al. (2004), Cronbach (1989), and Kane (2001), reliance on patterns of correlations to make inferences about hypothetical constructs has substantial limitations. Most importantly, what's missing in such data is evidence that changes in a construct **cause** changes in a variable supposedly linked to that construct. Finding evidence of such causal changes adds significantly to the construct validity case.

Thus, one might also gain confidence in the validity of the Klingon-prejudice scale if laboratory manipulations produce data that fits with the logical consequences of prejudice. For example, a person who scores high in anti-Klingon prejudice might evaluate an ambiguous-looking social target more negatively after being exposed to information that might lead one to believe that the target is a Klingon than when no such information is conveyed. Similarly, a person who scores high in anti-Klingon prejudice might evaluate an actor's task performance more negatively when led to believe that the task was performed by a Klingon than when the person is not led to that belief. These experimental manipulation/treatment studies have the advantage of suggesting that the anti-Klingon prejudice construct is not only *linked to* measures, but can *causally influence* these measures (see Borsboom, Mellenbergh, & van Heerden, 2003; Spencer, Zanna, & Fong, 2005).

Some validity-oriented scholars have overlooked the validity implications of this logical causal relation between manipulations and measures. For example, Ferguson and Kilburn (2009) published a meta-analysis examining the effects that observed media violence has on the observers. They omitted from their meta-analysis many laboratory studies that did not use measures for which traditional psychometric evidence of measure reliability had been provided. Ferguson and Kilburn questioned the validity of these unstandardized measures, later labeling these as examples of "poor methodology" (2009, p. 762). However, they ignored the fact that many of these measures had been repeatedly used in studies of violence, and the measures often "behaved" in ways that were consistent with the idea that

exposure to media violence causes increases in aggressiveness, which causes expressions of violence (for an example, see Crouch, Irwin, Wells, Shelton, Skowronski, & Milner, 2012). In our view, Ferguson and Kilburn (2009) committed a serious error because they discarded studies whose results did indeed provide evidence for measure validity. The validity evidence did not come from the usual psychometric sources but, consistent with the Borsboom et al. (2003) perspective, came from the fact that the measures behaved in a manner that was consistent with manipulation of the underlying construct.

However, the fact that a measure is responsive to manipulations is not itself a guarantee that the measure assesses the construct. Manipulations are often *not* surgical scalpels that work selectively – they are often like shotguns that influence many things all at once. The bottom line for us is that *no* single indicator or method is sufficient to establish the construct validity of a measure. Construct validity is established by the creation of a pattern of responses across many different kinds of studies (both observational and experimental), suggesting that if the measure works "like it should," the measure validly measures a construct. For us, the key ideas lie in **replication, triangulation, and convergence**. Confidence in the construct validity of a measure happens across many, many studies that replicate relations between the measure of interest and other measures, and it is especially important when those other measures differ in method (they attempt **triangulation**). When studies assessing a construct exhibit replication and triangulation, then one becomes increasingly convinced of the construct validity of the measure if the results of those studies exhibit **convergence** (e.g., the "expected" results emerge). See Soderberg and Errington, Chapter 14 for a detailed exposition on replication in the social and behavioral sciences.

2.3 Coda

Good science requires good measurements. The measures ought to reflect characteristics of the "thing" being measured (should be valid) and should be precise and repeatable (should have reliability). When all is said and done, it is probably better to have a measure that is a little sloppy but that still captures the influence of a construct than it is to have a repeatable measure that does not capture the construct. Nonetheless, researchers should strive for both characteristics.

Hence, one of the tasks that a behavioral or social science researcher has is to convince themselves, and to convince research consumers, that their measure of a construct validly measures the construct (and not some other construct). Doing so is *not* easy – often making such a case takes a lot of time and many studies. Achieving these goals is especially difficult in the social and behavioral sciences because social and behavioral scientists often try to measure hypothetical constructs. Hence, we don't know exactly what it is that is being measured. Moreover, in social and behavioral science research, people's responses may often be simultaneously influenced by multiple constructs, which complicates the compilation of reliability evidence and validity evidence. Despite these difficulties, the pursuit of good measures is worth the trouble. Good measures are necessary to the production of trustworthy results, and that is one of the most important pursuits of science.

KEY TAKEAWAYS

- Science prefers measurement tools that are valid (that measure attributes of what is supposed to be measured) and reliable (repeatable and precise).
- It is good practice for social and behavioral scientists to accumulate evidence supporting the reliability and the validity of the measurement tools employed, especially when exploring hypothetical constructs using tools of measurement that are not consensually accepted.
- This accumulation ought to be an ongoing process that involves replication and triangulation (replication with multiple methods) in the hopes of producing convergence.

- In the domain of reliability, one reason that it is good practice to use multiple methods to explore reliability is that different methods can produce different information. For example, exploration of test-retest reliability can provide information about the precision of a test that one cannot get from simple correlation-based statistics, such as Cronbach's alpha.
- Explorations of validity should similarly employ multiple methods to show that a construct is linked to things to which it is "supposed to" be linked (producing various forms of validity, such as predictive validity, convergent validity,

and concurrent validity), and responds to manipulations that supposedly alter the construct of interest. Research also needs to determine discriminant validity by showing that a measure is not linked to, or does not respond to manipulations of, alternative constructs.

IDEAS FOR FUTURE RESEARCH

- People often rely on existing measures because doing so is easy and everyone uses them. How can we increase the number of researchers that are interested in the refinement and improvement of such methods to enhance their reliability and validity?
- Do high-quality measures (valid, reliable) predict trustworthy, replicable results? We think that doing so would encourage researchers to employ the best possible measures in their research.
- Have the recent changes in journal policies and procedures led to improvements in how researchers approach reliability and validity?

SUGGESTED READINGS

Furr, R. M., & Bacharach, V. R. (2013). *Psychometrics: An Introduction* (2nd ed.). Thousand Oaks, CA: Sage.

Nunnally, J. (1978). *Psychometric Theory*. New York: McGraw-Hill.

Price, L. R. (2017). *Psychometric Methods: Theory into Practice*. New York: Guilford Press.

Watson, P. F., & Petrie, A. (2010). Method agreement analysis: A review of correct methodology. *Theriogenology, 73*, 1167–1179. doi:10.1016/j.theriogenology.2010.01.003

REFERENCES

Bland, J. M., & Altman, D.G. (1986). Statistical methods for assessing agreement between two methods of clinical measurement. *Lancet, 327*, 307–310. doi:10.1016/S0140-6736(86)90837-8

Blanton, H., Jaccard, J., Klick, J., Mellers, B., Mitchell, G., & Tetlock, P. E. (2009). Strong claims and weak evidence: Reassessing the predictive validity of the IAT. *Journal of Applied Psychology, 94*, 567–582. doi:10.1037/a0014665

Borsboom, D., Mellenbergh, G. J., & van Heerden, J. (2003). The theoretical status of latent variables. *Psychological Review, 110*, 203–219. doi:10.1037/0033-295X.110.2.203

Borsboom, D., Mellenbergh, G. J., & van Heerden, J. (2004). The concept of validity. *Psychological Review, 111*, 1061–1071. doi:10.1037/0033-295X.111.4.1061

Campbell, D. T., & Fiske, D. W. (1959). Convergent and discriminant validation by the multitrait-multimethod matrix. *Psychological Bulletin, 56*, 91–105. doi:10.1037/h0046016

Caspi, A., Houts, R. M., Belsky, D. W., Goldman-Mellor, S. J., Harrington, H., Israel, S., … Moffitt, T. E. (2014). The p factor: One general psychopathology factor in the structure of psychiatric disorders? *Clinical Psychological Science, 2*, 119–137. doi:10.1177/2167702613497473

Cortina, J. M. (1993). What is coefficient alpha? An examination of theory and applications. *Journal of Applied Psychology, 78*, 98–104. doi:10.1037/0021-9010.78.1.98

Cronbach, L. J. (1989). Construct validation after thirty years. In R. E. Linn (Ed.), *Intelligence: Measurement, Theory, and Public Policy* (pp. 147–171). Urbana, IL: University of Illinois Press.

Cronbach, L. J., & Meehl, P. E. (1955). Construct validity in psychological tests. *Psychological Bulletin, 52*, 281–302. doi:10.1037/h0040957

Crouch, J. L., Irwin, L. M., Wells, B. M., Shelton, C. R., Skowronski, J. J., & Milner, J. S. (2012).

The word game: An innovative strategy for
assessing implicit processes in parents at risk
for child physical abuse. *Child Abuse & Neglect*,
36, 498–509. doi:10.1016/j.chiabu.2012.04.004

Delambré, J. B. J. (1799). *Méthodes analytiques
pour la détermination d'un arc du méridien*.
Paris: Crapelet.

Dienes, Z., Scott, R. B., & Wan, L. (2011). The role
of familiarity in implicit learning. In P. A.
Higham & J. P. Leboe (Eds.), *Constructions of
Remembering and Metacognition: Essays in
Honour of Bruce Whittlesea*. Basingstoke:
Palgrave Macmillan.

Fazio, R. H., & Olson, M. A. (2003). Implicit
measures in social cognition research: Their
meaning and use. *Annual Review of Psychology*,
54, 297–327. doi:10.1146/annurev.
psych.54.101601.145225

Feldt, L. S., Woodruff, D. J., & Salih, F. A. (1987).
Statistical inference for coefficient alpha.
Applied Psychological Measurement, *11*,
93–103.

Ferguson, C. J., & Kilburn, J. (2009). The public
health risks of media violence: A meta-analytic
review. *Journal of Pediatrics*, *154*, 759–763.
doi:10.1016/j.jpeds.2008.11.033

Furr, R. M., & Bacharach, V. R. (2013).
Psychometrics: An Introduction (2nd ed.).
Thousand Oaks, CA: Sage.

Hamaker, E. L., Nesselroade, J. R., & Molenaar,
P.C. M. (2007). The integrated trait–state model.
Journal of Research in Personality, *41*, 295–315.
doi:10.1016/j.jrp.2006.04.003

Kane, M. T. (2001). Current concerns in validity
theory. *Journal of Educational Measurement*, *38*,
319–342. doi:10.1111/j.1745-3984.2001.
tb01130.x

Kelly, T. L. (1927). *Interpretation of Educational
Measurements*. New York: Macmillan Press.

Kenny, D.A., & Kashy, D. A. (1992). Analysis of the
multitrait-multimethod matrix by confirmatory
factor analysis. *Psychological Bulletin*, *112*,
165–172. doi:10.1037/0033-2909.112.1.165

Koo, T. K., & Li, M. Y. (2016). A guideline of
selecting and reporting intraclass correlation
coefficients for reliability research. *Journal of
Chiropractic Medicine*, *15*, 155–163.
doi:10.1016/j.jcm.2016.02.012

Krause, S., Back, M. D., Egloff, B., & Schmukle,
S. C. (2011). Reliability of implicit self-
esteem measures revisited. *European Journal
of Personality*, *25*, 239–251. doi:10.1002/
per.792

Lamers, F., van Oppen, P., Comijs, H. C., Smit,
J. H., Spinhoven, P., van Balkom, A. J. L.
M. ... Pennix, B. W. J. H. (2011). Comorbidity
patterns of anxiety and depressive disorders in a
large cohort study: The Netherlands Study of
Depression and Anxiety (NESDA). *Journal of
Clinical Psychiatry*, *72*, 341–348. doi:10.4088/
JCP.10m06176blu

Lin, L. I.-K. (1989). A concordance correlation
coefficient to evaluate reproducibility.
Biometrics, *45*, 255–268. doi:10.2307/2532051

Mount, M. K. (1984). Supervisor, self- and
subordinate ratings of performance and
satisfaction with supervision. *Journal of
Management*, *10*, 305–320. doi:10.1177/
014920638401000304

National Institute of Standards and Technology.
(2017). Retrieved from www.nist.gov/pml/
weights-and-measures/si-units-length

Nichols, A. L., & Webster, G. D. (2013). The
single-item need to belong scale. *Personality
and Individual Differences*, *55*, 189–192.
doi:10.1016/j.paid.2013.02.018

Nunnally, J. (1978). *Psychometric Theory*. New
York: McGraw-Hill.

Olson, M. A., & Fazio, R. H. (2004). Reducing
the influence of extra-personal associations on
the Implicit Association Test: Personalizing the
IAT. *Journal of Personality and Social
Psychology*, *86*, 653–667. doi:10.1037/0022-
3514.86.5.653

Price, L. R. (2017). *Psychometric Methods: Theory
into Practice*. New York: Guilford Press.

Sijtsma, K. (2009). On the use, the misuse, and the
very limited usefulness of Cronbach's Alpha.
Psychometrika, *74*, 107–120. doi:10.1007/
s11336-008-9101-0

Spencer, S. J., Zanna, M. P., & Fong, G. T. (2005).
Establishing a causal chain: Why experiments

are often more effective than mediational analyses in examining psychological processes. *Journal of Personality and Social Psychology, 89,* 845–851. doi:10.1037/0022-3514.89.6.845

Watson, P. F., & Petrie, A. (2010). Method agreement analysis: A review of correct methodology. *Theriogenology, 73,* 1167–1179. doi:10.1016/j.theriogenology.2010.01.003

3 Performing Research in the Laboratory

Austin Lee Nichols

John E. Edlund

The laboratory provides reliable and valid data for many social and behavioral scientists for whatever phenomenon they are studying (for a detailed discussion of reliability and validity, see Wagner & Skowronski, Chapter 2). For some, this is due to the convenience of student samples and space allocated by their institution/organization to performing one's research. For others, the laboratory is a place to eliminate confounds so that the data obtained is free from statistical "noise" (i.e., unexplained variance in the data). Whatever the reason for using a lab, one must understand the benefits and potential drawbacks of doing so. To this end, the following chapter discusses what a laboratory is, the general benefits and drawbacks of using a laboratory to conduct research, the types of research often conducted in a lab, and specific issues that one should consider when preparing to conduct lab research.

3.1 What Makes a Laboratory?

Before we get too far into the discussion of lab research, we must first set the stage for what is or is not a lab. Having worked in many different institutions around the world, we've adopted a very broad definition of what a lab is. However, we believe there are still certain characteristics that are necessary to make some space a lab. The most important thing to understand is that the often relied upon stereotype of a laboratory is only a small subset of the kinds of labs that are currently in use. That is, a lab need not contain sophisticated contraptions, be staffed by people in white coats, or be in the basement of an old university building. In fact, many of the labs we have worked in have limited equipment, are staffed by college students wearing typical college clothes (see Edlund, Hartnett, Heider, Perez, & Lusk, 2014), and are housed in bright, airy spaces.

In general, a laboratory need only be a space dedicated to performing some kind of research. For the purposes of this chapter, we further define a lab as an enclosed space that minimizes distractions unrelated to the study (i.e., is controlled). One important distinction here is that research performed outdoors would not be lab research but would likely be field research (see Elliott, Chapter 4). However, it is worth noting that, over the years, some research has blurred the lines between field and laboratory research (e.g., Goldstein,

Cialdini, & Griskevicius, 2008). Other than these simple criteria, a lab can come in any number of shapes, sizes, and uses. This includes a room housing chemistry experiments, a vacant classroom used to administer surveys, a mock trading floor, or even the International Space Station – arguably the world's largest laboratory. One of our colleagues even turned an old closet into a functional comparative cognition lab. These are only a very small subset of the many types of spaces that are used as laboratories. With a little creativity, you can turn just about anything into a productive, functional lab.

3.2 Advantages and Disadvantages

As we stated from the beginning, people perform research in a lab for a variety of reasons. Because lab research is not for everyone nor every project, it is important that you understand both why you should or should not use a lab to help achieve your research goals. We begin this discussion below and will end the chapter with a more detailed discussion of potential issues to understand and protect against when performing research in a lab.

3.2.1 Advantages

One benefit worth highlighting, as well as something that defines a lab, is the minimization of "noise" in your data (Kerlinger & Lee, 2000). Simply put, the more "noise" in your data, the harder it is to detect a relationship between two variables. As our main goal in research is to detect a relationship (if one exists), using a lab is extremely useful and beneficial. Labs also tend to eliminate most, if not all, relevant distractions. When performing research with humans, the last thing we want is to take their attention away from what we are trying to focus it on. Therefore, labs are great tools for doing research without distraction and with minimal "noise."

Related to the minimization of "noise," labs are ideal places to perform experiments. Since experiments are the only research method by which one is able to truly examine and support cause and effect relationships, labs offer the unique opportunity to claim that one variable causes another. It should be noted, however, that research outside of a lab can also be experimental and result in cause and effect conclusions (e.g., Goldstein et al., 2008). However, since a controlled environment is the ideal in any experimental research, labs are one of the most powerful tools for conducting research. This control provides two sets of benefits. First, as a controlled environment, a lab offers researchers the opportunity to manipulate some variable (i.e., the independent variable – IV) and measure how a second variable (i.e., the dependent variable – DV) changes in response to that manipulation without worrying that the IV is simultaneously being affected by other variables. Second, this same controlled environment minimizes the likelihood that any additional variables are affecting the DV, which thus increases the confidence that any change in this variable is due solely to the IV. This being said, unintended factors can still influence research outcomes in the lab (e.g., the influence of gender on persuasiveness, Eagly & Carli, 1981).

What these two advantages also lend a hand to is the increased reliability of effects. Due to the lack of extraneous variables affecting any part of the research process, the likelihood that any future attempt is able to replicate these effects is higher. Restated in terms of validity, a lab is the most effective place to perform research if your concern is internal validity (i.e., cause and effect), but this advantage often comes at the expense of external validity (i.e., the ability to apply findings to the real world). We discuss this more next.

3.2.2 Disadvantages

For many social and behavioral scientists, the goal is to one day be able to apply one's research to the "real world." As such, external validity is extremely important. Unfortunately, due to the often artificial nature of lab research, it is possible that many effects discovered in the lab will not replicate or even exist outside of the lab. However, it is worth noting that some research has compared lab and field studies and found similar effects and effect sizes between the two (Anderson, Lindsay, & Bushman, 1999) – we will discuss this more in the last section of this chapter. Nonetheless, lab research is often the first step in being able to apply one's findings. Once an effect is found in the lab, it is generally best for a researcher to next examine the key relationship(s) in the field (see Elliott, Chapter 4).

What happens in the lab does not always stay in the lab. That is, despite efforts to eliminate confounding variables from entering or exiting the lab, participants can still change the lab environment. This happens in several ways, including bringing expectations or information about the research into the lab or leaving with and disseminating information about the research to other, possibly future, participants (e.g., Edlund, Sagarin, Skowronski, Johnson, & Kutter, 2009). In addition,

due to the sterility of the lab environment, participants are often highly attentive to even the most minor details of the researcher or research, and this can often affect the data (see Beilock, Carr, MacMahon, & Starkes, 2002 for a discussion on how paying attention affects novices and experts). In Chapter 7 Papanastasiou covers many of these effects in detail.

3.3 Types of Lab Research

When people think about research conducted in the social and behavioral sciences, many think of it being conducted in a lab and they think of it being an experiment. However, just as is the case for field research, labs house many types of research. We will describe some of the main types of research that are often carried out in a lab next.

3.3.1 Experiment

As you are likely well aware by now, the experiment is the ideal research method in many respects (such as eliminating noise and controlling confounds), and experiments are most commonly carried out in labs. Again, the lab is the ideal location for an experiment since it provides an opportunity for the researcher to further minimize external influences and allows one to manipulate something without the fear of other variables affecting the IV or DV. What this looks like, in practice, varies from lab to lab, but often follows a similar procedure.

In a lab experiment, a researcher or experimenter often enters the lab before anyone else, sets up whatever hard or electronic materials are necessary, and prepares for the arrival of the participant(s). This could include opening a computer program that automatically randomizes participants to conditions and displays the corresponding questions and instructions or setting up an elaborate deception with confederates and props. For example, the second author's Master's thesis involved an experimenter that would administer a series of questionnaires to a small group of participants (which included one confederate). During the course of the study, each of the participants would leave to a side room to complete a computerized task which would allow the confederate to have a scripted series of interactions with the participants (Edlund, Sagarin, & Johnson, 2007).

3.3.2 Quasi-Experiment

A quasi-experimental design is basically an experiment minus the random assignment. This often occurs when it is difficult or even impossible to randomly assign participants to conditions. One situation that commonly results in a quasi-experimental design is when researchers use some meaningful variable as a means by which to assign participants to either the control or experimental condition. Commonly used quasi-experimental variables of this sort include sex, ethnicity, and age. Another common design is the pre-test/post-test design where a researcher assigns all participants to a single group and obtains a pre- and post-test measure for each. As all participants are in the same condition, without a control group, there is no random assignment, and it is not fully experimental (e.g., McCarthy et al., 2018).

3.3.3 Surveys

Researchers also may use the lab simply as a place to conduct survey research. Considering that much of the research in social and behavioral science is questionnaire based, it should be no surprise that many studies currently conducted in a lab are either paper and pencil or computerized surveys. Since many of these studies do not manipulate or randomize anything, we categorize them as non-experimental research. These take many different forms, including a questionnaire designed to establish the relationship between people's salaries and their job satisfaction (see Judge, Piccolo, Podsakoff, Shaw, & Rich, 2010 for a summary of such studies) or a measurement of risk-taking and one's position in a company or political organization.

3.4 *Issues to Consider*

Although we are huge proponents of lab research, one can only do such research with a full understanding of the pitfalls that pave the way. Some of these are fairly minor and require only a certain level of care in the design and execution of the research. Other issues are such that a simple misstep can doom the research and make the data useless. Since much of these more serious issues are presented in detail throughout this book, we focus in the following section of this chapter on describing and providing advice on those issues that often plague the novice lab researcher.

3.4.1 Participants

Human research is not possible without having participants to engage in it. Therefore, they are arguably the most important consideration when preparing and conducting research. As such, it is important to consider many aspects of recruiting and interacting with participants.

In particular, one must understand how to get people to participate, what to expect when they enter the lab, and the proper way to end the research relationship.

Recruitment. The first consideration with participants is how to recruit them. The strategies for recruitment in the social and behavioral sciences are as diverse as the fields of study. With that said, there are several commonly used strategies. The simplest of these is found at many large research-based universities. In many of these settings, students of some courses are required to participate in a certain number of research studies as part of their course requirement. As with any issue relevant to participants, ethical considerations abound. Although we won't discuss those in this chapter, Ferrero and Pinto address many of these issues in detail in Chapter 15.

Within these large university settings, as with any participant recruitment context, it is important to have a good service or program to manage recruitment. One of the most popular participant management systems is Sona Systems® (www.sona-systems.com/participant-recruitment.aspx). Like similar programs, it allows you to post research opportunities for registered users to see. Users are then able to select and register for a study by choosing a particular date and time that the researcher has made available. The researcher then has the option to receive notifications of these sign ups as well as reminders prior to the specified date and time. Finally, researchers can provide credit for participation that can then be used to fulfill whatever course requirement or other compensation agreed upon.

Recruitment even at large universities is not always so easy and straightforward. In fact, university research pools often get depleted, or you might face challenges in recruiting a sub-population (such as deaf and/or hard of hearing participants). We have found that a raffle opportunity for gift cards is a very cost-effective way to increase participant recruitment. Rather than paying all participants for their time, you have a small number of raffle prizes that can be just as effective in increasing participation. However, this is an issue that you must first work out with the appropriate ethical review board (ERB), as it will often have specific requirements for using raffles or lotteries as compensation.

Another option for student or non-student populations is the snowball data collection technique. In the snowball data collection technique, you ask volunteers (perhaps students in an upper-level university class) to recruit participants for you that meet some demographic requirement (typically friends, family, or coworkers). In exchange, you offer the volunteers some compensation (perhaps one point of extra credit or one dollar for each participant successfully recruited). This technique is best done for simple and short studies where participants would be most likely to agree to take part for limited compensation. In fact, this technique was used in the second author's first published paper (Edlund, Heider, Scherer, Farc, & Sagarin, 2006). A version of the snowball technique is also often used in medical studies where doctors or nurses recruit their patients as potential participants (e.g., Wohl et al., 2017).

Another way to recruit participants is to advertise in areas that your desired population is likely to frequent. This may include a university campus, in a local grocery store, or possibly in a certain doctor's office or coffee shop. The place of choice should directly reflect your decision regarding the type and size of sample that you desire. Whatever place you choose, it is often necessary to get your advertisement approved by your local ERB and to obtain permission from the owner of the establishment to post the material.

As Guadagno focuses on in Chapter 5, the internet is also a convenient research tool. As far as participants go, there are several recruitment strategies that utilize this technology. First, a researcher can advertise online and would want to consider the same things as discussed in the previous paragraphs. Another popular way to use the internet for recruitment is through email. As such, a researcher can either email out to an entire predefined list of people or may target a subset of email addresses that fit the defined parameters of the desired sample. As with all recruitment efforts, we always suggest that you first refer to your local ERB before beginning.

Biases. Once a researcher recruits participants to the lab, it is important to understand how to deal with them when they arrive. Beyond treating them in an ethical manner, one must understand the many biases and contextual factors that affect participant responses. We discuss a few of these next but encourage you to refer to Chapters 6–9 for a much more detailed discussion.

The first general concern that must be addressed is the state in which participants arrive. This includes basic factors such as time of day, how rested/hungry the participant is, or their interest in the research itself (e.g., Mischel, Shoda, & Rodriguez, 1989). However, other influential factors such as

participant mood, the weather outside, and the moments immediately prior to arrival must also be considered. One should also take into account history effects. For example, a national or local event could have altered participant responses (e.g., a disaster). Next, researchers must consider how the lab environment may influence participants, including the presence and effect of the experimenter(s). At times, we also may have to deal with participants who show up to the lab in altered states of consciousness. With all of these considerations, it is up to a researcher to understand these external factors and ideally to control for or measure them. However, it is also very important for researchers to understand the potential influence on their specific measures and research question, as the same factors may or may not have significant effects on various research paradigms.

Debriefing. When you have recruited participants and run them through the study, your job is not yet over. In fact, what researchers do after the research is complete can significantly affect both participants and the research. The main job of the researcher is to ensure that the participant has not been nor will be harmed due to his/her participation in the research. Since most of these issues relate to ethical review boards and general research ethics, we refer you to Ferrero and Pinto, Chapter 15 for more discussion. See also Blackhart and Clark, Chapter 9 for a discussion of debriefing participants.

Another important consideration is how each participant will potentially affect your research in the future. Part of this is the spread of information that you desire to be kept secret until the research is completed.

This is presented in both Papanastasiou, Chapter 7 and Strohmetz, Chapter 8, so we will not elaborate here. One related consideration (not discussed in future chapters) is lab reputation. For a variety of reasons, certain laboratories can gain either a positive or negative reputation, and this is likely to affect your ability to conduct successful research in the future. As such, researchers should take care to make sure participants are both satisfied and willing not to discuss the research with others. This is perhaps especially important when doing online studies (see also Guadagno, Chapter 5) as there are many resources available to potential participants to investigate research labs. For example, there are innumerable resources for mTurk workers (mTurk is a commonly used online subject pool; www.mturk.com) such as TurkOpticon and the HitsWorthTurkingFor reddit forum, and research suggests that mTurk workers consult these forums (Edlund, Lange, Sevene, Umansky, Beck, & Bell, 2017).

3.4.2 Research Assistants

Another important consideration when performing lab research is how you will staff the lab. Many lead investigators do not have the ability to run every participant personally, so they often employ one or more research assistants (RAs) for help. Finding the right people and providing the right training and resources for them are integral to the success of your research.

Recruitment. The first decision to make when choosing research assistants is the amount of resources that you have available to hire them. For many researchers, there are limited to no resources, so looking for volunteers is the best option. For those with more resources and/or grant money to pay RAs, the options are much greater, and you can be much more selective in the process.

When resources are tight, and volunteers are needed, one of the most important considerations is motivation. At one of the first author's previous institutions, he simply put out an advertisement to current undergraduates in the psychology department stating that he was interested in volunteers to help run some of his research. Since these people are not being compensated monetarily, and there are few to no consequences for mistakes or misbehavior, it is necessary to find people who value the experience and what it might bring them in the future. To this end, he first used markers that he believed represented a degree of commitment and responsibility (e.g., GPA), as well as past lab experience, to narrow down the list of potential candidates from several dozen to roughly a handful. He then interviewed each one independently and asked them about their past experience, current obligations, and future goals, to get an idea of each student's motivation and ability to make a positive contribution to his lab. After providing clear expectations for the upcoming semester, he then chose those students who seemed to fit with these expectations.

Another free option that may be available to you as a researcher is to have a student or intern sign up as an independent study/ cooperative education student. In these cases, the students are being paid in a valuable currency (course credit) that is basically free to you. Although there are usually no direct monetary costs to the experimenter for this arrangement, significant costs are still accrued in the time spent mentoring them and in potentially assessing the student for the assigned course.

Having resources to pay RAs provides a much greater opportunity to hire a successful experimenter. First, you are likely to get many more individuals who are interested in the opportunity. Second, you have the ability to be much more selective. For example, when seeking volunteers, you may have to accept five hours a week from a sophomore student who has no familiarity with research. In contrast, offering compensation may afford you a full-time graduate student with extensive experience in lab research. Although this helps in the recruitment process, it remains important to understand a candidate's motivation and to pick someone who will be as interested in and passionate about your research as you are.

Training. Getting the right people for the job is important, but it is useless if they don't learn how to do exactly what you need them to in your lab. This is where the often-arduous process of adequate training becomes essential. Many researchers take this process for granted or fail to allocate adequate time and resources to training, and they often pay for this in the long run.

The first type of training that many RAs need, before beginning, is some kind of human research training mandated by the institution, state, or even nation in which you are conducting research. If you are not aware of any requirements for RA training at your location, it is worth asking your ERB to prevent any problems in the future. Even if it is not required from a higher authority, many researchers choose to have their RAs complete something similar to ensure a basic understanding of participants' rights and other ethical considerations. One of the most popular of these options is CITI training for human subjects (https://about.citiprogram.org/en/series/human-subjects-research-hsr/).

Once RAs are prepared with the basics for dealing with human participants, it is necessary to train them on their role in your research. Although these roles may include anything from performing literature reviews to coding data to running actual participants, we focus on the latter of these here. In particular, many lab studies have quite complex procedures and cover stories. Even the experienced experimenter will often need training and practice to ensure each session goes without a hitch.

As each study is very different in its complexity and requirements of the experimenter(s), it will ultimately be up to the lead researcher to determine what and how to train each RA. To give two competing examples, we will use two very different studies from the first author's lab experience. One of the simplest lab studies he conducted simply required the experimenter to greet the participant, have them sign an informed consent form, provide them with a questionnaire to complete, then briefly debrief them. As you can imagine, training for this study was quite quick, and even the most novice RA was able to conduct the study. On the other hand, a different study consisted of a detailed script in which the experimenter had to memorize dozens of lines to recite to three different participants in the same session over a well-choreographed hour. This study required hours of training and practice and only the most senior RAs were considered and used for the study.

3.4.3 Deception

One very common aspect of lab research is the use of deception to prevent participants from acting irregularly due to knowledge of the true purpose of the research. Although this certainly is not used in all lab research (or solely in lab research), its use and the controversy surrounding it are common enough to warrant some discussion of it here. It is also helpful to refer to Blackhart and Clark, Chapter 9, as it focuses on the debriefing process as a way to alleviate the possible risks and negative effects of deception.

Tradeoffs. In the social and behavioral sciences, there is little that is more controversial than the use of intentional deception in research. Many believe that any potential risks are outweighed by the need and benefit that are realized from its use (e.g., Christensen, 1988). Others believe that these benefits are overstated and that deception has no place in science (e.g., Baumrind, 1985; Kelman, 1967). Some have even examined the assertion that participants' trust in research and those that conduct it is eroded as a result of deception (Sharpe, Adair, & Roese, 1992). However, even early on, when ethical concerns were at their highest regarding deception, views of research were not tainted by experiencing deception. Despite this, the moral argument still largely favors eliminating deception. However, even an examination of different ethical perspectives reveals different conclusions regarding deception that are dependent on what viewpoint you prescribe to (e.g., utilitarianism vs. duty-based). For more on ethics in research, refer to Ferrero and Pinto, Chapter 15. Also, for an alternate framework by which to decide on the use of deception, refer to Kimmel, Smith, and Klein (2011).

For the purposes of understanding deception's use in lab research, we want to share some thoughts and examples of our own to help you decide when and if you use deception in your own research. Let's begin by referencing a main research focus of the first author of this chapter – impression management. In most impression management research, a participant is asked to report an impression of another person. In its simplest form, deception likely brings no benefit to the research, as the participant can simply interact with someone or something then answer questions regarding the impression perceived. However, many times this involves a much more complicated interaction where researchers would likely use deception. Take, for example, the multiple audience problem. This involves several people interacting at the same time with at least one person desiring to convey different impressions to (at least some of) the different people present (e.g., Nichols & Cottrell, 2015). The procedure for this research often requires the "actor" to conceal the fact that s/he is attempting to convey a particular impression to a particular person and for the researcher to create a cover story regarding the reason for the interaction. Because the lack of information as well as incorrect information are often both thought of as deception, we move on to discussing the different types of deception next.

Types of Deception. No discussion surrounding deception is sufficient without understanding that there are different types of deception. The most important distinction to make here is deception by commission versus by omission. Deception by commission is what most people agree is the more serious of the two and the one almost all people would call deception. This is when you provide information to the participant that you know to be false. One of the most common examples of this is the use of a cover story in which the experimenter will tell the participant that the purpose of the study is different than the true purpose. Other examples include telling a participant that his/her actions will result in something that they won't, incorrectly stating the outcome or purpose of a specific task or measure, providing bogus feedback on an experimental task, or asking one participant to deceive another participant. We will discuss the use of these in more detail in the next section.

The other type of deception is that of omission or failing to provide information to the participant. The conundrum here is that similar to interpersonal lies. Many people think that a lie of commission is worse than that of omission (Haidt & Baron, 1996). Similarly, many researchers believe that deception by omission is less harmful, and some would even claim that omission itself is not deceptive. Examples of this form of deception abound and theoretically range from not informing participants of the hypotheses of the research to allowing them to believe something is true that is not (without actually telling them the untruth). Rather than provide our own opinion on the right or wrong associated with the use of deception, we prefer to simply mention that we both have used deception in our own research. As such, we will provide some details and advice for those who choose to use deception in their research.

Effectively Using Deception. Once the decision has been made to use deception, there are many things to consider to ensure that one is effective in doing so. The first and most common use of deception is concealing the purpose and hypotheses of the study. The reasons for doing so are numerous and are detailed in Papanastasiou, Chapter 7 (see also Nichols & Edlund, 2015). Relevant here is the best way to go about doing so. The first way is to simply omit the purpose and/or hypotheses from the information provided to participants. Since most ethical review boards require a stated purpose during the consent process, simple omission of the purpose is often not possible. However, it is usually acceptable (and recommended) to omit or generalize information regarding the hypotheses.

Since complete omission of the study's purpose is often prohibited, researchers frequently use one of two approaches. The first, and simplest, approach involves generalizing the title and purpose of a study to the point that any real understanding of what the researcher is interested in or his/her expectations is very unlikely. For example, if one is interested in examining how the personality of employees' supervisors affects their home life, one could state that the purpose of the study is "to understand information regarding your work and home life." Alternatively, one could choose to create a cover story using deception by commission.

In the above example, simply generalizing the study's purpose may be enough to avoid participant demand. However, let us instead take a more complex example, one that is best served by a detailed cover story. The first author's first ever independently run study examined the effect of participants' knowledge of the researcher's hypothesis on the results of the study (Nichols & Maner, 2008). Both researchers agreed that deception was necessary since informing participants of the study's purpose would inhibit the ability to examine the research question. Instead, an elaborate procedure and cover story, designed to keep participants from knowing the true purpose and hypothesis, started with the first author (i.e., the confederate – we will discuss this next) pretending to be the previous participant. The experimenter opened the lab door and, in front of the next participant, thanked the confederate for participating, asked the participant to wait a moment while s/he prepared the lab for the next session, then closed the door. The confederate then told the participant that the experimenter would show the participant pairs of pictures and that the researcher expected people to choose the ones on the right more than those on the left. The participant then entered the room and completed the task. In this example, there were effectively two levels of deception. The first was what was told to the participants prior to entering the lab. The second was what the experimenter told them when conducting the study. In fact, neither of those stated purposes was the true purpose. The use of deception in this example was extremely effective as not a single participant, when asked after the conclusion of the study, was able to correctly guess the purpose of the study. In fact, some even guessed that it was about something other than the true, experimenter-stated, or confederate-stated purpose.

So, what exactly is a confederate and why/ how should one use him/her? Essentially, a confederate is anyone who the experimenter has asked to help with the study, but the participants don't know is working with the experimenter. One of the most well-known examples of a confederate was in the Asch conformity research (Asch, 1951) where Solomon Asch had several confederates pretend to be participants and give the same wrong answer to see if the real participant would "conform" to that wrong answer. Although many confederates pose as real participants, they can also serve other roles in or outside of the laboratory to help get participants into a certain state of mind prior to entering the experiment or to study their behavior without the demands of being inside the lab. For example, confederates will often pretend to be passersby and will either act in front of or interact with the participant prior to the participant's arrival at the lab (e.g., en route to or just outside of the lab).

3.4.4 Applying Results to the Real World

One issue that receives extra attention when discussing lab research is that of applying the results obtained in the lab to the real-life phenomena that we are seeking to understand. To be clear, this is not necessarily unique to lab research since most field research also creates a context that is not identical to what people normally encounter. In addition, some argue that generalization to the real world is not necessary or even desired (e.g., Mook, 1983). However, the lab environment is purposefully modified to eliminate the parts of the real world that one would expect to confound the participant's behavior. As such, it is important to understand the concern of external validity as well as learn some ways to strengthen it when conducting in lab research.

Does Lab Research Generalize? Unfortunately, there is not a clear answer to this question. Frankly, people who want to suggest that lab research is as generalizable to the real world as is field research will not find difficulty supporting their position (e.g., Anderson et al., 1999); the same is true for those suggesting the opposite (e.g., Sears, 1986). In reality, a more textured view is likely the most accurate and unbiased (Mitchell, 2012). That is, depending on the research question and the participants used, research in the lab may or may not be likely to generalize to contexts outside of the lab. We will discuss this next while providing some recommendations for increasing the likelihood that your lab research will generalize outside of the lab.

Tips for Increasing Generalizability. As you likely know by now, all decisions leading up to and including the research process must be carefully considered as each option often has definite advantages and disadvantages. After all, this is one of the main purposes of this book. When conducting lab research, in particular, generalizability is one consideration that should weigh into your methodological decisions. In particular, there are two factors that are most likely to affect your study's ability to generalize outside of the lab: 1) the research question; 2) the participants.

The first thing to consider when thinking of generalizability is your research question. Some questions and phenomena are just more likely to generalize outside of the lab than others. As such, if you are interested in understanding how being outside in the fresh air and sunlight affects people's moods, it is unlikely that results obtained in a lab will represent what happens when someone is outside, and a lab study may not even make sense to conduct. On the other hand, when the editors of this book examine how different aspects of the research process affect participants and the data they provide, the lab is the "real-world" equivalent, and there is no reason to believe that the results obtained in lab experimentation do not generalize to the target phenomena. Therefore, you must truly understand your research question to understand how much of a concern external validity likely is when examining it in a lab.

You often decide whether or not to conduct research in the lab after you know your research question. Thus, the only things that you can do to improve external validity are to make the lab as similar to your target external environment as possible. However, doing so is likely to minimize the benefits of doing the research in a lab in the first place. In contrast, the decision regarding what and how many participants to use often follows that of research setting. Unlike the research question, participants are something that you can purposefully tailor to the lab setting. As such, you can and should choose the sample that you believe best represents your target population.

We realize that this may sound obvious, and we hope that it does to you. However, although sampling should always be an extremely well thought out decision, people often neglect this when engaging in lab research. The reason for this goes back to the fact that labs are usually in large institutions that also provide easy access to a very specific set of participants. In much research, these are people from Western, Educated, Industrialized, Rich and Democratic (WEIRD) societies (Henrich, Heine, & Norenzayan, 2010). Instead, researchers should take care to choose and recruit participants relevant to the research question. As an example of what not to do, the first author once (as a first-year graduate student) recruited a sample of freshman college students to examine the effect of leadership experience on the personality that one desires in his/her leader. As you can imagine, this didn't go over well with the reviewers of the first journal the paper was submitted to. After realizing the narrow-minded decision, the author recruited an international working sample (Nichols, 2016).

3.5 Conclusion

This chapter only scratches the surface of what issues exist and should be considered when performing research in a laboratory. Nonetheless, we think that it is a good start especially for those new to lab research, and trust that the following chapters will provide whatever additional information and details you will need in your journey to the lab. The one thing that you must always remember is that every single decision is important in research, and fully realizing the benefits of lab research requires that every single step, word, and fly on the wall should be attended to in a manner that is consistent with the goals of the research and the considerations that we have detailed above.

KEY TAKEAWAYS

- Although lab research provides many benefits over field research, one must understand the inherent risks and potential confounds to ensure that the rewards outweigh the risks.
- When recruiting both participants and research assistants, it is important to ensure you are doing so based on the research question and resulting needs, as well as the motivation of those being considered.
- The use of deception is a hotly debated topic and should only be done with a complete understanding of the pros and cons specific to your research.

IDEAS FOR FUTURE RESEARCH

- Are there ways to alter the lab environment to achieve an ideal balance between internal and external validity?
- What aspects of the lab environment may change participant biases and demand?
- How can we manage both participant interest and motivation so as not to let them affect the results on a study?

SUGGESTED READINGS

Baumrind, D. (1985). Research using intentional deception: Ethical issues revisited. *American Psychologist, 40*(2), 165–174.

Mitchell, G. (2012). Revisiting truth or triviality: The external validity of research in the psychological laboratory. *Perspectives on Psychological Science, 7*(2), 109–117.

Nichols, A. L., & Edlund, J. E. (2015). Practicing what we preach (and sometimes study): Methodological issues in experimental laboratory research. *Review of General Psychology, 19*(2), 191–202.

REFERENCES

Anderson, C. A., Lindsay, J. J., & Bushman, B. J. (1999). Research in the psychological laboratory: Truth or triviality? *Current Directions in Psychological Science, 8*(1), 3–9.

Asch, S. E. (1951). Effects of group pressure upon the modification and distortion of judgment. In H. Guetzkow (Ed.), *Groups, Leadership and Men* (pp. 177–190). Pittsburgh, PA: Carnegie Press.

Baumrind, D. (1985). Research using intentional deception: Ethical issues revisited. *American Psychologist, 40*(2), 165–174.

Beilock, S. L., Carr, T. H., MacMahon, C., & Starkes, J. L. (2002). When paying attention becomes counterproductive: Impact of divided versus skill-focused attention on novice and experienced performance of sensorimotor skills. *Journal of Experimental Psychology: Applied, 8*(1), 6–16. http://dx.doi.org/10.1037/1076-898X.8.1.6

Christensen, L. (1988). Deception in psychological research: When is its use justified? *Personality and Social Psychology Bulletin, 14*(4), 664–675.

Eagly, A. H., & Carli, L. L. (1981). Sex of researchers and sex-typed communications as determinants of sex differences in influenceability: A meta-analysis of social influence studies. *Psychological Bulletin, 90*(1), 1–20. http://dx.doi.org/10.1037/0033-2909.90.1.1

Edlund, J. E., Hartnett, J. H., Heider, J. H., Perez, E. J. G., & Lusk, J. (2014). Experimenter characteristics and word choice: Best practices when administering an informed consent. *Ethics & Behavior, 24*(5), 397–407.

Edlund, J. E., Heider, J. D., Scherer, C. R., Farc, M. M., & Sagarin, B. J. (2006). Sex differences in jealousy in response to actual infidelity experiences. *Evolutionary Psychology, 4*, 462–470.

Edlund, J. E., Lange, K. M., Sevene, A. M., Umansky, J., Beck, C., & Bell, D. (2017). Participant crosstalk: Issues when using the Mechanical Turk. *The Quantitative Methods in Psychology, 13*(3), 174–182.

Edlund, J. E., Sagarin, B. J., & Johnson, B. S. (2007). Reciprocity and the belief in a just world. *Personality and Individual Differences, 43*, 589–596.

Edlund, J. E., Sagarin, B. J., Skowronski, J. J., Johnson, S. J., & Kutter, J. (2009). Whatever happens in the laboratory stays in the laboratory: The prevalence and prevention of participant crosstalk. *Personality and Social Psychology Bulletin, 35*(5), 635–642.

Goldstein, N. J., Cialdini, R. B., & Griskevicius, V. (2008). A room with a viewpoint: Using social norms to motivate environmental conservation in hotels. *Journal of Consumer Research, 35*(3), 472–482. https://doi.org/10.1086/586910

Haidt, J., & Baron, J. (1996). Social roles and the moral judgement of acts and omissions. *European Journal of Social Psychology, 26*(2), 201–218.

Henrich, J., Heine, S. J., & Norenzayan, A. (2010). The weirdest people in the world? *Behavioral and Brain Sciences, 33*(2–3), 61–83.

Judge, T. A., Piccolo, R. F., Podsakoff, N. P., Shaw, J. C., & Rich, B. L. (2010). The relationship between pay and job satisfaction: A meta-analysis of the literature. *Journal of Vocational Behavior, 77*(2), 157–167. http://dx.doi.org/10.1016/j.jvb.2010.04.002

Kelman, H. C. (1967). Human use of human subjects: The problem of deception in social psychological experiments. *Psychological Bulletin, 67*(1), 1–11.

Kerlinger, F. N., & Lee, H. B. (2000). *Foundations of Behavioral Research.* Fort Worth, TX: Harcourt.

Kimmel, A. J., Smith, N. C., & Klein, J. G. (2011). Ethical decision making and research deception in the behavioral sciences: An application of social contract theory. *Ethics & Behavior, 21*(3), 222–251.

McCarthy, B., Trace, A., O'Donovan, M., O'Regan, P., Brady-Nevin, C., O'Shea, M., … & Murphy, M. (2018). Coping with stressful events: A pre-post-test of a psycho-educational intervention for undergraduate nursing and midwifery students. *Nurse Education Today*, *61*, 273–280.

Mischel, W., Shoda, Y., & Rodriguez, M. I. (1989). Delay of gratification in children. *Science*, *244*(4907), 933–938. doi:10.1126/science.2658056

Mitchell, G. (2012). Revisiting truth or triviality: The external validity of research in the psychological laboratory. *Perspectives on Psychological Science*, *7*(2), 109–117.

Mook, D. G. (1983). In defense of external invalidity. *American Psychologist*, *38*(4), 379–387.

Nichols, A. L. (2016). What do people desire in their leaders? The effect of leadership experience on desired leadership traits. *Leadership & Organization Development Journal*, *37*(5), 658–671.

Nichols, A. L., & Cottrell, C. A. (2015). Establishing versus preserving impressions: Predicting success in the multiple audience problem. *International Journal of Psychology*, *50*(6), 472–478.

Nichols, A. L., & Edlund, J. E. (2015). Practicing what we preach (and sometimes study): Methodological issues in experimental laboratory research. *Review of General Psychology*, *19*(2), 191–202.

Nichols, A. L., & Maner, J. K. (2008). The good-subject effect: Investigating participant demand characteristics. *The Journal of General Psychology*, *135*(2), 151–166.

Sears, D. O. (1986). College sophomores in the laboratory: Influences of a narrow data base on social psychology's view of human nature. *Journal of Personality and Social Psychology*, *51*(3), 515.

Sharpe, D., Adair, J. G., & Roese, N. J. (1992). Twenty years of deception research: A decline in subjects' trust? *Personality and Social Psychology Bulletin*, *18*(5), 585–590.

Wohl, A. R., Ludwig-Barron, N., Dierst-Davies, R., Kulkarni, S., Bendetson, J., Jordan, W., … Pérez, M. J. (2017). Project Engage: Snowball sampling and direct recruitment to identify and link hard-to-reach HIV-infected persons who are out of care. *Journal of Acquired Immune Deficiency Syndromes*, *75*(2), 190–197. doi:10.1097/QAI.0000000000001312

4 Field Research

Sinikka Elliott

"People say one thing and do another." It's a common enough aphorism to suggest there is some truth to it. In fact, research documents a wide gap between what people say (about their beliefs, values, and behaviors) and what they do (their actual behaviors) (Jerolmack & Khan, 2014). Fundamentally, social and behavioral scientists want to understand and explain human behavior. Field research, which involves observing and participating in people's lives and keeping a detailed written record of those experiences and observations, is a powerful method for uncovering people's behaviors in situ. Some social and behavioral scientists even argue that field research is the *only* way we can truly understand what people do and why they do it (Jerolmack & Khan, 2014). Although I believe other forms of data collection also offer important insight into group and individual behavior (see Pugh, 2013 on in-depth interviewing), field research is a potent way to capture and explain people's situated doings. In what follows, I discuss the practice of field research for social and behavioral scientists. This chapter is largely intended as a primer on field research for the uninitiated.

4.1 Getting Started

The goal of field research is to generate descriptions of and explanations for people's behavior. Research with human subjects conducted in laboratory settings may be hampered by the fact that this type of research removes people from their natural environment and thus potentially changes how they would otherwise behave (please see Nichols & Edlund, Chapter 3 for a detailed exposition on laboratory research). Field research, in contrast, follows people into their natural settings in an effort to document group and individual behavior in place. Researchers enter the field with basic epistemological assumptions. Fundamentally, field researchers hold that the best way to understand a phenomenon is to get close to it and uncover the meanings people give to it. Field researchers value the experiential aspects of human existence. They strive to feel and experience the social world to gain insight into it. To do so requires getting out into the field. But what is "the field"?

The field may be a bounded group(s) and location(s), such as a workplace office, a school, a farm, a university dormitory, or a rural town. It can be a bounded location(s) with some regulars and numerous people coming and going, such as a hospital, a soup kitchen, a street corner, or a retail store. It may also be a group of people whom the researcher follows through space and time, such as a group of families whose days might take them to a number of places, including schools, daycares, workplaces, stores, and doctor's offices. It could also be a group of activists who meet in a variety of locations, travel to coordinate action with other activist groups, and engage in protests in a number of locales. The researcher may stay in touch with members of the group and revisit them over the course of several years. Thus, in this chapter and in field research in general, the field is a broad moniker for describing research that involves a researcher (also known as a fieldworker) entering a setting or settings bounded by the parameters of the research question.

The first step in conducting a field project is to consider what we hope to learn in the field. What are the questions anchoring the research? Although this may seem presumptuous (how can we know what we want to learn prior to conducting the research?), the idea that researchers enter the field as blank slates, with no preconceptions about what they might find in the field has largely given way to the understanding that we cannot wipe clean or ignore our training, background, and interests in conducting our research. Even research that begins with a highly open-ended approach, with the goal of building theory from the ground up (Charmaz, 2014; Corbin & Strauss, 2008) will ultimately be informed by the researcher's disciplinary training and interests. In addition, research that begins with a very narrow and focused approach may develop into a very different study over time – one the researcher would not have anticipated at the outset. Writing out research questions and what we hope to achieve by conducting the research at the beginning of a project helps us to see how we initially approached the study and how our ideas developed over time. The research question

should be broad and should encourage the researcher to attend to processes (i.e., examining *how* people act, interact, feel, and so on) and explanations (i.e., *why* people act, interact, feel). For example, a field researcher studying low-income households with limited access to food might pose the following research questions at the study's onset: How do family members experience and cope with food shortages? What are the meanings they give to their experiences and coping efforts?

Field research sometimes comes about in a serendipitous way – we happen upon an interesting scene or issue and realize there is an important story to be told that connects to topics of broad concern in the social and behavioral sciences. Sometimes researchers have theoretical questions in mind and use these questions to anchor their selection of a case or cases for field research. For example, in her classic study of men and women in non-traditional jobs (i.e., men in jobs defined as "for women" and women in jobs seen as "for men"), Williams (1989, 1991) was inspired by existing theories of gender differences. Specifically, she was troubled by how the main theoretical explanations accounting for differences between men's and women's behaviors at the time ignored social context. As she put it, "I wondered *how* these differences are constructed and, furthermore, *why* they are constructed" (Williams, 1991, p. 231, emphasis in original). Her theoretical interests led her to choose cases highly associated with masculinity and femininity with close to the same proportion of non-traditional workers in them – the Marine Corps and the nursing profession. By engaging in participant observation, conducting in-depth interviews, and doing archival research on these two cases,

Williams (1991) reached the conclusion that "gender differences are constructed by subjecting men and women to different social expectations and environments" (p. 233). By taking theory as a starting point, Williams was positioned to accomplish the fundamental aim of field research – to build theoretical explanations for how and why people behave as they do. Choosing exemplary and contrasting empirical cases, as Williams did, also helps researchers gain theoretical purchase.

One of the many benefits of doing field research is that it puts us at the center of human action, where we often discover that what we expected to find was wrong or at least not the full story. Things may not go as anticipated in the field. High degrees of uncertainty and unpredictability characterize fieldwork. Field researchers can quickly feel overwhelmed by the amount of data they collect in a short amount of time. They can feel lost in a sea of information and interesting observations with little idea of how it all fits together. They can struggle to balance the equally important and time-consuming tasks of being in the field and writing fieldnotes (more on fieldnotes below). Returning to our original research questions, even if they require reframing or rewriting (which usually they do), this helps remind us of our initial interests and how our experiences on the ground match, extend, or challenge them.

Although I focus primarily on qualitative field research in this chapter, it is worth noting that much field research is more quantitative in approach and content and may focus primarily on the collection of numeric data. For example, when I worked for a market research firm, one of my field assignments involved unobtrusively "hanging out" in the dairy aisle

of a grocery store taking count of how many people came through the section in each fifteen-minute segment. I also recorded how long each shopper spent in a particular part of the dairy aisle, how many items shoppers selected from that area and, if they picked up an item to examine it, how long they spent doing so. This kind of numeric information gives marketers insight into how best to market products based on the amount of time people are devoting to shopping.

4.2 Locating the Self in Fieldwork

Fieldworkers need to be highly thoughtful about who we are, why we want to study a particular group or phenomenon, and how we go about doing so. Fieldworkers are a part of the setting, and, moreover, they are the research instrument. It is through the field researcher's decisions, conduct, and identity that the data are produced. Thus, far from being extraneous or an act of navel gazing, researchers must necessarily know themselves and pay attention to their reactions and decisions throughout the research and analysis.

From the outset, and on an ongoing basis, fieldworkers should ask themselves regularly what it is they are trying to accomplish and how they fit into the picture. How are our motivations, expectations, dispositions, and identities playing a role in the research (Elliott, McKelvy, & Bowen, 2017)? For example, a middle-class white male fieldworker conducting an ethnography of a call center who experiences call center work as highly competitive and game-like should be attuned to how his perceptions are informed by the intersections of race, class, and gender in his biography. The ways we approach and experience the field, the decisions we make in the field, the questions we ask, the ways we characterize goings-on in the field, and our analyses are shaped by our upbringing and training. As I discuss elsewhere (Elliott et al., 2017), even something as simple as how researchers experience and characterize time in the field can be shaped by unconscious temporal perceptions and preferences stemming from our background and the disciplinary pressures we are under. The increasing pressure on academics to be productive and efficient, for example, can lead field researchers to approach the field with a highly instrumental mindset, attempting to maximize their time in the field while minimizing the time actually spent in the field. Paying attention to these dynamics can help researchers spot moments when the constraints and expectations they hold around time shape how they conceive and organize the research project as well as the ways temporal value judgements seep into the data (Elliott et al., 2017).

Kleinman and Copp (1993) also encourage field researchers to examine their emotional investments in studying a particular topic. People may choose a fieldsite because they feel drawn to it. It is important to examine why that might be. This process of self-examination helps fieldworkers uncover what might otherwise remain unconscious and unarticulated. It is by examining our goals and assumptions (our biases, if you will) that fieldworkers can be attuned to how these underlying factors may be shaping our fieldwork. The aim is not to control or "control for" our emotions but rather to be highly attuned to and thoughtful about them. The crucial role that field researchers play as a conduit for the research underpins the necessity of examining

our commitments, beliefs, and identity investments on an ongoing basis.

Although a field researcher's presence always influences the field, some field research is purposefully interventionist. Field experiments involve manipulating people in natural settings to observe how they will behave under certain conditions. This manipulation can be as minor as subjecting hotel guests to different signage about reusing towels and observing whether a given message results in people being more (or less) likely to reuse their towels (Goldstein, Cialdini, & Griskevicius, 2008). Some field experiments may involve significant deception or manipulation, such as Stanley Milgram's 1960s studies of obedience (Milgram, 1974) and Zimbardo's Stanford prison experiment (Zimbardo, 2007). These studies subjected research participants to conditions of authority and control to observe how they would behave. Akin to research done in a laboratory setting, field experiments may involve randomly assigning one group of individuals to receive an intervention while the control group does not. The key difference between field and laboratory experiments is that individuals participating in field experiments are typically in natural environments, which can improve the validity of the findings since people may be more likely to behave in ordinary ways when they are not in contrived research settings. However, in field experiments researchers have less control over their subjects than they do in a lab setting and the likelihood of data contamination is greater.

4.2.1 Fitting In: On Insiders and Outsiders

One major element of locating the self in the field is figuring out how researchers' characteristics and experiences are similar to and/or different from those they wish to study. For simplicity's sake, I will refer to this as whether researchers are insiders or outsiders to the groups they study, although I question this dichotomy below. Holding status as an insider to the group under study – that is, someone who is part of the group because they share key characteristics, values, and/or experiences – can help the researcher gain access to the fieldsite and establish relationships of openness and trust with field participants. Insiders may also have a deep understanding of the dynamics in the field, which can aid in sense-making. However, the researcher's intimate familiarity with the group can obscure just as it can illuminate. An insider researcher, for example, may struggle to notice and document certain dynamics because they seem so commonplace. Being an insider can thus mean neglecting to interrogate taken-for-granted practices and meanings that should be subjected to scrutiny. In her study of traditional healers in southern Thailand, for instance, Suwankhong, a cultural insider, reflects on how her pre-existing knowledge meant she did not always probe deeply enough because the information felt so familiar and transparent to her. Similarly, some participants were not as detailed in explaining certain beliefs or practices because they assumed no explanation was required given Suwankhong's insider knowledge (Suwankhong & Liamputtong, 2015).

Outsiders to a fieldsite – researchers who do not share in common experiences, values, or characteristics with participants in the field – may also miss or, more often, misunderstand dynamics in the field. But because everything is new and unfamiliar, outsiders may be better positioned to uncover elements that insiders might take for granted by meticulously documenting and asking questions about them. Being an outsider can make access to the field harder to accomplish though. The researcher may not know how best to approach the group to gain approval for the study and may be greeted with suspicion due to being different. Outsider researchers must spend time establishing that they are trustworthy in a bid to gain trust and rapport. Outsiders who wish to study stigmatized or oppressed groups should be prepared to demonstrate that they are aware of the stakes, such as how others, like policymakers, may use their findings to further marginalize the group. Outsiders must also be prepared to offer an account of their interests in conducting the research that resonates with those they wish to study. Even so, they should be prepared to have their commitments and values tested. For example, in her research on the heated debates over foie gras in the United States, DeSoucey (2016) had to convince both those for and against foie gras that she was trustworthy. She did this in part by forming key relationships with individuals who vouched for her and by honestly explaining her interest in foie gras as an exemplary case to examine political fights over food, not because she had a stake in one side or the other prevailing.

In short, insiders must work to make the familiar strange whereas outsiders must work to make the strange familiar. In both instances, researchers must strive for understanding. Outsiders will have to uncover and articulate that which feels highly familiar whereas outsiders may be completely flummoxed by goings-on in the field and have their work cut out for them as they try to gain close, intimate knowledge of the processes at play. Both lenses come with benefits and drawbacks that the researcher must strategize around, such as asking questions of a diverse array of field participants to check our interpretations and asking others (e.g., members of dissertation committees, peers, colleagues) to read our fieldnotes and alert us to elements we are insufficiently unpacking. When researchers participate in their own field experiments, they may also get caught up in the field dynamics, as Philip Zimbardo did by playing the role of the prison's superintendent in his now famous 1971 Stanford prison experiment. It was only when an outsider entered the experiment six days in and raised concerns about the conditions the "guards" were subjecting the "prisoners" to that Zimbardo terminated the experiment (Zimbardo, 2007). Insiders may also have to work harder within academic and non-academic communities to establish the legitimacy of their work because of the prevalent misconception that people who study what they are or do are inherently biased and cannot be objective. One response is to remind others that all research involves researchers making choices that ultimately result in data and thus it is not objectivity but rather a continuous process of interrogating our decisions, behaviors, and interpretations that is the gold standard of research.

But it is also facile to assume one is either a complete insider or outsider. Unless the researcher is studying a site for which their characteristics are a perfect match, it is likely that insiders will share some but not all similarities with the "insider" group (Suwankhong & Liamputtong, 2015; Zinn, 2001). Only rarely

do researchers study those whose identities perfectly match that of the researcher. And our researcher status makes us somewhat of an outsider, even to a group we are closely aligned with. In recognition of this, long-standing debates about field researchers as insiders or outsiders to those they study – in particular whether insiders have privileged insight (Collins, 1986; Merton, 1972) – have largely given way to the understanding that who we are and what we believe matter but not in such simplified ways as the terms "insider" and "outsider" suggest (Corbin Dwyer & Buckle, 2009). Whether one belongs or does not belong to the group under study, researchers now recognize how our identities shape the access we gain, the trust and rapport we develop, and the insights we have. Knowing this, researchers need to pay attention to and carefully document how our location shapes the ways we navigate the field, the ways others perceive and act toward us in the field, and the understandings we develop therein.

4.3 Getting In

Once field researchers decide on a topic, develop research questions, and establish the case or cases they wish to examine, the next step is to figure out how to gain access to the field. A beautifully crafted research project will fall apart if researchers are unable to get in, that is, if they cannot get close to the people and interactions they wish to understand. How we gain access to the field will also shape what we are, and are not, privy to. A study of workplace interactions will have different vantage points, for example, depending on whether the researcher gains access through management to observe employees at work and to sit in on managers' meetings or whether the researcher gains access by getting a job and joining the workforce. There are many ways to gain access to a setting. Some researchers hang around and become known to those in the field before asking for permission to study the site. Some gain access through a friend or colleague (or a friend of a friend of a friend) who acts as a key informant and gatekeeper, making introductions in the field and smoothing our entrance to it. Depending on the type of fieldsite a researcher has chosen, they may need to take a more formal route, writing letters of introduction, phoning, and requesting face-to-face meetings in the hopes of finding individuals willing to accept a researcher in their midst.

For a novice fieldworker, covert research – entering the field *not* as a known researcher but as a (seemingly) full-fledged participant or member of the setting – may seem to solve a lot of problems. The novice fieldworker may think people won't change their behavior if they don't know they are being observed and that it solves the thorny issue of gaining access. However, covert research comes with important ethical implications and downsides. Covert researchers may not ask questions of those in the field out of fear of being "found out" and thus may lack important information about people's motivations and sense-making. Covert researchers may feel so deeply conflicted about the ethics of lying to people in the field that it affects the ways they view the field and write about it (on how emotional responses matter for fieldwork, see Kleinman & Copp, 1993). Once out of the field, covert researchers may rightly worry that others will discover their research and feel betrayed, which can shape their ability to analyze the data cogently. Also, even if they are not "out" as a researcher in the field, merely by their presence, covert researchers influence and change the dynamics they witness. In short, while it may seem appealing to bypass the whole effort of gaining access through overt and legitimate channels, covert research should be treated as a tactic of last resort – when there is no possible way to gain access to an important and worthy fieldsite other than by concealing one's identity as a researcher. Covert researchers should be prepared to offer a compelling explanation to justify this choice to the public. In addition, most universities require ethics approval for research. Researchers who wish to conduct covert research will need to provide a strong rationale for going covert and be able to demonstrate that the benefits of the research outweigh the risks and drawbacks (Calvey, 2017). For more discussion on ethical considerations, see Ferrero and Pinto, Chapter 15.

4.4 Writing Fieldnotes

The goal of field research is to develop what Geertz (1973) called "thick description." Ultimately, we want to bring readers of our research into the topic and scene to help them understand what people do in specific settings and contexts, and why they do it. How do we accomplish this? The simple answer is we write detailed fieldnotes which form the basis for our eventual analysis. Writing fieldnotes requires us to recall and record with as much precision as possible what we witnessed in the field. Fieldworkers must use all of their senses in the field to offer an accurate account in their fieldnotes of what they not only saw and heard but also smelled, touched, and perhaps even tasted, as well as what they felt and perceived. In writing fieldnotes, field researchers try at all times to document, in detail, the information that led them to form certain conclusions. Rather than write that someone "seems tired," for example, the fieldworker should provide the empirical evidence they used to form this conclusion (e.g., bags under eyes, stooped posture, slow gait). If there is a chair in the room, what kind of chair is it? If someone makes a face, what does the expression look like? If we characterize the atmosphere a certain way (e.g., gloomy, upbeat), what information led us to this impression? Did we check our perception by asking others in the field how they perceived the atmosphere?

Fieldnotes document, in rich detail, how the setting and people in it appear, what people say, how they say it, and who they say it to, as well as what people do, how they do it, and with whom they do it. In short, fieldnotes must capture the environment – this includes spatially mapping out the setting and where people and objects are located in it – and the actions and interactions of individuals in the setting. How do they look, act, and interact? How are people expected to behave in this setting? Where do these expectations come from? Do all people behave this way? What happens when someone doesn't behave accordingly? Field researchers should also pay close attention to absences in the field. Who isn't present? What isn't said? Who doesn't talk or interact with whom? What doesn't appear in the fieldsite (e.g., what isn't represented in the artwork on the walls?)? Where don't participants go? What don't they do? Who don't they spend time with? Social and behavioral scientists want to understand human behavior and interaction. Questions such as these force us to drill down into the social world – what is and is not permissible to say, do, or be in a particular setting.

Before launching into a field research project, researchers should practice and hone their fieldnote writing skills. This might involve spending an hour or two at a public venue, such as a coffee shop, library, bar, or grocery store, and writing up our observations afterwards. In writing fieldnotes, researchers should pay attention to what they remember and what they struggle to recall. Returning to the scene with fieldnotes in hand can help us check our observations. Although the people in the setting may not all be the same, we can determine the extent to which we correctly described the environment. Fieldworkers should continue in this vein, extensively practicing and sharpening their observational, recall, and writing skills prior to conducting fieldwork.

Our memories are notoriously unreliable. It is helpful to bear this in mind. If we forget or neglect to document a detail, we may be tempted to make it up. Thus, field researchers must develop techniques to help with recall, such as writing jottings in the field of snippets of conversations, sequences of events, characteristics of the setting, and other important details (for a helpful guide to jottings and other aspects of fieldnote writing, see Emerson, Fretz, & Shaw, 1995). Some fieldworkers train themselves to record the scene as if their minds were video cameras which they play back as they write fieldnotes afterwards. I prefer to write brief jottings in the field and copious handwritten notes as soon as I've left the field. Given the type of fieldwork I do, this typically involves pulling my car over to the side of the road after leaving the field and scribbling notes for thirty minutes or more. Once home, I type up comprehensive field-notes using my jottings and the notes I took after leaving the field as guides. Some field-workers orally record their observations into a digital recorder upon leaving the field that they play back as they write fieldnotes. What-ever practice fieldworkers adopt, it is crucial to write a complete set of fieldnotes as soon as possible after spending time in the field, with-out talking to others, checking our email, or doing anything else that might cloud our recall.

Writing fieldnotes can be exhilarating and cathartic. It can feel good to get our observa-tions down on paper. But writing fieldnotes can be equally tedious and exhausting. It can be the last thing we want to do after being in the field (which is tiring itself). Yet it is essen-tial to prioritize fieldnote writing from the outset and throughout our time in the field. For every hour in the field, a field researcher can expect to spend at least two to three hours writing fieldnotes and notes-on-notes. Notes-on-notes, discussed in more detail below, are essentially reflections, ideas, and questions we have about what is going on in the field.

How much researchers include themselves in their fieldnotes is an ongoing question for the practice of field research. Field researchers are part of the field. Our presence fundamen-tally changes the field. From loaning people money, offering them rides, helping with chores or work duties, eating meals together, offering opinions, asking questions, sharing our stories, and simply being there, our pres-ence in the field means that what happens is different than if we were not there. Moreover, what we see, how we feel, and what we do are key to what we learn. Because of this, we must necessarily write ourselves into our fieldnotes. This does not mean that everything in our fieldnotes should be about us. We must be equally invested in recording the details of the scene and what others do and say. As with much of fieldwork, figuring out how to include ourselves in our fieldnotes without sacrificing other important details is an ongoing practice and balancing act.

In addition to writing fieldnotes, many field researchers keep other written records, such as a methods journal in which they record the methodological decisions they make. It is also advisable to keep a log that records basic information, such as hours spent in the field, the number of times the researcher visited a particular location, the number of people pre-sent during each field visit, and so on. Field researchers benefit when they develop good recall and writing techniques as well as strong organizational practices. I also use a spread-sheet to track numeric information such as how many times I observe a particular inci-dent or see a particular individual in the field. This might involve, for example, recording each time I observe racism in the field and providing a cross-reference so that I can easily access this information in my fieldnotes.

4.5 A Note on Ethics

Embroiling ourselves in the lives of others can be ethically "sticky." For example, whereas trust and rapport are important to field research, what happens when partici-pants come to trust the researcher so much they forget they are being researched? What happens when a participant divulges some-thing that may put another participant in harm's way? Do we breach promises of confidentiality by sharing that information with others? What happens when people in our field experiments begin behaving in ways that potentially pose a danger to themselves or others? The dynamics and entanglements of fieldwork demand that fieldworkers be prepared to encounter and deal with many murky ethical dilemmas as they emerge (see also Ferrero & Pinto, Chapter 15). The field researcher's ethical code should involve the following basic precepts: Do no harm to those in the field, to one's profession and institutional affiliation(s), and to oneself. However, there are good reasons for not forming absolutist principles around our conduct in the field. Often, the best course of action will depend on the context. Field researchers must take seriously the ethical implications of their work and be prepared to confront and deal with ethical issues on the ground in ways that minimize harm to self, participants, and the discipline. However, because our behaviors must mesh with the demands and context of the situation, a clear blueprint for such ethical conduct does not exist. It is worth bearing in mind at all times that field researchers must be able to defend the ethical decisions they made in the field to others in the academic community and general public. If we cannot come up with a credible defense for doing something that others might view as unethical (e.g., lying to participants or engaging in illegal activity in the field), then we should not do it.

Our ethical obligations extend from our conduct in the field to analyzing and publishing the data. For example, researchers should be attuned to how others might use our research in unanticipated ways and in ways that could negatively impact the people we researched (Wilk, 1997). In writing and publishing, the fieldworker also needs to carefully consider issues around participants' privacy and confidentiality. If we promise research participants that their identities will be concealed such that no one will identify them in publications, then we must take seriously the ways we protect participants' anonymity by obscuring their identities while still providing an accurate rendering of them. Social media, professional profiles, and a plethora of other online information mean that those who wish to unmask the identities of our research participants have powerful tools available to them, intensifying both the importance of and challenges around our obligation to protect participants' confidentiality.

Ethical conduct is crucial for our discipline and the continued efficacy and vibrancy of field research. When researchers violate codes of ethical conduct, they don't just risk causing harm to those in the field or to themselves, they also risk damaging the profession. If people develop a general opinion of field researchers as untrustworthy or unethical, they may not agree to participate in field research. Additionally, if researchers don't feel they have handled well the ethical dilemmas they have encountered in the field, they may experience writer's block or what Kleinman and Copp (1993) describe as analysis avoidance – the unwillingness or inability to analyze the data. Researchers who feel deeply concerned about their conduct in the field may not be able to think and write cogently about their research because of lingering feelings of unease. Hence, if a researcher is not convinced that ethical conduct is important from a purely moral or even professional standpoint, it may still influence our insight and block our analytic process.

4.6 Developing an Analysis

Analysis involves an ongoing process of asking questions of the data, forming tentative ideas, checking those ideas by gathering more data and asking questions in the field, and gradually developing a coherent argument. Analysis is not a linear or straightforward process. It requires us to spend a great deal of time reading and coding our fieldnotes, writing notes-on-notes about ideas we are forming or hunches we may have that guide us to collect more data, and writing memos that deepen the analytic process by forcing us to clarify our insights and link them to specific pieces of data. If I notice a pattern occurring in the field, I will search for key terms related to the pattern in my fieldnotes, but I also pay attention to exceptional occurrences as they may be rare in the field but nevertheless potentially highly illustrative. The process of identifying themes and patterns is often a precursor for coding the fieldnotes, which involves reading them line-by-line and assigning thematic codes to begin to sort and organize the data into analytic categories (Emerson et al., 1995).

In addition to identifying patterns and themes in the data, analysis also entails sense-making and interpretation. A note-on-note might set the stage for further exploration of why a particular dynamic is occurring in the field. For example, in a brief note-on-note, I might observe: "Rereading my fieldnotes, I notice boredom keeps coming up. Why? Are people feeling bored in the field? Am I feeling bored? What does it mean to be bored in this setting? I need to pay closer attention to processes around boredom." A memo about boredom would involve a much more elaborate analysis of when boredom arises, who expresses feelings of being bored (and who doesn't), and, crucially, how boredom is characterized and why (Elliott et al., 2017).

Asking "how" (how do people act?) and "why" (why do people behave as they do?) questions of our data primes us to think of our research in explanatory terms, which is the goal of analysis. In offering an explanation, field researchers are not attempting to provide data generalizable in a demographic sense (i.e., we are not trying to say that all people who are similar to those we observed act in a certain way), but rather to offer explanations that connect to broader processes, such as the construction of gender differences (Williams, 1989; 1991) or the reproduction of race and class privilege. A field study of a handful of white fathers and their families, for example, cannot definitively claim that all affluent, progressive-minded white fathers racially socialize their children in one particular way, but it can illuminate "*how* these fathers understand their role as a white father, *how* their attempts to raise antiracist children both challenge and reinforce hegemonic whiteness, and *what* role race and class privilege play in this process" (Hagerman, 2017, p. 60, emphasis added).

Field research starts from the premise that we must get close to something to understand it, but what happens when we get so close that we take on the perspectives of those in the field as our own? In an effort to gain intimate familiarity with the setting, people, and processes, fieldworkers want to and, indeed, *must* immerse themselves in field research. At the same time, the analytic process requires fieldworkers to critically assess what is going on and why. When we unconsciously adopt the attitudes and concerns of our participants, it can stymie our analysis. For example, in his study of a domestic violence and sexual assault service organization, Kolb (2014) became a volunteer and found himself, over time, adopting and embracing at face value the terminology frequently used in the organization, such as the concept of empowerment. Rather than subjecting the term to critical evaluation, he writes, "my field notes and notes-on-notes were filled with speculation about whether their services were 'empowering.' And if they were not, I wondered how they could be changed to better 'empower' clients" (Kleinman & Kolb, 2011, p. 432). Kolb essentially got caught up in the questions that were of concern to his research participants rather than subjecting them to analytic interrogation. Thus, field researchers should pay attention to how their interests may shift over time in ways that reflect the worldviews of those in the field and potentially cloud researchers' perception of events in the field as well as their analytic process. Although analysis is necessarily an ongoing process, time away from the field can sometimes help us to better see our imbrication in field dynamics and provide us with important insight that enriches the analysis.

4.7 Getting Out

A field study can last a few weeks to a few years and can involve sporadic trips to the field or a prolonged stay. How much time a researcher spends in the field depends on the type of fieldwork (e.g., whether it is intermittent or immersive) and the research questions, as well as things beyond the researcher's control, like funding for fieldwork (or lack thereof), time for fieldwork (or lack thereof), and whether the field continues to exist (e.g., a volunteer group might dissolve or a debate might be resolved). The goal should be to stay in the field long enough to gain a deep understanding of it, which generally means no new information is arising and the researcher knows *what* is happening and has developed explanations for *why* it is happening.

Even after leaving the field, researchers often maintain ties to individuals in the field, allowing them to check in and get updates on an ongoing basis. Some researchers enter the field for an immersive period, leave, and come back at a later date. Leaving the field and returning to it (i.e., recurring fieldwork) can be helpful for providing researchers with distance from the field, time for analysis, and the opportunity to test and expand our ideas with further fieldwork. Elsewhere, I discuss how the professional pressures of academia, such as time constraints for data collection and publishing imperatives, may require fieldworkers to carve out bounded time periods for doing fieldwork, with consequences for how they approach and may characterize the field (Elliott et al., 2017). Whatever way we leave, or get out of, the field, we should use grace and tact, showing our appreciation for being granted the privilege of sharing in the lives of others.

4.8 Looking Ahead

As field research continues to improve and expand, there are several issues to keep an eye on. First, field researchers are increasingly calling attention to temporal and spatial dynamics in the field as well as to how time and space are interconnected and overlapping (what Patil, 2017 calls "webbed connectivities"). Being aware of and dealing with these will be important. In addition, new technologies continue to enhance the field researcher's toolkit (Hallett & Barber, 2014; Pink, Horst, Postill, Hjorth, Lewis, & Tacchi, 2016), while posing ethical and practical conundrums. Field researchers will have to find the right balance between doing ethical work and collecting rich data. Community-based and participatory-action projects – research by and for participants and their communities (Hollowell & Nicholas, 2009) – have also begun to raise important questions about the long-term responsibilities of field researchers to those we study. Finally, field researchers are starting to question the ways human action and subjectivity have been privileged in field accounts (see Todd, 2014, 2017 for an exposition of human–fish relationships, for example). It will be important to consider these contemporary issues when conducting field research. For those interested in investigating these issues, I provide some questions for future research at the end of this chapter.

4.9 Conclusion

Human behavior is fascinating and complex. Field research allows social and behavioral scientists to observe and participate in social action with an eye to explaining how and why people behave as they do. Fieldwork can be exciting, but it can also be time-consuming, emotionally and physically draining, and unpredictable. Fieldworkers must use all of their senses, documenting in rich detail in fieldnotes what they see, hear, taste, smell, and touch, as well as what they feel and perceive in the field. They must check their perceptions and interpretations by gathering more data and asking questions of those in the field. We write ourselves into the research because our presence shapes the data we collect and our training, interests, and background influence the orientations we hold and the interpretations we draw. This imbrication means researchers must be very thoughtful about the choices they make in and out of the field. Field researchers must also be prepared to encounter and deal with the many complex ethical conundrums that accompany observing and participating in the lives of others. We build ideas about what we think is going on through reading and coding our fieldnotes and through the practice of writing notes-on-notes and analytic memos. Finally, we develop a written account that brings readers into the setting(s) we have studied and offers an explanation for an aspect of social life and human behavior. Field research is intense and demanding, but highly rewarding. Moreover, done carefully and thoughtfully, it is a powerful method for advancing understanding of the social world.

KEY TAKEAWAYS

- Cultivate strong recall and descriptive writing skills.
- Pay attention to and document one's reactions and social location in the field.
- Capture detailed descriptions of settings, actions, and interactions *and* broader observations and reflections about what is happening in the field.

- Be mindful of the ethical commitments of field researchers.
- Focus on *how* and *why* questions in analysis.

IDEAS FOR FUTURE RESEARCH

- How do team-based field studies differ from individual-based field research?
- How do we situate and historicize our fieldsites in ways that transcend simple dichotomies such as past and present, local and global, online and offline?
- How do new technologies (digitally driven) meaningfully change the way field research is conducted?
- How can we document and theorize human, land, and non-human relations as mutually reciprocal?
- What are the social justice obligations of field researchers?

SUGGESTED READINGS

Calvey, D. (2017). *Covert Research: The Art, Politics and Ethics of Undercover Fieldwork*. London: Sage.

Corbin Dwyer, S., & Buckle, J. L. (2009). The space between: On being an insider-outsider in qualitative research. *International Journal of Qualitative Methods*, 8(1), 54–63.

Emerson, R., Fretz, R., & Shaw, L. (1995). *Writing Ethnographic Fieldnotes*. Chicago, IL: University of Chicago Press.

Kleinman S., & Copp, M. A. (1993). *Emotions and Fieldwork*. Thousand Oaks, CA: Sage.

Kleinman, S., & Kolb, K. H. (2011). Traps on the path of analysis. *Symbolic Interaction*, 34(4), 425–446.

REFERENCES

Calvey, D. (2017). *Covert Research: The Art, Politics and Ethics of Undercover Fieldwork*. London: Sage.

Charmaz, K. (2014). *Constructing Grounded Theory: A Practical Guide Through Qualitative Analysis* (2nd ed.). Thousand Oaks, CA: Sage.

Collins, P. H. (1986). Learning from the outsider within: The sociological significance of black feminist thought. *Social Problems*, 33, 14–32.

Corbin, J., & Strauss, A. (2008). *Basics of Qualitative Research* (3rd ed.). Thousand Oaks, CA: Sage.

Corbin Dwyer, S., & Buckle, J. L. (2009). The space between: On being an insider-outsider in qualitative research. *International Journal of Qualitative Methods*, 8(1), 54–63.

DeSoucey, M. (2016). *Contested Tastes: Foie Gras and the Politics of Food*. Princeton, NJ: Princeton University Press.

Elliott, S., McKelvy, J. N., & Bowen, S. (2017). Marking time in ethnography: Uncovering temporal dispositions. *Ethnography*, 18(4), 556–576.

Emerson, R., Fretz, R., & Shaw, L. (1995). *Writing Ethnographic Fieldnotes*. Chicago, IL: University of Chicago Press.

Geertz, C. (1973). *The Interpretation of Cultures: Selected Essays*. New York: Basic Books.

Goldstein, N. J., Cialdini, R. B., & Griskevicius, V. (2008). A room with a viewpoint: Using social norms to motivate environmental conservation in hotels. *Journal of Consumer Research*, 35(3), 472–482.

Hagerman, M. A. (2017). White racial socialization: Progressive fathers on raising "antiracist" children. *Journal of Marriage and Family*, 79, 60–74.

Hallett, R. E., & Barber, K. (2014). Ethnographic research in a cyber era. *Journal of Contemporary Ethnography*, 43(3), 306–330.

Hollowell, J., & Nicholas, G. (2009). Using ethnographic methods to articulate community-based conceptions of cultural heritage management. *Public Archaeology*, *8*(2–3), 141–160.

Jerolmack, C., & Khan, S. (2014). Talk is cheap: Ethnography and the attitudinal fallacy. *Sociological Methods & Research*, *43*(2), 178–209.

Kleinman S., & Copp, M. A. (1993). *Emotions and Fieldwork*. Thousand Oaks, CA: Sage.

Kleinman, S., & Kolb, K. H. (2011). Traps on the path of analysis. *Symbolic Interaction*, *34*(4), 425–446.

Kolb, K. H. (2014). *Moral Wages: The Emotional Dilemmas of Victim Advocacy and Counseling*. Berkeley, CA: University of California Press.

Merton, R. (1972). Insiders and outsiders: A chapter in the sociology of knowledge. *American Journal of Sociology*, *78*(1), 9–47.

Milgram, S. (1974). *Obedience to Authority: An Experimental View*. New York: HarperCollins.

Patil, V. (2017). Sex, gender and sexuality in colonial modernity: Towards a sociology of webbed connectivities. In J. Go & G. Lawson (Eds.), *Global Historical Sociology* (pp. 139–155). Cambridge: Cambridge University Press.

Pink, S., Horst, H., Postill, J., Hjorth, L., Lewis, T., & Tacchi, J. (2016). *Digital Ethnography: Principles and Practice*. London: Sage.

Pugh, A. J. (2013). What good are interviews for thinking about culture? Demystifying interpretive analysis. *American Journal of Cultural Sociology*, *1*(1), 42–68.

Suwankhong, D., & Liamputtong, P. (2015). Cultural insiders and research fieldwork: Case examples from cross-cultural research with Thai people. *International Journal of Qualitative Methods*, *14*(5), 1–7.

Todd, Z. (2014). Fish pluralities: Human–animal relations and sites of engagement in Paulatuuq, Arctic Canada. *Etudes/Inuit/Studies*, *38*(1–2), 217–238.

Todd, Z. (2017). Fish, kin, and hope: Tending to water violations in amiskwaciwâskahikan and Treaty Six Territory. *Afterall: A Journal of Art, Context and Inquiry*, *43*(1), 102–107.

Wilk, R. (1997). *Household Ecology: Economic Change and Domestic Life Among the Kekchi Maya of Belize (with a new preface)*. DeKalb, IL: Northern Illinois University Press.

Williams, C. L. (1989). *Gender Differences at Work: Women and Men in Non-Traditional Occupations*. Berkeley, CA: University of California Press.

Williams, C. L. (1991). Case studies and the sociology of gender. In J. Feagin, A. Orum, & G. Sjoberg (Eds.), *A Case for the Case Study* (pp. 224–243). Chapel Hill, NC: University of North Carolina Press.

Zimbardo, P. (2007). *The Lucifer Effect: Understanding How Good People Turn Evil*. New York: Random House.

Zinn, M. B. (2001). Insider field research in minority communities. In R. M. Emerson (Ed.), *Contemporary Field Research: Perspectives and Formulations* (2nd ed., pp. 159–166). Long Grove, IL: Waveland Press.

5 Using the Internet for Research

Rosanna E. Guadagno

The headlines were shocking, questioning, and resigned: "Facebook's Unethical Experiment" (Waldman, 2014); "Do You Consent? If tech companies are going to experiment on us, they need better ethical oversight" (Grimmelmann, 2015). These were representative of the wide-ranging responses to news that Facebook, in conjunction with Cornell University scholars, had published a study in which nearly 700,000 Facebook users' news feeds were manipulated to examine emotional contagion – whether people caught the emotions of their Facebook friends' posts (Kramer, Guillory, & Hancock, 2014). Specifically, people's news feeds were manipulated to reduce either the number of positive or negative words expressed by their Facebook friends. The results revealed that emotional contagion did indeed occur on Facebook – people exposed to fewer negative emotions expressed significantly more positive and fewer negative emotions in their posts, and people exposed to fewer positive emotions expressed more negative and fewer positive emotions.

Once this study was published and subsequently publicized, the backlash from consumers and the media was swift and severe, as the only statement in the paper addressing ethical considerations of this experiment stated that: "consistent with Facebook's Data Use Policy, to which all users agree to prior to creating an account on Facebook, constituting informed consent for this research" (Kramer et al., 2014, p. 8789). This study also ignited intense debate within the social and behavioral science community as to whether the study was ethical, prompting the outlet that published the study to issue an "Editorial Expression of Concern" written by the Editor-in-Chief (Verma, 2014). Specifically, the statement suggested that, while it was unclear whether participants provided informed consent and had the choice to opt out of the study, the statement suggested that because Facebook is a corporation, it was not obligated to follow the same ethical guidelines for data collection that scholars at federally funded institutions must follow. For social and behavioral scientists, this example highlights some of the key ethical considerations that should be taken into account when conducting research via the internet (see Ferrero & Pinto, Chapter 15 for a detailed exposition on ethical issues in the social and behavioral sciences).

5.1 Overview of Present Chapter

The internet has existed since the 1960s, having initially been developed as a means for communication between US scientists. Since then, it has grown exponentially, both in the number of users as well as in the number of uses for the technology. Today's predominant internet technologies – social media, smartphones, and apps – emerged in the mid-2000s and have been widely adopted globally. Where the people go, the social and behavioral scientists follow, and, to date, thousands of books and journal articles include research conducted on the internet using a wide range of research methodologies and theoretical perspectives. The present chapter will review this literature, exploring issues such as: What does it mean to conduct research on the internet? What topics do social and behavioral scientists study on the internet? And what are the different approaches for conducting research on the internet? A discussion of ethical considerations for internet research and suggestions for improving these methodologies conclude this chapter.

5.2 Conducting Internet Research

Literature reviews on internet research have focused on the prevalence, concepts commonly examined, and research methodologies employed to conduct this research (e.g., Gosling & Mason, 2015; Peng, Zhang, Zhong, and Zhu, 2013). Peng et al. (2013) reported that, between 2000 and 2009, there were 27,340 published papers examining social behavior on the internet. Similarly, a survey of Internal Review Board (IRB) members indicated that 94 % of respondents reported that an online study was the most common methodology reviewed by the IRB (Buchanan & Hvizdak, 2009). A February 2018 search of Google Scholar, conducted by this author for articles mentioning the internet, revealed over 94,000 articles published in psychology journals alone. Thus, it appears that conducting research on the internet has become increasingly common in the social and behavioral sciences.

5.2.1 Why Do Social Scientists Conduct Internet Research?

Scholars have pointed to several advantages for conducting research on the internet: 1) the discovery of novel findings; 2) the ability to collect large sets of data quickly and easily; 3) increased access to larger, more diverse, non-college student, non-WEIRD (Western, Educated, Industrialized, Rich, and Democratic; Henrich, Heine, & Norenzayan, 2010), and/or specialized samples, such as parents against vaccinating their children (Cheung, Burns, Sinclair, & Sliter, 2017; Gosling & Mason, 2015). Finally, hosting studies online saves resources (e.g., data entry and paper) and can provide immediate feedback to both researchers and participants.

A key issue explored in the early days of internet research was the question of whether data obtained from online samples would be comparable to traditional offline samples. For instance, a myriad of studies suggest that people are more likely to self-disclose in online contexts relative to offline, face-to-face contexts (e.g., Nguyen, Bin, & Campbell, 2012). Other research has shown that traditional versus online samples differ in other significant ways. Chang and Krosnick (2009) reported greater accuracy in responses from an internet sample compared with random-digit phone dialing. Similarly, Casler, Bickel, and Hackett (2013) sampled from two internet sources (Amazon's Mechanical Turk and

Social Media) versus a college student sample and reported that, other than a greater diversity in background among the online samples, the results were indistinguishable. They conclude that the anonymity afforded by the internet is largely beneficial for accurate survey responses. Overall, the literature suggests that data collected online often produce comparable results to traditional lab studies with some noticeable advantages of internet-based research in the diversity of samples and amount of self-disclosure observed.

5.2.2 What Topics Do Social Scientists Study on the Internet?

Peng et al. (2013) reviewed the research published on the internet between 2000 and 2009 and concluded that literature from this time span focused on four general topics: e-Health, e-Business, e-Society, and Human–Technology interactions. Similarly, Gosling and Mason (2015) provided examples of types of internet research. For instance, online personality quizzes (Guadagno, Okdie, & Eno, 2008) are an example of translational research. Virtual environments (Guadagno, Swinth, & Blascovich, 2011), cyberbullying (Wingate, Minney, & Guadagno, 2013), crowdsourcing (Miller, Crowe, Weiss, Maples-Keller, & Lynam, 2017), internet addiction (Anderson, Steen, & Stavropoulos, 2017), and rumor spreading (Kwon, Cha, & Jung, 2017) all fall under the rubric of phenomenological research on the internet. Finally, research on the viral spread of emotions, behaviors, and internet content (Guadagno, Rempala, Murphy, & Okdie, 2013) and the analysis of text in social media posts (Mitra, Counts, & Pennebaker, 2016) are illustrations of novel internet

research. Thus, the answer to this question is broad and has evolved along with internet-based technology.

Similarly, the types of methodologies employed by internet researchers have also developed and changed over time and in response to technological advances. Early research on the internet (i.e., pre-2009) generally fell into three categories: 1) simple surveys (Gosling & Bonnenburg, 1998); 2) experiments on computer-mediated communication (CMC) using computers networked to each other rather than the internet (Guadagno & Cialdini, 2002; 2007); and 3) simple experiments using custom-built computer programs (Fraley, 2004). These practices too have changed over time as easy to use point-and-click options for creating and deploying surveys and experiments became the norm (e.g., Qualtrics and Google Forms are two options widely used by social and behavioral scientists).

5.3 Crowdsourced Data Collection

Introduced in 2006, the phrase "crowdsourcing" refers to: "the practice of obtaining needed services, ideas, or content by soliciting contributions from a large group of people and especially from the online community rather than from traditional employees or suppliers" (Merriam-Webster's, n.d.). Examples of crowdsourcing include websites that aggregate movie ratings (e.g., rottentomatoes.com), reviews of businesses and services (e.g., Yelp), and applications that aggregate driver's input on road conditions to improve the quality of GPS directions (e.g., Waze). Some crowdsourcing applications have been used by researchers to recruit participants into their internet studies. The most widely used variant of this is Amazon's Mechanical Turk, so the next section will focus on it (although there are alternatives such as Crowdflower, Clickworker, and SurveyMonkey).

5.3.1 Amazon's Mechanical Turk (MTurk)

MTurk works because people (MTurk "workers" or MTurkers) compete HITs ("human intelligence tasks") for small amounts of money. Researchers ("requesters") seek assistance on their tasks by posting HITs – typically a link to a survey or experiment – for workers to complete. For new researchers, guides exist to instruct interested scholars on the ins and outs of using MTurk for data collection (Miller et al., 2017). Requesters can select participants based on demographic characteristics such as country of residence, age, gender, income, education, voting behavior, employment sector, internet use patterns, lifestyle habits, and more. Thus, researchers can use MTurk to recruit both diverse samples as well as more specialized samples relative to other options (e.g., university subject pool).

Mounting evidence suggests that MTurk is becoming increasingly common as a means of data collection across most social and behavioral science fields, particularly psychology. For instance, Chandler and Shapiro (2016) searched Google Scholar for papers published between 2006 and 2014, containing the phrase "Mechanical Turk," and found over 15,000 citations. Similarly, Zhou and Fishbach (2016) reviewed data on the percentage of papers published in three top tier psychology journals between 2012 to 2015 and reported a 10% to 30% increase in the inclusion of at

least one MTurk study in papers published in each journal across time.

Although MTurk provides access to research participants from more than 100 countries, most scholars limit their samples to the United States (Chandler & Shapiro, 2016). After the United States, India is the most prevalent source of MTurk workers (Paolacci, Chandler, & Ipeirotis, 2010). Similarly, other research has shown that MTurk participants are generally younger, poorer, more politically liberal, and less religious (Berinsky, Huber, & Lenz, 2012). Casler, Bickel, and Hackett (2013) and others have reported that MTurk provides access to more diverse samples with respect to age, ethnicity, and socio-economic status and yields research results that are typically comparable and/or superior. Furthermore, Buhrmester, Kwang, and Gosling (2011) surveyed MTurkers and found the following reasons why participants accept HITs: to 1) enjoy interesting tasks, 2) kill time, 3) have fun, 4) make money, and 5) gain self-knowledge. Thus, while crowdsourcing is a convenient way to collect data, understanding why people participate in crowdsourced research is essential for conducting quality research using the methodology.

Miller et al. (2017) proposed the following advantages of sampling from MTurk: it's efficient; it provides both diverse and specialized sample selection; the service is flexible in the types of research it can support; and it generally yields high-quality data. Evidence on the quality of data collected via MTurk relative to other, more traditional, offline sources generally demonstrates that the quality of data collected on MTurk is broadly comparable (Buhrmester et al., 2011). However, there have been notable exceptions with some evidence suggesting that collection via MTurk results in better- (e.g., Beymer, Holloway, & Grov, 2018) or poorer-quality data (e.g., Rouse, 2015) and that this outcome is affected by the way a study is posted on MTurk (Zhou & Fishbach, 2016). For instance, Rouse (2015) provided evidence that the use of attention checks – questions that are added into a survey to ensure that participants are paying attention (e.g., "please select the number 6 on the scale below") ameliorated the problem. Similarly, Buhrmester et al. (2011) reviewed data on the length of the task and compensation, finding that shorter surveys with higher compensation resulted in faster and larger recruitment of participants.

Limitations of sampling from MTurk include a small worker pool, non-naïve participants, awareness of attention checks, and high, differential attrition (Miller et al., 2017). Specifically, Cheung et al. (2017) noted that while attention checks, such as the one described above, are useful for data cleaning, non-naïve participants may become aware of attention checks. Attrition – when participants start an experiment but do not complete it – is also a concern for samples obtained from MTurk. Zhou and Fishbach (2016) reviewed studies published in the *Journal of Personality and Social Psychology* between 2014 and 2015, reporting the results of data collected on MTurk. Of 289 individual studies, only six reported the dropout rates from their samples and that the attrition rate was over 30%. The authors then replicated several classic experiments online and found markedly different attrition across conditions (e.g., 77.6% vs. 22.8%) and that this differential attrition affected the validity of results. The authors also found that, among other things, asking participants to commit to completing the study at the outset decreased attrition rates.

5.3.2 Snowball Sampling

Snowball sampling occurs when researchers use their initial participants to recruit others by sharing invitations to take part either via social media or other mediated communication methods or through verbal word of mouth. Research suggests that one way in which snowball sampling can be successful is through virality – a common way in which information spreads online (Ellis-Barton, 2016). For example, one experiment examined whether people could influence their friends to agree to join a social media group (Kwon, Stefanone, & Barnett, 2014). In this study, participants received one or more requests from friends in their social networks. Results revealed that the following characteristics increased participant responses: If they were women, if they received more than one group invitation, and if their social media friends had also joined the group. Thus, this study revealed that not only can snowball sampling be effective in recruiting participants, but also there are notable situational determinants on the extent to which snowball sampling effectively recruits new participants. Young, Shakiba, Kwik, and Montazeri (2014) used snowball samplings to examine the link between social media use and religious veil-wearing in Iran, finding that younger people and longer-term social media users were less likely to endorse wearing a veil. Thus, as these studies illustrate, snowball sampling can be an effective means of recruiting participants. However, snowball sampling does not randomly sample, and this can be a threat to the validity of research findings obtained using this method (see Wagner & Skowronski, Chapter 2 for a detailed exposition on validity issues).

5.4 Big Data: Collecting Data Directly from People's Internet Activity

Another trend that has emerged as the internet has transformed is the use of people's internet activities to predict their future behavior. An early example of this occurred in the mid-2000s when the retailer Target started employing statisticians, called Data Scientists, to predict life events from store purchases to better target (no pun intended!) their advertising. During this time, I was pregnant and was astounded by Target's ability to predict my pregnancy and child-rearing needs by promoting products just before I needed them. It was as if Target could read my mind. Instead, they were measuring my behavior and using this data to predict likely life events based on what I was purchasing at Target. This may seem like an innovative approach to customer service, but it turned into a public relations nightmare as parents started receiving pregnancy- and baby-related coupons in the mail addressed to their underage daughters (Hill, 2012), ending this early foray into Big Data.

Big Data refers to datasets that are far larger than contemporary data-analysis software can analyze within a reasonable timeframe (Snijders, Matzat, & Reips, 2012). While there is no clear-cut distinction between "big" data and "small" data (e.g., it makes no sense to arbitrarily designate that a dataset with 17,206 participants is a small dataset, while a dataset with 17,207 participants is a big dataset), Big

Data generally describes datasets that have so many participants and/or variables that traditional data-analysis programs (e.g., SPSS, SAS) cannot analyze the data due to the number of variables or the magnitude of the file storing the data. Data Scientists instead use computer programming languages such as Python and R to analyze large datasets.

At its core, Big Data is about predicting future outcomes from past behavior (Mayer-Schönberger & Cukier, 2013). Although Big Data is often presented as a subset of computer science referred to as **machine learning**, many social and behavioral scientists use these methods as well. Machine learning processes large amounts of data to infer probable outcomes such as the likelihood that a person would enjoy a specific song based on what previous songs they have listened to and/or liked or selecting recommended products based on people's previous online shopping history. This method reflects three shifts in traditional approaches to research: 1) entire populations being studied; 2) a reduction in sampling error with a corresponding increase in measurement error; 3) shifting away from the search for causality. Overall, this area of internet research is so new that there is little consensus among scholars on the detailed definition and parameters of Big Data (Cheung & Jak, 2016).

Social and behavioral scientists employing Big Data methods gather data on people's online behavior by downloading or "webscraping" people's posts from different online platforms, such as Facebook (Guadagno, Loewald, Muscanell, Barth, Goodwin, & Yang, 2013) and online dating services (Rudder, 2014). Illustrative of the type of insights obtained from webscraping, Bakshy,

Eckles, Yan, and Rosenn (2012) reported that when a person's Facebook friends "like" a product, they are more likely to follow their friends' lead and click on the ad as well. In a study of online movie ratings, Lee, Hosanagar, and Tan (2015) reported that, in the absence of friend ratings, people's movie preferences are influenced by a combination of the movie's popularity and the general positivity or negativity of strangers' ratings.

5.5 Linguistic Analysis of People's Posts

Scholars have also begun to apply methods of text analysis to gain social and behavioral scientific insights from people's online posts. One of the most widely used forms of text analysis – cited in over 4900 papers and patents (Boyd & Pennebaker, 2017) – is called the Linguistic Inquiry Word Count (LIWC, pronounced "Luke"; Pennebaker, Booth, Boyd, & Francis, 2015). Based on a solid foundation of empirical evidence supporting the notion that people's psychological states can be understood by examining their written words (Pennebaker, 2011), the LIWC is a software package bundled with a dictionary of commonly used words (see Pennebaker, Boyd, Jordan, & Blackburn, 2015).

The current version of the LIWC categorizes text into over eighty different language dimensions and also provides summary data, such as the total number of words in the passage, the average number of words per sentence, the percentage of words in the passages that were identified by the dictionary, and the percentage of long words (> 6 letters) (Boyd & Pennebaker, 2015). The LIWC also provides classification of text into categories

that map onto psychological states, express personal concerns, and punctuation. Research using the LIWC to analyze text has revealed relationships between the linguistic characteristics of people's writing and their health, personalities, and thought processes (Boyd & Pennebaker, 2015). Researchers can also create custom dictionaries to capture specific themes such as privacy concerns (Vasalou, Gill, Mazanderani, Papoutsi, & Joinson, 2011), civility in online discussions (Ksiazek, 2015), and task engagement in negotiations (Ireland & Henderson, 2014).

An early study using the LIWC examined changes in people's online posts during the two months before versus the two months after the 9/11 terrorist attacks (Cohn, Mehl, & Pennebaker, 2004). To examine this, the authors analyzed over 1000 posts, reporting significant psychological changes among participants after the attacks. In the first two weeks after 9/11, participants expressed more negative emotions, were more social and cognitively engaged, yet their writing showed significantly more social distance. By six weeks following 9/11, participants' posts indicated that they had psychologically returned to their pre-attack states. Thus, these results provided unique insight into how people cope with and recover from traumatic events.

Other research has shown that by analyzing people's social media posts, the LIWC can predict depression (De Choudhury, Gamon, Counts, & Horvitz, 2013), people's personality characteristics (Park et al., 2015), and, presaging the #MeToo movement, improve women's psychological well-being after Tweeting about sexual harassment (Foster, 2015). Thus, the application of linguistic analysis via software programs such as the LIWC to analyze people's online posts has been shown to reveal genuine psychological states and has tremendous potential for improving people's psychological well-being.

5.6 Ethical Considerations

In conducting internet research, ethical concerns fall into four broad themes: 1) participant anonymity and confidentiality, 2) privacy issues, 3) non-naïve participants and the veracity of their responses, 4) and fair compensation. To address these issues, scholars have made specific recommendations and proposed specific guidelines for ethics in internet research. For instance, the Association of Internet Researchers (AoIR) has established guidelines for ethics in internet research (Markham & Buchanan, 2012). These guidelines present examples and provide insights specific to internet research, examining questions such as: Is an avatar a person? What are the long-term ethical considerations of directly quoting someone's online posts? What are the risks for participants when a technology prevents the removal of identifying information (e.g., facial recognition data)? Members of university IRBs also have expressed ethical concerns specific to internet research. For instance, in one study, IRB members expressed the following concerns on the ethics of internet research: Security of data, participant anonymity, data deletion/loss and recovery, access to electronic data, validity of data collected online, the impact of anonymity on participants' responses, age verification, consent, and perception of the university (Buchanan & Hvizdak, 2009).

5.6.1 Participant Anonymity, Confidentiality, and Consent

Other key issues in internet research relate to the ability of researchers to properly consent and debrief participants recruited from the internet while also maintaining their confidentiality and anonymity. For instance, one ethics concern pertaining to internet research indicates, "researchers have less control over and knowledge of the research environment and cannot monitor the experience of participants or indeed their true identities" (Buchanan & Williams, 2010, p. 894). In a traditional lab experiment, participants who withdraw from an experiment prematurely are still debriefed. With participants recruited from the internet using the various methods reviewed above, those who drop out early will not get debriefed and it is also difficult to confirm that a participant reads and/or comprehends the debriefing (Miller et al., 2017). Similarly, when researchers do not know the identity of their participants, this raises concerns about proper consenting and debriefing but also concerns about double-participation, how to identify and reach participants to intervene if an experiment causes them distress, and how to protect a participant's confidentiality.

Illustrative of other concerns pertaining to participant anonymity, evidence suggests that people can be identified, from text collected from their online posts, by someone skilled using triangulation methods such as an internet search (Dawson, 2014). In this study, the researcher identified the participant-author of text-based postings in 8.9% of (10/112) papers. Of them, only one paper reported receiving IRB approval to publish identifiable text. In the remaining nine studies, five failed to anonymize the data or discuss ethics implications, and one unsuccessfully attempted to anonymize the data. Thus, while it is important to safeguard participants' anonymity and confidentiality, the anonymity of participants can also limit certain research in which it would be informative or useful to know the identities of the research participants.

5.6.2 Privacy Concerns

One of the most interesting (in this author's opinion) kinds of contemporary technology startups are the companies that extract and analyze DNA from a saliva sample provided via mail by customers. People receive details on their genetic predispositions (e.g., likelihood of flushing after consuming alcohol, lactose intolerance, being under- or overweight), DNA ancestry, and if they are carriers for congenital diseases (e.g., Tay-Sachs, Alzheimer's). Thus, in addition to having access to our words and actions, people's genetic makeup and other sensitive information are stored on the internet. This type of information, if revealed, would be a huge violation of privacy and may lead to unexpected negative consequences such as discrimination based on genetic predispositions. Thus, privacy is an important ethical consideration for research on the internet and will only become more pressing as technology becomes more personalized and intrusive.

Indeed, scholars suggest that certain ethical concerns pertaining to internet research result from the paucity of guidelines covering it as these guidelines largely pre-date the research practice (Gosling & Mason, 2015). Regarding privacy, scholars suggest that, along with poor consenting/debriefing guidance, the definition of "public behavior" should be considered in determining whether data collected from the internet should be considered private. This is particularly an issue for techniques such as webscraping that is used for direct data collation of people's texts from social media, blogs, and other online discussions.

Furthermore, as individuals' internet activities expand and change across time, they unwittingly create "digital dossiers" – profiles comprised of their posts, purchases, photos, and other contributions. People inadvertently create such records while using the internet. These profiles have also grown through the use of social media features allowing people to use their social media accounts to access other applications. This practice expands the amount of information that can be gleaned from people's behavior online as their activities cross more than one internet application. While this information can enhance the convenience and customization of people's internet experiences, the cost is privacy – both online and off. As far as this author can tell, no consensus exists among internet researchers on this issue. Scholars have provided only thought-provoking guidelines on considerations when determining if a certain action or context should be considered public or private and eligible to be observed/recorded without participant permission (Buchanan & Williams, 2010).

5.6.3 Non-Naïveté and Honesty in Participants

While participant non-naïveté and honesty are also issues with traditional research methods (see Papanastasiou, Chapter 7), these concerns are intensified by the relative anonymity afforded by the internet. For instance, in the samples recruited via MTurk, there is a specific subset of MTurkers that complete most psychological studies (Chandler & Shapiro,

2016). One study reported that, at any given time, there are approximately 7300 people in the worker pool, and it takes nearly three months to replenish the pool (Stewart et al., 2015). This suggests that there are a small number of people participating in quite a bit of social science research and that MTurkers are generally aware of the common procedures in psychological research. Furthermore, Chandler and Shapiro (2016) reported a link between the non-naïveté of human research participants and decreases in effect sizes over time. They also suggest that MTurk is another variation on convenience sampling with all the associated limitations (e.g., non-representative sample, non-naïve participants, and participants self-select into studies and therefore are not randomly sampled). The non-naïveté of MTurk workers can easily be observed on the various discussion forums that MTurkers use to communicate with each other about different studies, researchers, and labs (e.g., turkernation.com; mturkcrowd.com; mturkforum.com; turkerhub.com). This type of participant crosstalk can be damaging in traditional participant pools (Edlund et al., 2014) and likely is more damaging by the public nature of these forums. Edlund, Lange, Sevene, Umansky, Beck, and Bell (2017) examined participant crosstalk on MTurk and found it to be quite high – key details for 33% of MTurk studies were openly discussed on MTurker discussion forums and the majority of MTurk workers reported being exposed to crosstalk. Edlund and colleagues further reported that only 53% of MTurk researchers whose work had been discussed by participants reported being aware of this, few were concerned that this crosstalk would affect their results, and none reported taking any measures to reduce crosstalk. To address this limitation of crowdsourced research, scholars suggest that researchers employ careful data screening and customized samples built into the system and/or available for individual researchers to set up themselves (Cheung et al., 2017). Additionally, Edlund et al. (2017) also reported that asking participants to refrain from crosstalk significantly reduces its occurrence.

5.6.4 Fair Compensation

There are three ethical concerns regarding fair compensation for participants: The use of people's online comments without compensation, payment for participants recruited through crowdsourcing services, and compensation for participants who provide blanket consent to participate in research as part of an application's end user agreement. Specifically, when researchers webscrape online message boards and social media, the unwitting participants are not compensated for their contribution to the study. Further complicating this issue, it may not be possible for researchers to contact these participants to acknowledge their participation and/or compensate them. This is, in part, owing to the anonymity afforded by some types of online forums and, in part, because people's online activities and behavior are stored by the internet indefinitely. For instance, when a person dies, much of their online activity remains archived on the internet. Similarly, people change email addresses, user names, and avatars on a regular basis so it may be impossible to contact them about their posts.

Specific to crowdsourcing, many MTurkers report that financial compensation is one of their primary motivations for participating in internet research (Buhrmester et al., 2011).

Chandler and Shapiro (2016) note that studies with higher wages generally complete data collection faster and they further suggest that, for the sake of ethical treatment of human research participants, researchers pay them a fair wage rather than a few cents for as much as thirty minutes of work. Furthermore, compensation for participation in studies posted on MTurk varies widely: from $0.50 per hour on the low end to $50 per hour on the high end, averaging $8.56 per hour of work (Edlund et al., 2017). Finally, as evidenced by Kramer et al. (2014), the corporate practice of embedding blanket consent to participate in research in exchange for using a "free" application is ethically problematic. Thus, providing better compensation for participants would improve the ethics of internet research.

5.7 Conclusion

Using the internet for research is becoming increasingly common in the social and behavioral sciences and, despite the many challenges, it has proven to be a useful tool. Although there are many issues to consider when using the internet for research, in terms of practicality, ethics, and quality of data collected online, the existing research largely indicates that using the internet for research can provide scholars with access to more diverse and more specialized samples than typically available using other methods. I hope that my thoughts and experiences studying behavior and collecting data online can help readers of this chapter benefit from these tools while avoiding the many pitfalls described throughout the chapter.

KEY TAKEAWAYS

- Social and behavioral scientists use a wide variety of research methodologies to study people's behavior on the internet and also use the internet itself as a means of data collection (e.g., online surveys, webscraping).
- The advances of internet research include the ability to: 1) recruit more diverse samples, thereby increasing the generalizability of research findings; 2) broadly study language and behavior on the internet in its natural environment; and 3) use Big Data to draw large-scale inferences about human behavior.
- The drawbacks of internet research largely pertain to three broad areas: 1) the ethical treatment of participants recruited from the internet and/or whose language and behaviors are used for research purposes; 2) the veracity of participants' responses and their self-reported identities; and 3) the high rates of participant crosstalk and differential attrition. All three factors may impact the integrity of research findings yet are difficult for researchers to control.

IDEAS FOR FUTURE RESEARCH

- What can be done to develop more detailed and widely accepted rules for protecting participants' privacy while also ensuring proper consent, debriefing, and ensuring that internet research participants are protected from harm?

- What steps can be taken to encourage online data collection across the various social and behavioral sciences?
- How replicable are the results of studies that are adapted to online technologies across the various social and behavioral sciences?

SUGGESTED READINGS

Boyd, R. L., & Pennebaker, J. W. (2015). A way with words: Using language for psychological science in the modern era. In Claudiu V. Dimofte, Curtis P. Haugtvedt, and Richard F. Yalch (Eds.), *Consumer Psychology in a Social Media World* (pp. 222–236). London: Routledge.

Chandler, J., & Shapiro, D. (2016). Conducting clinical research using crowdsourced convenience samples. *Annual Review of Clinical Psychology, 12*, 1253–81. doi:10.1146/annurev-clinpsy-021815-093623

Edlund, J. E., Lange, K. M., Sevene, A. M., Umansky, J., Beck, C. D., & Bell, D. J. (2017). Participant crosstalk: Issues when using the Mechanical Turk. *The Quantitative Methods for Psychology, 13*(3), 174–182. doi:10.20982/tqmp.13.3

Gosling, S. D., & Mason, W. (2015). Internet research in psychology. *Annual Review of Psychology, 66*, 877–902. doi:10.1146/annurev-psych-010814-015321

Miller, J. D., Crowe, M., Weiss, B., Maples-Keller, J. L., & Lynam, D. R. (2017). Using online, crowdsourcing platforms for data collection in personality disorder research: The example of Amazon's Mechanical Turk. *Personality Disorders: Theory, Research, and Treatment, 8*(1), 26–34.

REFERENCES

Anderson, E. L., Steen, E., & Stavropoulos, V. (2017). Internet use and problematic internet use: A systematic review of longitudinal research trends in adolescence and emergent adulthood. *International Journal of Adolescence and Youth, 22*(4), 430–454.

Bakshy, E., Eckles, D., Yan, R., & Rosenn, I. (2012, June). Social influence in social advertising: Evidence from field experiments. In *Proceedings of the 13th ACM Conference on Electronic Commerce* (pp. 146–161). Valencia, Spain: ACM.

Berinsky, A. J., Huber, G. A., & Lenz, G. S. (2012). Evaluating online labor markets for experimental research: Amazon. com's Mechanical Turk. *Political Analysis, 20*, 351–368.

Beymer, M. R., Holloway, I. W., & Grov, C. (2018). Comparing self-reported demographic and sexual behavioral factors among men who have sex with men recruited through Mechanical Turk, Qualtrics, and a HIV/STI clinic-based sample: Implications for researchers and providers. *Archives of Sexual Behavior, 47*, 133. https://doi.org/10.1007/s10508-016-0932-y

Boyd, R. L., & Pennebaker, J. W. (2015). A way with words: Using language for psychological science in the modern era. In Claudiu V. Dimofte, Curtis P. Haugtvedt, and Richard F. Yalch (Eds.), *Consumer Psychology in a Social Media World* (pp. 222–236). London: Routledge.

Boyd, R. L., & Pennebaker, J. W. (2017). Language-based personality: A new approach to personality in a digital world. *Current Opinion in Behavioral Sciences, 18*, 63–68.

Buchanan, E. A., & Hvizdak, E. E. (2009). Online survey tools: Ethical and methodological concerns of human research ethics committees. *Journal of Empirical Research on Human Research Ethics, 4*(2), 37–48. doi:10.1525/jer.2009.4.2.37

Buchanan, T., & Williams, J. E. (2010). Ethical issues in psychological research on the Internet. In S. D. Gosling & J. A. Johnson (Eds.), *Advanced Methods for Conducting Online Behavioral Research* (pp. 255–271).

Washington, DC: American Psychological Association.

Buhrmester, M., Kwang, T., & Gosling, S. D. (2011). Amazon's Mechanical Turk: A new source of inexpensive, yet high-quality, data? *Perspectives on Psychological Science, 6*(1), 3–5.

Casler, K., Bickel, L., & Hackett, E. (2013). Separate but equal? A comparison of participants and data gathered via Amazon's MTurk, social media, and face-to-face behavioral testing. *Computers in Human Behavior, 29*(6), 2156–2160.

Chandler, J., & Shapiro, D. (2016). Conducting clinical research using crowdsourced convenience samples. *Annual Review of Clinical Psychology, 12*, 53–81. doi:10.1146/annurev-clinpsy-021815-093623

Chang, L., & Krosnick, J. A. (2009). National surveys via RDD telephone interviewing versus the Internet: Comparing sample representativeness and response quality. *Public Opinion Quarterly, 73*(4), 641–678.

Cheung, J. H., Burns, D. K., Sinclair, R. R., & Sliter, M. (2017). Amazon Mechanical Turk in organizational psychology: An evaluation and practical recommendations. *Journal of Business and Psychology, 32*(4), 347–361.

Cheung, M. W. L., & Jak, S. (2016). Analyzing Big Data in psychology: A split/analyze/meta-analyze approach. *Frontiers in Psychology, 7*, 1–13. doi:10.3389/fpsyg.2016.00738

Cohn, M. A., Mehl, M. R., & Pennebaker, J. W. (2004). Linguistic markers of psychological change surrounding September 11, 2001. *Psychological Science, 15*(10), 687–693.

Dawson, P. (2014). Our anonymous online research participants are not always anonymous: Is this a problem? *British Journal of Educational Technology, 45*(3), 428–437. doi:10.1111/bjet.12144

De Choudhury, M., Gamon, M., Counts, S., & Horvitz, E. (2013). Predicting depression via social media. *ICWSM, 13*, 1–10.

Edlund, J. E., Lange, K. M., Sevene, A. M., Umansky, J., Beck, C. D., & Bell, D. J. (2017).

Participant crosstalk: Issues when using the Mechanical Turk. *The Quantitative Methods for Psychology, 13*(3), 174–182. doi:10.20982/tqmp.13.3

Edlund, J. E., Nichols, A. L., Okdie, B. M., Guadagno, R. E., Eno, C. A., Heider, J. D., ... Wilcox, K. T. (2014). The prevalence and prevention of crosstalk: A multi-institutional study. *The Journal of Social Psychology, 154*(3), 181–185.

Ellis-Barton, C. (2016). Ethical considerations in research participation virality. *Journal of Empirical Research on Human Research Ethics, 11*(3), 281–285. doi:10.1177/1556264616661632

Foster, M. D. (2015). Tweeting about sexism: The well-being benefits of a social media collective action. *British Journal of Social Psychology, 54*(4), 629–647.

Fraley, R. C. (2004). *How to Conduct Behavioral Research Over the Internet: A Beginner's Guide to HTML and CGI/Perl.* New York: Guilford Press.

Gosling, S. D., & Bonnenburg, A. V. (1998). An integrative approach to personality research in anthrozoology: Ratings of six species of pets and their owners. *Anthrozoös, 11*(3), 148–156.

Gosling, S. D., & Mason, W. (2015). Internet research in psychology. *Annual Review of Psychology, 66*, 877–902. doi:10.1146/annurev-psych-010814-015321

Grimmelmann, J. (2015, May 27). Do you consent? If tech companies are going to experiment on us, they need better ethical oversight. *Slate Magazine.* Retrieved from https://slate.com/technology/2015/05/facebook-emotion-contagion-study-tech-companies-need-irb-review.html

Guadagno, R. E., & Cialdini, R. B. (2002). Online persuasion: An examination of gender differences in computer-mediated interpersonal influence. *Group Dynamics: Theory, Research, and Practice, 6*(1), 38.

(2007). Persuade him by email but see her in person: Online persuasion revisited. *Computers in Human Behavior*, *23*(2), 999–1015.

Guadagno, R. E., Loewald, T. A., Muscanell, N. L., Barth, J. M., Goodwin, M. K., & Yang, Y. L. (2013). Facebook history collector: A new method for directly collecting data from Facebook. *International Journal of Interactive Communication Systems and Technologies*, *3*, 57–67.

Guadagno, R. E., Okdie, B. M., & Eno, C. A. (2008). Who blogs? Personality predictors of blogging. *Computers in Human Behavior*, *24*, 1993–2004.

Guadagno, R. E., Rempala, D. M., Murphy, S. Q., & Okdie, B. M. (2013). Why do Internet videos go viral? A social influence analysis. *Computers in Human Behavior*, *29*, 2312–2319.

Guadagno, R. E., Swinth, K. R., & Blascovich, J. (2011). Social evaluations of embodied agents and avatars. *Computers in Human Behavior*, *6*, 2380–2385.

Henrich, J., Heine, S. J., & Norenzayan, A. (2010). Beyond WEIRD: Towards a broad-based behavioral science. *Behavioral and Brain Sciences*, *33*(2–3), 111–135.

Hill, K. (2012, February 12). How Target figured out a teen girl was pregnant before her father did. *Forbes*. Retrieved from www.forbes.com/ sites/kashmirhill/2012/02/16/how-target-figured-out-a-teen-girl-was-pregnant-before-her-father-did/#72d076516668

Ireland, M. E., & Henderson, M. D. (2014). Language style matching, engagement, and impasse in negotiations. *Negotiation and Conflict Management Research*, *7*(1), 1–16.

Kramer, A. D., Guillory, J. E., & Hancock, J. T. (2014). Experimental evidence of massive-scale emotional contagion through social networks. *Proceedings of the National Academy of Sciences of the United States of America*, *111*(24), 8788–8790.

Ksiazek, T. B. (2015). Civil interactivity: How news organizations' commenting policies explain civility and hostility in user comments. *Journal of Broadcasting & Electronic Media*, *59*(4), 556–573.

Kwon, K. H., Stefanone, M. A., & Barnett, G. A. (2014). Social network influence on online behavioral choices exploring group formation on social network sites. *American Behavioral Scientist*, *58*, 1345–1360.

Kwon, S., Cha, M., & Jung, K. (2017). Rumor detection over varying time windows. *PLoS One*, *12*(1), e0168344.

Lee, Y. J., Hosanagar, K., & Tan, Y. (2015). Do I follow my friends or the crowd? Information cascades in online movie ratings. *Management Science*, *61*, 2241–2258.

Markham, A., Buchanan, E., & AoIR Ethics Working Committee (2012). Ethical decision-making and Internet research: Version 2.0. Association of Internet Researchers. Retrieved from www.expertise121.com/curriculum/3 .pdf

Mayer-Schönberger, V., & Cukier, K. (2013). *Big Data: A Revolution that Will Transform How We Live, Work, and Think*. Boston: Houghton Mifflin Harcourt.

Merriam-Webster's (n.d.). Crowdsourcing. *Merriam-Webster's Dictionary Online*. Retrieved from www.merriam-webster.com/dictionary/ crowdsourcing

Miller, J. D., Crowe, M., Weiss, B., Maples-Keller, J. L., & Lynam, D. R. (2017). Using online, crowdsourcing platforms for data collection in personality disorder research: The example of Amazon's Mechanical Turk. *Personality Disorders: Theory, Research, and Treatment*, *8*(1), 26–34.

Mitra, T., Counts, S., & Pennebaker, J. W. (2016, March). Understanding anti-vaccination attitudes in social media. In *Proceedings of the Tenth International Association for the Advancement of Artificial Intelligence (AAAI) Conference on Web and Social Media (ICWSM 2016)* (pp. 269–278) Palo Alto, CA: AAAI Press.

Nguyen, M., Bin, Y. S., & Campbell, A. (2012). Comparing online and offline self-disclosure: A systematic review. *Cyberpsychology, Behavior, and Social Networking*, *15*(2), 103–111.

Paolacci, G., Chandler, J., & Ipeirotis, P. G. (2010). Running experiments on Amazon

Mechanical Turk. *Judgment and Decision Making*, 5, 411–419.

Park, G., Schwartz, H. A., Eichstaedt, J. C., Stillwell, D., Ungar, L. H., & Seligman, M. E. P. (2015). Automatic personality assessment through social media language. *Journal of Personality and Social Psychology*, *108*(6), 934–952.

Peng, T. Q., Zhang, L., Zhong, Z. J., & Zhu, J. J. (2013). Mapping the landscape of internet studies: Text mining of social science journal articles 2000–2009. *New Media & Society*, *15*(5), 644–664.

Pennebaker, J. W. (2011). *The Secret Life of Pronouns: How Our Words Reflect Who We Are*. New York: Bloomsbury.

Pennebaker, J. W., Booth, R. J., Boyd, R. L., & Francis, M. E. (2015). Linguistic Inquiry and Word Count: LIWC 2015 [Computer software]. Pennebaker Conglomerates.

Pennebaker, J. W., Boyd, R. L., Jordan, K., & Blackburn, K. (2015). *The Development and Psychometric Properties of LIWC2015*. Austin, TX: University of Texas at Austin.

Rouse, S. V. (2015). A reliability analysis of Mechanical Turk data. *Computers in Human Behavior*, *43*, 304–307.

Rudder, C. (2014). *Dataclysm: Love, Sex, Race, and Identity – What Our Online Lives Tell Us About Our Offline Selves*. New York: Random House.

Snijders, C., Matzat, U., & Reips, U. D. (2012). "Big Data": Big gaps of knowledge in the field of internet science. *International Journal of Internet Science*, *7*(1), 1–5.

Stewart, N., Ungemach, C., Harris, A. J., Bartels, D. M., Newell, B. R., Paolacci, G., &

Chandler, J. (2015). The average laboratory samples a population of 7,300 Amazon Mechanical Turk workers. *Judgment and Decision Making, 10*(5), 479.

Vasalou, A., Gill, A. J., Mazanderani, F., Papoutsi, C., & Joinson, A. (2011). Privacy dictionary: A new resource for the automated content analysis of privacy. *Journal of the Association for Information Science and Technology*, *62*(11), 2095–2105.

Verna, I. M. (2014). Editorial expression of concern. *PNAS Early Edition*, *111*(29), 10779.

Waldman, K. (2014, June 28). Facebook's unethical experiment: It intentionally manipulated users' emotions without their knowledge. *Slate Magazine*. Retrieved from www.slate.com/articles/health_and_science/science/2014/06/facebook_unethical_experiment_it_made_news_feeds_happier_or_sadder_to_manipulate.html

Wingate, V. S., Minney, J. A., & Guadagno, R. E. (2013). Sticks and stones may break your bones, but words will always hurt you: A review of cyberbullying. *Social Influence*, *8*, 2–3, 87–106.

Young, S. D., Shakiba, A., Kwok, J., & Montazeri, M. S. (2014). The influence of social networking technologies on female religious veil-wearing behavior in Iran. *Cyberpsychology, Behavior, and Social Networking*, *17*(5), 317–321. doi:10.1089/cyber.2013.0338

Zhou, H., & Fishbach, A. (2016). The pitfall of experimenting on the web: How unattended selective attrition leads to surprising (yet false) research conclusions. *Journal of Personality and Social Psychology*, *111*(4), 493–504. doi:10.1037/pspa0000056

Part Two: Understanding Issues Present Throughout the Research Process

6 Issues in Informed Consent

David S. Festinger

Karen L. Dugosh

Esther Choi

Chloe Sierka

Informed consent is a process that optimally occurs in the context of an investigator–participant relationship characterized by trust and honesty. Importantly, as policy has evolved over the years, *informed consent is no longer viewed as a one-time event but rather as an ongoing process*. Active and dynamic informed consent may involve re-obtaining informed consent from participants when appropriate, welcoming and answering questions from participants, sharing new information that may affect an individual's decision to continue participation, and ensuring that participants are continuously aware of study procedures, risks, benefits, and their rights as participants (Gupta, 2013; US Department of Health and Human Services, n.d.).

Unfortunately, the philosophical and legal doctrine of informed consent emerged from celebrated advancements with dark histories. Revelations about the horrors of World War II (e.g., Nazi medical experiments), unethical investigations in the United States (e.g., Tuskegee syphilis study; Centers for Disease

Control and Prevention, 2011), Milgram's obedience and individual responsibility study (Milgram, 1971), human radiation experiments (Faden, 1996), and international instances of unethical conduct (e.g., the trovafloxacin trials in Nigeria, nevirapine PMTCT trials in Uganda, streptokinase trials in India, VGV-1 trials in China, the cariporide trial in Argentina; Weyzig & Schipper, 2008) heightened public awareness of the potential for research misconduct.

Over the past seventy years, the global medical community has taken numerous steps to protect people who take part in research and has derived three seminal international documents on human research ethics: the Nuremberg Code, the Declaration of Helsinki, and the Belmont Report. In response to the Nuremberg trials of Nazi doctors who performed unethical experimentation during World War II, the United Nations General Assembly adopted the Nuremberg Code as the first major international document to provide guidelines on research ethics. It made voluntary consent a requirement in research studies, emphasizing that consent can be voluntary only if individuals:

1. are able to consent;
2. are free from coercion (i.e., outside pressure); and
3. comprehend the risks and benefits involved.

The Nuremberg Code also clearly states that researchers should minimize risk and harm, ensure that risks do not significantly outweigh potential benefits, use appropriate study designs, and guarantee participants' freedom to withdraw from the research at any time.

The next major development in human research protections came in 1964 at the 18th World Medical Assembly in Helsinki, Finland. With the establishment of the Helsinki Declaration, the World Medical Association adopted twelve principles to guide physicians on ethical considerations related to biomedical research. The Declaration emphasized the distinction between medical care that directly benefits the patient and research that may or may not provide direct benefit.

In 1979, the National Commission for the Protection of Human Subjects of Biomedical and Behavioral Research issued "The Belmont Report: Ethical Principles and Guidelines for the Protection of Human Subjects of Research." The report sets forth three principles underlying the ethical conduct of research:

1. *respect for persons* – recognizing the autonomy and dignity of individuals, and the need to protect those with diminished autonomy (i.e., impaired decision-making skills), such as children, the aged, and the disabled;

2. *beneficence* – an obligation to protect persons from harm by maximizing benefits and minimizing risks; and
3. *justice* – fair distribution of the benefits and burdens of research.

Biomedical and legal developments promulgated three key elements of ethical and valid informed consent based on the Belmont Report's key principles:

1. *intelligence* – the individual has the cognitive capability to make a rational and informed decision based on the consent- and study-related information provided to them;
2. *knowingness* – the individual fully understands the information shared with them and the implications of participation on their well-being; and
3. *voluntariness* – the individual decides, free from coercion and undue influence, whether or not to participate in the research.

6.1 Concerns About and Strategies to Improve Informed Consent

Informed consent is an essential prerequisite to an individual participating in any type of research, including social and behavioral studies, biomedical trials, and therapeutic interventions. The three elements of informed consent are important to uphold, whether the study is examining a life-altering condition in high-risk clinical trials or social interactions in low-risk psychological research, because they recognize and protect individuals' rights and welfare. However, decades of research on informed consent and human subjects protections have revealed troubling findings regarding potential participants' compromised intelligence, knowingness, and voluntariness. Issues relating to these three elements of informed consent and strategies to improve each are discussed here.

6.1.1 Intelligence

Within the context of informed consent, intelligence, also referred to as competence, describes an individual's capacity and capability to make decisions about their participation in research. This definition of intelligence is most often used for legal purposes and is particularly relevant when developing safeguards for populations with diminished decision-making capacity (e.g., the elderly, the severely mentally ill).

Following the mistakes of past research, researchers began to systematically exclude vulnerable populations from research trials. As a result of this overcompensation, fewer members of vulnerable populations were represented in and could benefit from important scientific advances. For example, individuals with diminished cognitive capacity or severe mental illness were often excluded from research because of the difficulty associated with obtaining their informed consent and potentially for purposes of risk management (National Institutes of Health, 2009). This exclusionary practice violates the principle of distributive justice that ensures that the risks and benefits of research are equally distributed across all individuals. Importantly, distributive justice is bidirectional and can be violated by both overinclusion and underinclusion of certain groups of people, such as those with diminished capacity.

For these reasons, it is important to find ways to conduct research with individuals who may have diminished capacity or intelligence. The first step in the process is to evaluate using standardized methods whether potential participants have the capacity to make an informed decision about their involvement in the study. There is much debate about how, and even if, individuals who are deemed to have diminished capacity can participate in research (Carlson, 2013; Iacono & Carling-Jenkins, 2012).

Assessment. As discussed, it is important to evaluate the decisional capacity of any potential research participant. According to National Institutes of Health (NIH) guidelines (NIH, 2009), researchers may use their discretion about suitability for participation after briefly talking with a potential participant. However, this subjective determination lacks standardization, and the decisions made regarding intelligence may vary from researcher to researcher. This points to the need for standardized and validated measures of decisional capacity that can be incorporated into the consent process.

One such measure is the MacArthur Competency Assessment Tool for Clinical Research (MacCAT-CR; Appelbaum & Grisso, 2001), a well-validated fifteen- to twenty-minute interview that can be customized to different research studies. Unfortunately, the administration time of this measure often limits its utility in the consent process. The University of California Brief Assessment of Capacity to Consent (UBACC; Jeste et al., 2007) is a suitable alternative. This brief ten-item tool to measure decisional capacity has been validated for use in a range of vulnerable populations, including individuals who have schizophrenia (Jeste et al., 2007; Jeste et al.,

2009), individuals who have substance use disorders (Martel et al., 2017), and individuals who have Alzheimer's disease (Seaman, Terhorst, Gentry, Hunsaker, Parker, & Lingler, 2015). The UBACC can be easily incorporated into the consent process and provide researchers with the necessary information to make a determination about the individual's capacity. Despite the brevity of the ten-question UBACC, researchers may prefer to use the MacCAT-CR for its thorough exploration of specific aspects of consent. This in-depth information may be legally and ethically necessary with participants at a high risk of diminished capacity in a high-risk study (Appelbaum & Grisso, 2001).

Surrogates and Legally Authorized Representatives. Some individuals, including those with severe mental disabilities, may fail to meet the required level of decisional capacity to provide informed consent. However, excluding these individuals would violate the principle of distributive justice. The scientific and legal communities have attempted to address this issue though the use of surrogates (NIH, 2009) and legally authorized representatives (LARs; Wynn, 2014). Surrogates and LARs serve as representatives for potential participants in making decisions about participation in the study. Policies surrounding the use of surrogates in research can vary from state to state and organization to organization (Wynn, 2014). Many state and institutional policies have a hierarchy of potential surrogates. In most cases, close relatives (e.g., parent, spouse, adult child) may serve as a surrogate for an individual with diminished capacity (e.g., California Code, Health and Safety Code Section 24178, 2003). In some cases, they may be appointed by the potential participant if the latter anticipates having diminished capacity in the future (NIH, 2009; Wendler & Prasad, 2001; Wynn, 2014).

It is imperative that surrogates understand all aspects of the study and rely on their knowledge and experience to decide whether the individual whom they are representing should participate in the research. Surrogates must follow two primary decision-making principles: substituted judgement and the best interest standard. Substituted judgement requires that the surrogate make the decision in the same manner that they believe the person with diminished capacity would. The best interest standard requires that the surrogate make a decision that would result in the optimal outcome for the individual (NIH, 2009; Wendler & Prasad, 2001). It is important to note that the decision of the surrogate cannot supersede that of an individual with diminished capacity; if the person with diminished capacity expresses dissent toward participating in a study, they cannot be forced to participate even if the research may help their condition. Ideally, the participant with diminished capacity will still be able to express assent in some manner (NIH, 2009).

Diminished capacity or intelligence may present an ethical barrier to conducting research with vulnerable individuals. Nevertheless, it is critical to include such individuals to help ensure that these populations benefit from scientific research and advancements. Researchers should evaluate capacity using standardized assessments during the consent process and consider the use of surrogates as needed.

6.1.2 Knowingness

The second element of informed consent is knowingness, which refers to an individual's understanding and retention of study-related information that is presented during the consent process. The increasing complexity of research protocols has introduced a growing list of new risks and benefits participants may experience, which, in turn, has lengthened and complicated the informed consent process (Morán-Sánchez, Luna, & Pérez-Cárceles, 2016; Sonne et al., 2013). Some research suggests that current consent procedures do not facilitate optimal participant understanding, especially in vulnerable populations with diminished decision-making capacities (Morán-Sánchez et al., 2016; Neilson, Chaimowitz, & Zuckerberg, 2015; Westra & De Beaufort, 2015) or in populations from developing countries (Mandava, Pace, Campbell, Emanuel, & Grady, 2012). As a result, individuals may not fully understand or remember important aspects of the study procedures, risks and benefits, and human subject protections (Festinger, Ratanadilok, Marlowe, Dugosh, Patapis, & DeMatteo, 2007; Madeira & Andraka-Christou, 2016).

Comprehending and Recalling Consent and Study Information. Studies have demonstrated that research participants are often unaware that they are participating in a research study rather than receiving treatment, have poor understanding of study information, do not comprehend important risks and benefits of participation, do not know the meaning of randomization or placebo/control conditions, and are not aware that they can withdraw from the study at any time (Appelbaum, Roth, & Lidz, 1982; Cassileth, Zupkis, Sutton-Smith, & March, 1980; Edwards, Lilford, Thornton, & Hewison, 1998; Festinger et al., 2007; Festinger, Marlowe, Crift, Dugosh, Arabia, & Benasutti, 2009; Levine, 1992; Muss et al., 1979; Robinson & Merav, 1976; Silva, Sorell, & Busch-Rossnagel, 1988; Sugarman, McCrory, & Hubal, 1998; Verheggen & van Wijmen, 1996). In developing countries, participants are less likely to understand that they can refuse or withdraw their participation at any time without penalty. One particular concern is the removal of healthcare they normally do not have access to (Krogstad et al., 2010; Mandava et al., 2016). Moreover, participants often fail to retain much of the information presented during the informed consent process even only a few days after the initial date of consent (Cassileth et al., 1980; Festinger et al., 2007; Festinger et al., 2009; Miller, Searight, Grable, Schwartz, Sowell, & Barbarash, 1994; Rosique, Pérez-Cárceles, Romero-Martín, Osuna, & Luna, 2006). These findings pose a challenge to the assumption that consent is truly informed and highlight the need to modify the consent process to ensure that individuals understand and remember the critical aspects of their study participation that are provided during the consent process.

Therapeutic Misconception. Some individuals may choose to participate in a clinical trial under a false assumption that the treatment offered to them as part of the research will be better than their current treatment or alternative treatments. This phenomena is referred to as the therapeutic misconception (Appelbaum et al., 1982). Therapeutic misconceptions most commonly manifest in three ways: 1) an incorrect belief that treatment will be tailored to a participant's needs; 2) the failure to recognize that the primary aim of the research is to advance scientific knowledge as opposed to benefitting participants; and 3) an unrealistic expectation of benefits (Christopher, Stein, Springer, Rich, Johnson, & Lidz, 2016).

Therapeutic misconceptions can prevent individuals from accurately and objectively evaluating the risks and benefits of participation (Appelbaum et al., 1982; Christopher et al., 2016; Lidz, 2006). Therefore, it is paramount to diffuse these misconceptions by ensuring that participants fully understand the distinction between experimental and alternative treatments. Preliminary evidence suggests that educating individuals on the rationale behind clinical trials and the differences between research and ordinary clinical care during the consent process can significantly reduce therapeutic misconception (Christopher, Appelbaum, Truong, Albert, Maranda, & Lidz, 2017).

Strategies for Improving Knowingness. A number of strategies have been developed to improve understanding and recall of consent information. These innovations focus on modifying either the *structure* of the consent document or the *process* of obtaining informed consent.

Modifying the Structure of Consent Documents. Adjusting the reading level of consent documents is a common strategy designed to make content more understandable. Typically, consent documents are written at a grade level that is commensurate with that of the average adult or, in the case of children, at their specific grade level. Improving the readability of consent forms is important given that approximately 14% of adults in the United States have marginal literacy skills and another 29% can only perform simple literacy activities (White & Dillow, 2005). Decreasing the reading level alone, unfortunately, does not necessarily increase participants' comprehension and retention of the most significant aspects of the consent document (Sonne et al., 2013).

Shorter consent forms with simpler language have been shown to be as effective (Enama, Hu, Gordon, Costner, Ledgerwood, & Grady, 2012; Stunkel et al., 2010) or better (Beardsley, Jefford, & Mileshkin, 2007; Matsui, Lie, Turin, & Kita, 2012) than long consent forms at providing relevant information and enhancing comprehension. Supplementing simpler language with revised layouts, text styling, and diagrams is effective at significantly increasing participants' comprehension of consent information (Kim & Kim, 2015; Nishimura, Carey, Erwin, Tilburt, Murad, & McCormick, 2013). Modifying the structure and visual presentation of consent documents by using shorter sentences, larger font size, wider line spacing, diagrams and pictures, and bullet points may ensure that the information most relevant to the prospective participant is prioritized and help facilitate prospective participants' or their LARs' understanding of the reasons why they may or may not want to participate in research.

Modifying the Consent Process. Beyond improving the readability of consent forms, there is a growing body of research on improving the consent process itself. The intervention that has received the most empirical support is corrected feedback. Corrected feedback entails assessing a potential participant's comprehension of informed consent information following review of the consent form and then interacting with the participant to address incorrect responses. Corrected feedback consistently improves both initial comprehension and longer-term recall of consent information (Carpenter et al., 2000; Coletti et al., 2003; Festinger, Dugosh, Croft, Arabia, & Marlowe, 2010; Stiles, Poythress, Hall, Falkenbach, & Williams, 2001; Taub & Baker, 1983; Taub, Kline, & Baker, 1981; Wirshing, Wirshing, Marder, Liberman, & Mintz, 1998). Moreover, combining corrected feedback and structural interventions significantly improves participants' comprehension, especially on questions related to randomization (Kass, Taylor, Ali, Hallez, & Chaisson, 2015). Researchers should consider using corrected feedback throughout a study's duration as part of an ongoing informed consent process.

Enhancing Motivation. The structural and procedural interventions discussed above are remedial in nature as they are designed to simplify cognitive tasks or accommodate cognitive deficits. While cognitive variables, such as IQ, education level, and neuropsychological measures of memory and attention, are positively correlated with an individual's capacity to recall consent information (Dunn & Jeste, 2001; Flory & Emanuel, 2004; Taub & Baker, 1983), these variables statistically account for less than half of the variance in recall (Festinger et al., 2007). This suggests that remediation interventions do not address all of the potential factors that affect comprehension and retention, including motivation.

It is likely that some individuals may not understand or recall consent information because they do not pay adequate attention to the information presented. A monetary incentive such as $5 for each correct answer on a consent quiz can significantly increase motivation to commit consent information to memory and result in moderate gains in consent recall (Festinger et al., 2009; Festinger, Dugosh, Marlowe, & Clements, 2014). The scarcity of research on the effects of participant-level variables such as motivation warrants a shift in focus from simply modifying the structure and delivery of informed consent information alone.

Consent Quizzes. Despite the critical importance of ensuring that research participants understand and recall consent information, a survey of 220 researchers who received NIH grant funding in 2012 revealed that only 18% of studies utilized a consent quiz to evaluate comprehension of consent information at any time during the consent process (Ryan et al., 2015). Indeed, although documentation of informed consent is required, there exists no mandate for evidence that participants were truly informed before consenting to participate. It is, therefore, possible that participants with limited understanding of consent information are enrolled in research. Incorporating consent quizzes as a prerequisite to enrollment provides an efficient means to detect adequate understanding of consent information and an opportunity for researchers to clarify participants' confusion around any consent- or study-related component. Furthermore, consent quizzes can help investigators identify the areas of their consent form that need the most clarity and attention

(Allen et al., 2017). If a prospective participant fails to demonstrate adequate knowledge of consent and study information despite repeated corrected feedback, that individual should not be enrolled in the research. Unfortunately, there are no official standards for consent quizzes and many use short true or false or multiple choice instruments that rely on recognition rather than recall. Nevertheless, over the years, many researchers have cited the importance of using more comprehensive measures to ensure that the information covered in the consent procedure is fully comprehended (e.g., Allen et al., 2017; Appelbaum & Grisso, 2001; Festinger et al., 2009; Festinger et al., 2014; O'Brien, 2002).

Multimedia Consent Approaches. Changing the length or format of consent documents may not be enough to convey critical information to potential participants, especially to those who are illiterate or understand a different language. For example, translations often change the meaning and content of the original message, especially when some words – such as randomization – do not exist in local languages (Krogstad et al., 2010). Hence, it may be best to obtain informed consent without written documentation for illiterate or diverse populations.

With the advent of novel technologies, there has been increasing interest in incorporating multimedia communication modalities to facilitate the informed consent process. These strategies incorporate video, computer, and web-based elements into the consent process. A recent meta-analysis demonstrated that 31% of studies that used multimedia consent approaches resulted in significant increases in understanding and retention of study-related information (Nishimura et al., 2013). Since this systematic review, several studies have supported these positive results and provided new evidence of the efficacy of multimedia consent tools in improving participants' comprehension (Kraft et al., 2017; Rothwell et al., 2014; Siu, Rotenberg, Franklin, & Sowerby, 2016; Spencer, Stoner, Kelleher, & Cohen, 2015; Tipotsch-Maca, Varsits, Ginzel, & Vescei-Marlovits, 2016; Winter et al., 2016).

Computer-animated video presentations with audio narratives have been shown to increase comprehension (Kraft et al., 2017; Rothwell et al., 2014; Spencer et al., 2015; Winter et al., 2016) and knowledge retention (Siu et al., 2016; Tipotsch-Maca et al., 2016) when used in conjunction with traditional written and face-to-face consent procedures. Some participants expressed a preference for the paper forms due to the ability to go back and reread the information, though self-controlled return and advancement through a video can easily be built in (Sonne et al., 2013). Interestingly, almost all participants reported that the videos improved their procedural understanding but far fewer reported increased conceptual understanding. This finding suggests that comprehension is multi-faceted and one type of intervention may not equally target or improve all elements of understanding of a research study (Rothwell et al., 2014; Sonne et al., 2013).

Mixed multimedia tools also have the potential to significantly increase recall and understanding of consent information in participants with low literacy (Afolabi et al., 2014; Afolabi et al., 2015) and psychiatric conditions (Jeste et al., 2009; Sonne et al., 2013). In a recent study involving patients diagnosed with psychotic, mood, and anxiety disorders, participants with diminished decision-making capacity and participants without diminished capacity were exposed to computer-enhanced consent procedures. The enhanced consent involved a PowerPoint slide projected on a 17-inch computer monitor that shared the same key information as the paper consent form provided at baseline. Enhanced features included a larger text size, one key piece of information per slide, titles for each slide, simple graphics, and several summary slides at the beginning, middle, and end of the presentation. More than half of the participants who scored as having diminished capacity pre-intervention later scored as having non-diminished capacity post-intervention. Furthermore, these individuals who regained capacity exhibited the highest understanding and appreciation of the informed consent process out of all participants (Morán-Sánchez et al., 2016).

Obtaining Informed Consent Remotely. As technological advances continue, many settings have incorporated computer-enabled audio-visual communications (e.g., telemedicine, online education) to remotely reach individuals. Despite the usefulness of remote communications, there is limited evidence on the utility and effectiveness of multimedia approaches in remotely consenting and recruiting participants into research studies. In the first trial to investigate telemedicine-based consent, Bobb et al. (2016) found no significant differences in objective understanding of informed consent, perceived understanding of consent, and accrual rates between participants approached by telemedicine versus face-to-face. This preliminary evidence suggests that remote consent does not worsen potential participants' comprehension and willingness to participate compared to standard consent. Despite the promising start, more research is needed to examine the quality of remote informed consent and its impact on participants' and study personnel's time and travel costs.

It is likely that multimedia consent procedures will become part of a new standard of the informed consent process as technology continues to evolve and play a bigger role in research. Although most of the available research has treated multimedia consent as a supplement to traditional informed consent processes, thereby lengthening the entire consent process, multimedia interventions offer the potential to provide meaningful gains in understanding and willingness to participate in research. While recent research favors multimedia consent formats, it is important to note that the heterogeneity of the interventions and assessments makes it difficult to compare studies and draw broad conclusions. This lack of standardization may, in part, be due to the lack of regulations or guidance on electronic signatures, email consent, or video conferencing (Neff, 2008). Therefore, future research on multimedia consent procedures is imperative to identify the most efficient elements of multimedia consent and optimize understanding of the key elements of informed consent.

6.1.3 Voluntariness

The third element of informed consent is voluntariness. Specifically, the decision to participate in a research study must be made in an autonomous fashion that is free from undue influence and coercion. The US Department of Health & Human Services defines undue influence as occurring when a researcher offers an excessive or inappropriate reward or other overture for study participation. For example, a professor may offer students extra credit for participating in her research study. If this is the only way that students may earn extra credit, the professor is unduly influencing the decision to participate. Offering alternative ways to earn extra credit may serve to minimize undue influence. Undue influence may also occur when a person of authority, particularly one who controls sanctions and rewards, urges a person to act in a particular way. In this example, students may feel like they have to agree to participate because of the power that the professor holds over them. Coercion, on the other hand, occurs when a researcher or some other related entity explicitly or implicitly threatens an individual with harm if they do not participate in a study. Coercion would occur if the professor indicated that only students who participated in the research study would be permitted to attend a class trip.

Individuals who have certain vulnerabilities may be more likely to have compromised autonomy when making the decision to participate in research. For instance, low-income individuals may be more likely to experience undue influence relative to those with high incomes when large monetary incentives are offered. In this case, it is possible that their negative financial situation could influence their decision-making regarding participation and expose them to a greater level of risk than they normally would accept. Similarly, individuals may be recruited from settings that are, by nature, more coercive. These settings include prisons and other criminal justice settings, inpatient mental health facilities, and hospitals. When individuals are recruited from these settings, they may believe (correctly or incorrectly) that study participation is essential to their well-being. For instance, a prisoner may agree to participate in research because they believe that the warden will punish them if they do not. Likewise, a cancer patient whose oncologist tells them about a research trial may feel like they have to agree to participate in order to continue to receive excellent care by the physician. These two examples illustrate an important point: Threats to voluntariness may be real or *perceived*. In fact, it is often the case that study participants believe that they have been pressured or threatened when they, in fact, have not (Appelbaum, Lidz, & Klitzman, 2009).

Strategies to Help Ensure Voluntariness. Although it has been less widely studied than the other elements of informed consent, several strategies have been identified to help researchers ensure the voluntariness of research participation. These strategies are described below.

Assessment. As is the case for knowingness and understanding, researchers can evaluate the extent to which a person's decision to participate in a study is autonomous and voluntary at the time of consent. Applebaum et al. (2009) suggest that the MacArthur Perceived Coercion Scale (PCS; Gardner et al., 1993) can be tailored for this purpose. The PCS is a brief, five-item true/false measure that was designed to assess perceptions of coercion in the context of inpatient psychiatric hospitalization. It has been modified in several studies (e.g., Festinger, Marlowe, Dugosh, Croft, & Arabia, 2008; Moser et al.,

2004) to measure perceived coercion in the context of research (e.g., "I chose to enter *the study*," "I felt like I could say no to entering *the study*"). Moser et al. (2004) developed the Iowa Coercion Questionnaire (ICQ), which incorporates the five modified PCS items along with new items, to address self-presentation concerns (e.g., "I entered the study to appear cooperative"). Although these instruments may be useful in identifying whether someone feels pressured or coerced, they may have limited utility to the research in that they do not identify the source or intensity of the influence.

The Coercion Assessment Scale (CAS; Dugosh, Festinger, Croft, & Marlowe, 2010; Dugosh, Festinger, Marlowe & Clements, 2014) was developed to identify and measure the intensity of specific sources of coercion for criminal justice clients who participate in research. Scale items were developed through focus groups and interviews with clients and stakeholders and were designed to identify possible sources of influence (both actual and perceived) in the decision to participate in research. Sample items include: "I entered the study because I thought it would help my court case" and "I felt the judge would like it if I entered the study." Respondents indicate their level of agreement with each item on a four-point Likert scale ranging from "not at all" to "a lot." The CAS may provide a template for developing context-specific measures, as different sources of pressure and coercion may exist in different types of trials and research contexts.

By identifying real or perceived sources of coercion, researchers can discuss these issues with the potential participants. This allows the researcher to correct any misperceptions

that the individual may have as well as address any real and existing issues that may compromise voluntariness. Importantly, these items can be built into existing consent quizzes and become a standard component in the informed consent process.

Research Intermediaries. Another strategy to help ensure the autonomy of research participants is the use of research intermediaries (Benson, Roth, Appelbaum, Lidz, & Winslade, 1988; Reiser & Knudson, 1993; Stroup & Appelbaum, 2003). A research intermediary, sometimes referred to as a research advocate, neutral educator, or ombudsmen, helps the potential participant understand what the study entails so that they can make an informed decision regarding research participation. Research intermediaries are advocates for the potential participant and are independent from the research, treatment, and related entities. Intermediaries can serve many functions, including explaining to the participant what the study will entail and the risks and benefits of participation, advocating for the participant, and monitoring adverse events that the participant may experience throughout the course of their study participation. In addition to improving understanding and recall of informed consent information (e.g., Coletti et al., 2003; Fitzgerald, Marotte, Verdier, Johnson, & Pape, 2002; Kucia & Horowitz, 2000), the use of research intermediaries can help to improve autonomy by reducing perceptions of coercion or undue influence (Festinger, Dugosh, Croft, Arabia, & Marlowe, 2011). In fact, several federal advisory panels have recommended the inclusion of research intermediaries to improve the informed consent process (e.g., Bioethics Interest Group,

2009; National Bioethics Advisory Commission, 1998; World Medical Association, 1997).

6.2 Cultural Context

It is important to recognize that these tenets and strategies may not apply when conducting research in developing countries (see Krogstad et al., 2010). International researchers often need to adopt different informed consent processes that consider and reflect the cultural context of the study setting. For instance, the need to document written consent may be irrelevant in societies that are largely illiterate. Similarly, developing countries often place less emphasis on individual autonomy and more on the role of the chief, elders, and community in determining social roles and participation in activities. When conducting research in these locales, researchers need to first seek the approval or consent of community leaders (rather than the individual) regarding research conduct and participation. As can be seen, the informed consent process may vary substantially for research conducted in developed and developing countries as well as for Western and non-Western cultures.

6.3 Conclusion

Emanating from a dark history of harms experienced by unwitting and vulnerable human subjects, the consent process continues to evolve as one of the most essential procedures for protecting individuals who participate in research. Now, conceptualized as an ongoing process rather than a one-time event, informed consent is designed to ensure that research participants: 1) have the capacity to make decisions about engaging in research, 2) are fully informed, understand, and recall the research process, the potential risks and harms of participation, and their protections, and 3) engage in the research voluntarily. A substantial body of research has identified numerous evidence-based strategies to improve the effectiveness of the consent procedure:

1. Use assessments such as the MacArthur Perceived Coercion Scale or Coercion Assessment Scale to evaluate the voluntariness of a prospective participant's decision to participate, the MacCAT-CR or UBACC to evaluate their decisional capacity, and consent quizzes and corrected feedback to screen for adequate comprehension.

2. Obtain informed consent from legally authorized representations for individuals with diminished decision-making capacity or community leaders for individuals who exercise communal decision-making, and/or use research intermediaries to minimize participants' perceptions of coercion or undue influence.

3. Clearly distinguish experimental and alternative treatments to diffuse therapeutic misconception.

4. Improve the readability of consent forms by using shorter sentences, text styling, and diagrams, or use multimedia consent formats such as computer-animated video presentations.

5. Provide incentives to enhance prospective participants' motivation to understand and recall consent information.

KEY TAKEAWAYS

- Informed consent should be conceptualized as an ongoing process, not a one-time event that occurs prior to study entry.
- For consent to be truly informed, individuals must demonstrate intelligence (i.e., capacity), knowingness (i.e., understanding and recall), and voluntariness (i.e., free from coercion or undue influence).
- Informed consent best practices include the systematic and standardized assessment of participants' intelligence, knowingness, and voluntariness prior to research participation.
- Researchers must be considerate of cultural factors when conducting research in international settings.

IDEAS FOR FUTURE RESEARCH

- How can we encourage researchers to implement much of what we have learned about effectively obtaining consent?
- How can we adapt our consent methods to serve diverse populations and cultures? *It should not be assumed that the informed process is one-size-fits-all.* Procedures that are effective and appropriate for cultures that emphasize independent decision-making may, for example, be less suitable among cultures that rely on communal decision-making.
- What are the best methods for obtaining consent in the new era of social media?

SUGGESTED READINGS

Appelbaum, P. S., Lidz, C. W., & Klitzman, R. (2009). Voluntariness of consent to research: A conceptual model. *Hastings Center Report*, *39*(1), 30–39.

Appelbaum, P. S., Roth, L. H., & Lidz, C. (1982). The therapeutic misconception: Informed consent in psychiatric research. *International Journal of Law and Psychiatry*, *5*, 319–329.

Carlson, L. (2013). Research ethics and intellectual disability: Broadening the debates. *The Yale Journal of Biology and Medicine*, *86*(3), 303–314.

Federal Policy for the Protection of Human Subjects, 45 C.F.R. § 46 (2017).

Krogstad, D. J., Dop, S., Diallo, A., Mzayek, F., Keating, J., Koita, O. A., & Toure, Y. T. (2010). Informed consent in international research: The rationale for different approaches. *American Journal of Tropical Medicine and Hygiene*, *83*(4), 743–747.

Weyzig, F., & Schipper, I. (2008). SOMO Briefing Paper on Ethics in Clinical Trials: #1: Examples of Unethical Trials. Retrieved from www .wemos.nl/wp-content/uploads/2016/07/ examples_of_unethical_trials_feb_2008.pdf

REFERENCES

Afolabi, M. O., Bojang, K., D'Alessandro, U., Imoukhuede, E. B., Ravinetto, R. M., Lardon, H. J., … Chandramohan, D. (2014). Multimedia informed consent tool for a low literacy African research population: Development and pilot testing. *Journal of Clinical Research and Bioethics*, *5*(3), 178.

Afolabi, M. O., McGrath, N., D'Alessandro, U., Kampmann, B., Imoukhuede, E. B., Ravinetto, R. M., … Bojang, K. (2015). A multimedia consent tool for research participants in the Gambia: A randomized controlled trial. *Bulletin of the World Health Organization*, *93*(5), 320–328A.

Allen, A. A., Chen, D. T., Bonnie, R. J., Ko, T. K., Suratt, C. E., Lee, J. D., ... O'Brien, C. P. (2017). Assessing informed consent in an opioid relapse prevention study with adults under current or recent criminal justice supervision. *Journal of Substance Abuse Treatment*, *81*, 66–72.

Appelbaum, P. S., & Grisso, T. (2001). *MacArthur Competence Assessment Tool for Clinical Research (MacCAT-CR)*. Sarasota, FL: Professional Resource Press/Professional Resource Exchange.

Appelbaum, P. S., Lidz, C. W., & Klitzman, R. (2009). Voluntariness of consent to research: A conceptual model. *Hastings Center Report*, *39*(1), 30–39.

Appelbaum, P. S., Roth, L. H., & Lidz, C. (1982). The therapeutic misconception: Informed consent in psychiatric research. *International Journal of Law and Psychiatry*, *5*, 319–329.

Beardsley, E., Jefford, M., & Mileshkin, I. (2007). Longer consent forms for clinical trials compromise patient understanding: So why are they lengthening? *Journal of Clinical Oncology*, *23*(9), e13–e14.

Benson, P. R., Roth, L. H., Appelbaum, P. S., Lidz, C. W., & Winslade, W. J. (1988). Information disclosure, subject understanding, and informed consent in psychiatric research. *Law and Human Behavior*, *12*(4), 455–475.

Bioethics Interest Group, National Institutes of Health. (2009). Research Involving Individuals with Questionable Capacity to Consent: Ethical Issues and Practical Considerations for Institutional Review Boards (IRBs). Retrieved from https://grants.nih.gov/grants/policy/questionablecapacity.htm

Bobb, M. R., Van Heukelom, P. G., Faine, B. A., Ahmed, A., Messerly, J. T., Bell, G., ... Mohr, N. M. (2016). Telemedicine provides noninferior research informed consent for remote study enrollment: A randomized controlled trial. *Academic Emergency Medicine*, *23*(7), 759–765.

California Code, Health and Safety Code, HSC § 24178 (2003).

Carlson, L. (2013). Research ethics and intellectual disability: Broadening the debates.

The Yale Journal of Biology and Medicine, *86*(3), 303–314.

Carpenter, W. T., Gold, J. M., Lahti, A. C., Queern, C. A., Conley, R. R., Bartko, J. J., ... Appelbaum, P. S. (2000). Decisional capacity for informed consent in schizophrenia research. *Archives of General Psychiatry*, *57*(6), 533–538.

Cassileth, B. R., Zupkis, R. V., Sutton-Smith, K., & March, V. (1980). Informed consent: Why are its goals imperfectly realized? *New England Journal of Medicine*, *302*, 869–900.

Centers for Disease Control and Prevention. (2011). US Public Health Service Syphilis Study at Tuskegee: The Tuskegee Timeline. Retrieved from www.cdc.gov/tuskegee/timeline.htm

Christopher, P. P., Appelbaum, P. S., Truong, D., Albert, K., Maranda, L., & Lidz, C. (2017). Reducing therapeutic misconception: A randomized intervention trial in hypothetical clinical trials. *PLoS One*, *12*(9), e018224.

Christopher, P. P., Stein, M. D., Springer, S. A., Rich, J. D., Johnson, J. E., & Lidz, C. W. (2016). An exploratory study of therapeutic misconception among incarcerated clinical trial participants. *AJOB Empirical Bioethics*, *7*(1), 24–30.

Coletti, A. S., Heagerty, P., Sheon, A. R., Gross, M., Koblin, B. A., Metzger, D. S., & Seage, G. R. (2003). Randomized, controlled evaluation of a prototype informed consent process for HIV vaccine efficacy trials. *Journal of Acquired Immune Deficiency Syndromes*, *32*(2), 161–169.

Dugosh, K. L., Festinger, D. S., Croft, J. R., & Marlowe, D. B. (2010). Measuring coercion to participate in research within a doubly vulnerable population: Initial development of the coercion assessment scale. *Journal of Empirical Research on Human Ethics*, *5*(1), 93–102.

Dugosh, K. L., Festinger, D. S., Marlowe, D. B., & Clements, N. T. (2014). Developing an index to measure the voluntariness of consent to research. *Journal of Empirical Research on Human Ethics*, *9*(4), 60–70.

Dunn, L. B., & Jeste, D. V. (2001). Enhancing informed consent for research and treatment. *Neuropsychopharmacology*, *24*(6), 595–607.

Edwards, S. J. L., Lilford, R. J., Thornton, J., & Hewison, J. (1998). Informed consent for clinical trials: In search of the "best" method. *Social Science and Medicine*, *47*(11), 1825–1840.

Enama, M. E., Hu, Z., Gordon, I., Costner, P., Ledgerwood, J. E., & Grady, C. (2012). Randomization to standard and concise informed consent forms: Development of evidence-based consent practices. *Contemporary Clinical Trials*, *33*, 895–902.

Faden, R. (1996). The Advisory Committee on Human Radiation Experiments. *Hastings Center Report*, *26*(5), 5–10.

Festinger, D. S., Dugosh, K. L., Croft, J. R., Arabia, P. L., & Marlowe, D. B. (2010). Corrected feedback: A procedure to enhance recall of informed consent to research among substance abusing offenders. *Ethics & Behavior*, *20*(5), 387–399.

Festinger, D. S., Dugosh, K. L., Croft, J. R., Arabia, P. L., & Marlowe, D. B. (2011). Do research intermediaries reduce perceived coercion to enter research trials among criminally involved substance abusers? *Ethics & Behavior*, *21*(3), 252–259.

Festinger, D. S., Dugosh, K. L., Marlowe, D. B., & Clements, N. (2014). Achieving new levels of recall in consent to research by combining remedial and motivational techniques. *Journal of Medical Ethics*, *40*(4), 264–268.

Festinger, D. S., Marlowe, D. B., Croft, J. R., Dugosh, K. L., Arabia, P. L., & Benasutti, K. M. (2009). Monetary incentives improve recall of research consent information: It pays to remember. *Experimental and Clinical Psychopharmacology*, *17*(2), 99–104.

Festinger, D., Marlowe, D., Dugosh, K., Croft, J., & Arabia, P. (2008). Higher magnitude cash payments improve research follow-up rates without increasing drug use or perceived coercion. *Drug and Alcohol Dependence*, *96* (1–2), 128–135.

Festinger, D. S., Ratanadilok, K., Marlowe, D. B., Dugosh, K. L., Patapis, N. S., & DeMatteo, D. S. (2007). Neuropsychological functioning and recall of research consent information among drug court clients. *Ethics & Behavior*, *17*(2), 163–186.

Fitzgerald, D. W., Marotte, C., Verdier, R. I., Johnson, W. D., & Pape, J. W. (2002). Comprehension during informed consent in a less-developed country. *Lancet*, *360*, 1301–1302.

Flory, J., & Emanuel, E. (2004). Interventions to improve research participants' understanding in informed consent for research: A systematic review. *Journal of the American Medical Association 292*, 1593–1601.

Gardner, W., Hoge, S. K., Bennett, N., Roth, L. H., Lidz, C. W., Monahan, J., & Mulvey, E. P. (1993). Two scales for measuring patients' perceptions for coercion during mental hospital admission. *Behavioral Sciences and the Law*, *11*(3), 307–321.

Gupta, U. C. (2013). Informed consent in clinical research: Revisiting few concepts and areas. *Perspectives in Clinical Research*, *4*(1), 26–32.

Iacono, T., & Carling-Jenkins, R. (2012). The human rights context for ethical requirements for involving people with intellectual disability in medical research. *Journal of Intellectual Disability Research*, *56*(11), 1122–1132.

Jeste, D. V., Palmer, B. W., Appelbaum, P. S., Golshan, S., Glorioso, D., Dunn, L. B., ... Kraemer, H. C. (2007). A new brief instrument for assessing decisional capacity for clinical research. *Archives of General Psychiatry*, *64*(8), 966–974.

Jeste, D. V., Palmer, B. W., Golshan, S., Eyler, L. T., Dunn, L. B., Meeks, T., ... Appelbaum, P. S. (2009). Multimedia consent for research in people with schizophrenia and normal subjects: A randomized controlled trial. *Schizophrenia Bulletin*, *35*(4), 719–729.

Kass, N., Taylor, H., Ali, J., Hallez, K., & Chaisson, L. (2015). A pilot study of simple interventions to improve informed consent in clinical research: Feasibility, approach, and results. *Clinical Trials*, *12*(1), 54–66.

Kim, E. J., & Kim, S. H. (2015). Simplification improves understanding of informed consent information in clinical trials regardless of health literacy level. *Clinical Trials*, *12*(3), 232–236.

Kraft, S. A., Constantine, M., Magnus, D., Porter, K. M., Lee, S. S., Green, M., … Cho, M. K. (2017). A randomized study of multimedia informational aids for research on medical practices: Implications for informed consent. *Clinical Trials*, *14*(1), 94–102.

Krogstad, D. J., Dop, S., Diallo, A., Mzayek, F., Keating, J., Koita, O. A., & Toure, Y. T. (2010). Informed consent in international research: The rationale for different approaches. *American Journal of Tropical Medicine and Hygiene*, *83*(4), 743–747.

Kucia, A. M., & Horowitz, J. D. (2000). Is informed consent to clinical trials an "upside selective" process in acute coronary syndromes. *American Heart Journal*, *140*, 94–97.

Levine, R. (1992). Clinical trials and physicians as double agents. *The Yale Journal of Biology and Medicine*, *65*, 65–74.

Lidz, C. W. (2006). The therapeutic misconception and our models of competency and informed consent. *Behavioral Sciences and the Law*, *24*(4), 535–546.

Madeira, J. L., & Andraka-Christou, B. (2016). Paper trials, trailing behind: Improving informed consent to IVF through multimedia approaches. *Journal of Law and the Biosciences*, *3*(1), 2–38.

Mandava, A., Pace, C., Campbell, B., Emanuel, E., & Grady, C. (2012). The quality of informed consent: Mapping the landscape. A review of empirical data from developing and developed countries. *Journal of Medical Ethics*, *38*(6), 356–365.

Martel, M. L., Klein, L. R., Miner, J. R., Cole, J. B., Nystrom, P. C., Holm, K. M., & Biros, M. H. (2017). A brief assessment of capacity to consent instrument in acutely intoxicated emergency department patients. *The American Journal of Emergency Medicine*. doi:10.1016/j.ajem.2017.06.043. [Epub ahead of print]

Matsui, K., Lie, R. K., Turin, T. C., & Kita, Y. (2012). A randomized controlled trial of short and standard-length consent for a genetic cohort study: Is longer better? *Journal of Epidemiology*, *22*, 308–316.

Milgram, S. (1974). *Obedience to Authority*. New York: HarperCollins.

Miller, C. M., Searight, H. R., Grable, D., Schwartz, R., Sowell, C., & Barbarash, R. A. (1994). Comprehension and recall of the informational content of the informed consent document: An evaluation of 168 patients in a controlled clinical trial. *Journal of Clinical Research and Drug Development*, *8*(4), 237–248.

Morán-Sánchez, I., Luna, A., & Pérez-Cárceles, M. D. (2016). Enhancing the informed consent process in psychiatric outpatients with a brief computer-based method. *Psychiatry Research*, *245*, 354–360.

Moser, D. J., Arndt, S., Kanz, J. E., Benjamin, M. L., Bayless, J. D., Reese, R. L., … Flaum, M. A. (2004). Coercion and informed consent in research involving prisoners. *Comprehensive Psychiatry*, *45*(1), 1–9.

Muss, H. B., White, D. R., Michielutte, R., Richards, F., Cooper, M. R., William, S., … Spurr, C. (1979). Written informed consent in patients with breast cancer. *Cancer*, *43*, 549–556.

National Bioethics Advisory Commission. (1998). *Research Involving Persons with Mental Disorders that May Affect Decision-Making Capacity* (Vol. 1). Rockville, MD: US Government Printing Office.

National Institutes of Health (NIH). (2009). Research Involving Individuals with Questionable Capacity to Consent: Points to Consider. Retrieved from https://grants.nih.gov/grants/policy/questionablecapacity.htm

Neff, M. J. (2008). Informed consent: What is it? Who can give it? How do we improve it? *Respiratory Care*, *53*(10), 1137–1341.

Neilson, G., Chaimowitz, G., & Zuckerberg, J. (2015). Informed consent to treatment in psychiatry. *Canadian Journal of Psychiatry*, *60*, 1–12.

Nishimura, A., Carey, J., Erwin, P. J., Tilburt, J. C., Murad, M. H., & McCormick, J. B. (2013). Improving understanding in the research informed consent process: A systematic review of 54 interventions tested in randomized control trials. *BMC Medical Ethics, 14*, 28.

O'Brien, C. P. (2002). Re: Enhancing informed consent for research and treatment [Letter to the editor]. *Neuropsychopharmacology, 26*(2), 273.

Reiser, S. J., & Knudson, P. (1993). Protecting research subjects after consent: The case for the research intermediary. *IRB, 15*(2), 10–11.

Robinson, G., & Merav, A. (1976). Informed consent: Recall by patients tested post-operatively. *The Annals of Thoracic Surgery, 22*, 209–212.

Rosique, I., Pérez-Cárceles, M. D., Romero-Martín, M., Osuna, E., & Luna, A. (2006). The use and usefulness of information for patients undergoing anaesthesia. *Medicine and Law, 25*(4), 715–727.

Rothwell, E., Wong, B., Rose, N. C., Anderson, R., Fedor, B., Stark, L. A., & Botkin, J. R. (2014). A randomized controlled trial of an electronic informed consent process. *Journal of Empirical Research on Human Research Ethics, 9*(5), 1–7.

Ryan, C., Seymour, B., Stephens, M., Comly, R., Sierka, C., Musselman, T. G., ... Festinger, D. S. (2015, June). Informed Consent Practices among NIH-funded Researchers. Poster presented at the 77th annual meeting of the College on Problems of Drug Dependence, Phoenix, AZ.

Seaman, J. B, Terhorst, L., Gentry, A., Hunsaker, A., Parker, L. S., & Lingler, J. H. (2015). Psychometric properties of a decisional capacity screening tool for individuals contemplating participation in Alzheimer's disease research. *Journal of Alzheimer's Disease, 46*(1), 1–9.

Silva, L., Sorell, G., & Busch-Rossnagel, N. (1988). Biopsychosocial discriminators of alcoholic and nonalcoholic women. *Journal of Substance Abuse, 1*(1), 55–65.

Siu, J. M., Rotenberg, B. W., Franklin, J. H., & Sowerby, L. J. (2016). Multimedia in the informed consent process for endoscopic sinus surgery: A randomized control trial. *Laryngoscope, 126*(6), 1273–1278.

Sonne, S. C., Andrews, J. O., Gentilin, S. M., Oppenheimer, S., Obeid, J., Brady, K., ... Magruder, K. (2013). Development and pilot testing of a video-assisted informed consent process. *Contemporary Clinical Trials, 36*(1), 25–31.

Spencer, S. P., Stoner, M. J., Kelleher, K., & Cohen, D. M. (2015). Using a multimedia presentation to enhance informed consent in a pediatric emergency department. *Pediatric Emergency Care, 31*(8), 572–576.

Stiles, P. G., Poythress, N. G., Hall, A., Falkenbach, D., & Williams, R. (2001). Improving understanding of research content disclosures among persons with mental illness. *Psychiatric Services, 52*, 780–785.

Stroup, S., & Appelbaum, P. (2003). The subject advocate: Protecting the interests of participants with fluctuating decision making capacity. *IRB, 25*(3), 9–11.

Stunkel, L., Benson, M., McLellan, L., Sinaji, N., Bedairda, G., Emanuel, E., & Grady, C. (2010). Comprehension and informed consent: Assessing the effect of a short consent form. *IRB, 32*(4), 1–9.

Sugarman, J., McCrory, D. C., & Hubal, R. C. (1998). Getting meaningful informed consent from older adults: A structured literature review of empirical research. *Journal of the American Geriatric Society, 46*, 517–524.

Taub, H. A., & Baker, M. T. (1983). The effect of repeated testing upon comprehension of informed consent materials by elderly volunteers. *Experimental Aging Research, 9*, 135–138.

Taub, H. A., Kline, G. E., & Baker, M. T. (1981). The elderly and informed consent: Effects of vocabulary level and corrected feedback. *Experimental Aging Research, 7*, 137–146.

Tipotsch-Maca, S. M., Varsits, R. M., Ginzel, C., & Vescei-Marlovits, P. V. (2016). Effect of a multimedia-assisted informed consent procedure on the information gain, satisfaction, and anxiety of cataract surgery patients. *Journal of Cataract and Refractive Surgery, 42*(1), 110–116.

US Department of Health and Human Services, Office of Human Research Protections. (n.d.). Informed Consent FAQs. Retrieved from www.hhs.gov/ohrp/regulations-and-policy/guidance/faq/informed-consent/index.html

Verheggen, F. W., & van Wijmen, F. C. (1996). Informed consent in clinical trials. *Health Policy, 36*(2), 131–153.

Wendler, D., & Prasad, K. (2001). Core safeguards for clinical research with adults who are unable to consent. *Annals of Internal Medicine, 135*(7), 587–590.

Westra, A. E., & de Beaufort, I. (2015). Improving the Helsinki Declaration's guidance on research in incompetent subjects. *Journal of Medical Ethics, 41*, 278–280.

Weyzig, F., & Schipper, I. (2008). SOMO Briefing Paper on Ethics in Clinical Trials: #1: Examples of Unethical Trials. Retrieved from www.wemos.nl/wp-content/uploads/2016/07/examples_of_unethical_trials_feb_2008.pdf

White, S., & Dillow, S. (2005). *Key Concepts and Features of the 2003 National Assessment of Adult Literacy (NCES 2006–471).* US Department of Education. Washington, DC: National Center for Education Statistics.

Winter, M., Kam, J., Nalavenkata, S., Hardy, E., Handmer, M., Ainsworth, H., … Louie-Johnsun, M. (2016). The use of portable video media vs standard verbal communication in the urological consent process: A multicenter, randomised controlled, crossover trial. *BJU International, 118*(5), 823–828.

Wirshing, D. A., Wirshing, W. C., Marder, S. R., Liberman, R. P., & Mintz, J. (1998). Informed consent: Assessment of comprehension. *American Journal of Psychiatry, 155*, 1508–1511.

World Medical Association. (1997). Declaration of Helsinki: Recommendations Guiding Physicians in Biomedical Research Involving Human Subjects. [Reprinted in the] *Journal of the American Medical Association, 277*, 925–926.

Wynn, S. (2014). Decisions by surrogates: An overview of surrogate consent laws in the United States. *Bifocal: A Journal of the Commission on Law and Aging, 36*(1), 10–13.

7 Participant Preknowledge and Attitudes in Research

Elena C. Papanastasiou

The rise of the internet during the last decades has significantly increased the ease and frequency with which people can access information on the web. At the same time, people are constantly bombarded with information through web pages, social media, emails, etc. A teacher may search online for teaching methods that are effective for classes with high percentages of refugee students. A college student may obtain information, through Facebook groups, on the best exercises for strengthening the abdominal muscles. A parent can ask other parents online about their beliefs as to whether they should immunize their children or not. At the same time, individuals have the option to consult scientific, peer-reviewed journal articles, which are now available to the general public more easily than ever and are frequently widely publicized online even before they are printed (Weingart & Faust, 2014). However, the ease of access to information on the web has not been coupled with relevant training on how to critically evaluate the validity of the claims that are being made from the various sources mentioned above (Papanastasiou, 2007).

The lack of training in research methodology might be one of the factors that contributes to the inability of individuals to evaluate claims that are made online and elsewhere. Another factor is that of confirmation bias, which is the tendency to interpret and recall information that confirms one's pre-existing views or knowledge (Plous, 1993). Both of these factors can lead to erroneous decision-making that can sometimes lead to unfortunate or life-threatening outcomes. For example, a teacher who chooses to consult a blog written by a racist community might end up using ineffective teaching strategies with his immigrant students. The college student with health issues might be physically injured while trying the exercises obtained from a Facebook group. A child's life might even be lost due to the advice offered by non-professionals, on the supposed links between immunization and autism development.

At the same time, social media is flooded with questionnaires that can predict your personality, based on your favorite TV shows, as well as identify the Harry Potter character you would get along with, based on your knowledge about Hogwarts. These "playful questionnaires" entail a number of risks (Better Business Bureau, 2016) and do not claim to be scientific. Moreover, many individuals do not have the basic knowledge of research to be able to distinguish the results of peer-reviewed research articles from unsubstantiated claims based on bad science. Therefore, efforts are currently being made to expose students to the research process from an early age. Although student research training and involvement has most often been associated with graduate students, undergraduate students are increasingly becoming more involved in the research process, and universities are investing substantial resources to teach research skills (Murtonen & Lehtinen, 2003).

There are many ways in which students can become involved in the research process: a) students can engage in research by participating in projects as part of the learning process; b) students can work on the projects of experienced researchers as research assistants; c) students can undertake research projects themselves as part of a thesis; or d) students can attend, and possibly present at, research conferences. In addition to the research opportunities mentioned above, students in many disciplines (e.g., the social sciences, education, humanities) have the opportunity to take specifically designed research courses during which they can formally learn about and engage in the research process. These courses are now offered at different academic levels, so students can take research methods courses either at the Associate's, Bachelor's, Master's, or the Doctoral levels. The overall goal of the "students as researchers" pedagogic approach is to help develop the research culture in universities by breaking the long-standing disconnect between teaching and research (Walkington, 2015).

The research studies that are performed by students in the social and behavioral sciences come in many different methodological forms and serve many types of research goals. A portion of the research is descriptive in nature, and, thus, aims to describe the study participants at a given point in time (see Elliott, Chapter 4). For example, how have successful businesspeople with learning disabilities managed to overcome their disabilities? Another large portion of social and behavioral studies research is experimental in nature (see Nichols & Edlund, Chapter 3) and tries to measure and explain the effects of certain variables or interventions on dependent variables of interest. For example, how does the development of children's creativity affect their overall achievement in school? Regardless of the goals of the research studies or the types of methodologies that they employ, a number of issues could "contaminate" the results of each study, thus decreasing the validity of the data obtained, which in turn affect the results and the conclusions that are reached.

To be able to ensure the high quality of the data that are obtained in research studies, it is essential for researchers to be aware of the factors that could negatively influence the validity of the data, which are therefore considered as threats to the internal validity of a study's results (see Wagner & Skowronski, Chapter 2 for a more general discussion of this issue). By being aware of the possible threats to internal validity, efforts can be made to minimize or control them during the research process. The goal of this chapter is to discuss how preknowledge, and pre-existing attitudes can impact the ways in which individuals respond to research studies as well as possible methods of minimizing these threats to the validity of a study's results.

7.1 The Role of Measurement in the Research Process

An essential requirement of any research study in the social and behavioral sciences includes the ability of the researcher to be able to obtain accurate measures of the variables of interest in the study. This is not always an easy or straightforward task in the social and behavioral sciences since many variables of interest are not directly observable (Crocker & Algina, 2006). Therefore, questionnaires, tests, and other similar tools are utilized so that study participants can externalize their opinions, attitudes, knowledge, beliefs, etc. (see Stassen & Carmack, Chapter 12 for a detailed exposition on this issue). However, a number of issues can interfere with this process of data collection, add error to results, and eventually compromise the quality of the conclusions that are reached in a research study. Two such issues, which are not always emphasized enough either in research courses or in research textbooks, are a) participant preknowledge, and b) the participant's pre-existing attitudes toward the research study or toward research in general.

7.1.1 Participant Preknowledge

Participant preknowledge occurs when study participants obtain foreknowledge of information relevant to the research study, which has the potential to contaminate the quality of the data obtained from the study. For example, when people who have already participated in a research study inform future participants about the study (Edlund, Sagarin, Skowronski, Johnson, & Kutter, 2009), the future participants possess preknowledge. The act of

informing future participants of the details of a study is termed crosstalk (Edlund et al., 2009).

Crosstalk can take many different forms (and have varying degrees of seriousness). Nine forms, in particular, have been identified by Edlund, Lange, Sevene, Umansky, Beck, and Bell (2017) in relation to research studies on Amazon's Mechanical Turk. According to Edlund et al. (2017), the most problematic type of crosstalk is "Key Crosstalk." Key Crosstalk involves the act of sharing important information about aspects of the research process or protocol, of which participants should not be aware beforehand. This practice is particularly problematic since it can influence the ways in which participants respond to a study, thus affecting the degree of validity of the data obtained. In an experimental study performed by Nichols and Maner (2008), a confederate was sent by the researchers to inform study participants that, despite what they will be told by the researchers, the true intent of the study was to test if people prefer items that are presented on the left side (this bogus information provided by the confederate would represent a form of crosstalk). The results of this study found that participants did act in ways to support the researchers' hypothesis by mostly choosing the items that were located to their left, thus affecting the validity of the data that were obtained from the purported study. The degrees to which this occurred varied among individuals. However, this effect was more pronounced with participants who held more positive attitudes toward the study and toward the experimenter and especially when the subjects were single females with a male researcher.

In contrast to Key Crosstalk, Low Crosstalk is not seen as threatening to the validity of the results (Edlund et al., 2017). Low Crosstalk occurs when the information that is shared does not interfere with the study in any way. For example, Low Crosstalk could involve notifying future participants of study details that the researchers plan to tell the participants at the start of the study anyway. Such information could include providing information about how a questionnaire will be administered (e.g., computer based) or how long their participation in the study will last. Knowing such information in advance should not interfere with the quality of the data that will be obtained and would not affect the study negatively in any way.

Beyond Key and Low Crosstalk, other types of participant crosstalk have the potential to influence the attitudes of participants in a research study, either positively or negatively. Crosstalk that has offered praise to the research study, either due to its topic or due to the incentives that are offered for study participation, is likely to elicit more positive reactions from participants. Such reactions are likely to result in more positive attitudes toward the experiment and the experimenter (Nichols & Maner, 2008), and they might also lead to higher participation rates in the study. This is not necessarily bad as long as this praise does not lead participants to act as "Good Subjects." The Good Subject Effect is when participants act in ways that they believe are likely to please the researchers and can occur either due to social desirability or due to crosstalk (Nichols & Maner, 2008). This is usually done by purposefully responding in ways that help support the researchers' hypotheses, as in the example presented above.

When crosstalk includes negative comments about a study, then negative attitudes can be invoked. Crosstalk that includes complaints of any source will potentially make participants less likely to take part in a study or to make an effort to try their best while responding to the research prompts. Thus, negative crosstalk may lead to careless responses that are also known as aberrant responses. This occurs when the way research participants respond to a study (e.g., while responding to a questionnaire) is not based on the item content but on other reasons. Some patterns of careless responses include providing random answers, consistently choosing the same extreme response (e.g., strongly disagreeing with all items on a scale), consistently choosing the midpoint of the scale, etc. Such actions are extremely problematic since they increase error, and thus pose a significant threat to the validity of the study results (Schmitt, Cortina, & Whitney, 1993).

To explore the effects of random responding, Osborne and Blanchard (2011) examined the data from a research study that examined the effects of two teaching methods on students' learning and growth. The results of their study showed that: a) the differences in the two teaching methods were masked when the participants' random responses were kept in the analyses, and b) growth was less likely to be observed in the cases of participants who were unmotivated to perform well in the study. Thus, it is clear that careless responding can undermine the validity of a study's results and diminish the researcher's ability to obtain a realistic representation of a situation from their data.

The frequency with which crosstalk takes place is difficult to measure. An experimental study by Lichtenstein (1970) had estimated that the rate of crosstalk was about 78%. It should be noted, however, that in that specific study, the students were prompted to reveal information about the study as part of the experimental design. Recently, researchers have suggested that these rates are less than 5% within university settings (Edlund et al., 2014) but potentially higher online (Edlund et al., 2017; see Guadagno, Chapter 5 for a discussion of the unique challenges of online research). Moreover, the extent to which crosstalk occurs varies depending on the settings in which the studies take place. University students, for example, might be less likely to participate in crosstalk when the methodology of the study involves randomly sampling students from various courses in a large university, as students will be less likely to know others who have participated in the study (compared to a school-based classroom study). Finally, crosstalk cannot occur when the researchers utilize whole classes who concurrently take part in the research study, since all participants learn about the study at the same time.

7.1.2 Attitudes Beyond Crosstalk

Pre-existing attitudes and attitudes toward research that might exist either due to extra-experimental communication (termed as Participant Crosstalk) or due to prior experiences with any specific aspects of the research process could also affect the research process and the quality of the research results. Attitudes, in general, according to the tripartite framework originally presented by McGuire (1969), may be defined as the cognitive, affective, and behavioral predispositions toward a concept, a situation, an object, a course, etc. Thus, the attitudes that individuals hold, regardless of their type or source, could potentially have an effect on the ways in which they engage in the research process, thus affecting the internal validity of the study results.

The most frequently cited attitude in relation to research methods is that of anxiety. Unfortunately, despite the positive effects of engaging students in research, research courses are also associated with negative attitudes toward research for some students (e.g., Murtonen, 2005; Murtonen & Lehtinen, 2003; Papanastasiou & Zembylas, 2008). Some students dislike research because they do not see themselves as researchers and because they do not believe that they will need research skills in their future (Murtonen, Olkinuora, Tynjala, & Lehtinen, 2008). Other students dislike research because they do not feel empowered to understand or use it (Lodico, Spaulding, & Voegtlee, 2004; Onwuegbuzie, Slate, Patterson, Watson, & Schwartz, 2000). Regardless of the reasons one might have the negative attitudes toward research, these attitudes might affect the quality of the research that is performed, either from a researcher or from a research subject perspective.

One consequence of research anxiety is associated with an individual's willingness to participate in research studies as well as with the quality of the data that might be provided by study participants. Individuals with high levels of research anxiety are least likely to choose to voluntarily participate in research studies as subjects. Moreover, such individuals are least likely to make an effort to respond truthfully to the research prompts, either because they do not see the value of research in general or because they would like to complete their participation in the study as quickly as possible. In contrast, participants who value research tend to have more favorable attitudes toward it (Papanastasiou, 2005), and might be more likely to make an effort to participate in research studies, as well as attempt to accurately convey their true values and opinions on the data collection instruments. People's differential levels of willingness to participate in studies can also influence the representativeness of the results that are obtained. If the people who refrain from participating in research studies are different from individuals who participate in these studies, then the results obtained will not be representative and will be less generalizable to the general population.

In a study performed by Papanastasiou (2005), it was found that research anxiety had a medium-sized correlation with research usefulness in the profession ($r = .36, p < .01$) and research relevance to life ($r = .32, p < .01$). Also, the students who perceived research to be useful and relevant to life and their profession were more likely to have lower levels of research anxiety. Although this was not examined as a cause–effect relationship, the negative relationships among these variables should not be ignored. People who do not see the usefulness of research and who have anxiety toward research might be more likely to avoid the research process overall and be least likely to participate in research studies. In turn, students who do not see the value of research are also more likely to avoid placing adequate effort while participating in a study, thus significantly affecting the power and the validity of the results. However, when required to take a research methods course, students learn to appreciate the topic since they end up with lower levels of research anxiety, as well as consider research more useful compared to their counterparts who have not

taken such courses (Szymanski, Whitney-Thomas, Marshall, & Sayger, 1994).

A different study, by Papanastasiou and Karagiorgi (2012), found that being exposed to research methodology courses is related to an individual's involvement in research. More specifically, teachers who had taken research courses were more frequently involved in various research-related activities, such as in participating in research studies. Also, teachers who had taken two research courses (one at the graduate and one at the undergraduate level) read research articles more frequently. This finding might indicate that by taking two research courses, teachers have more opportunities to consolidate their research knowledge, which, in turn, enables them to read research articles with greater ease. These results are in accord with Bard, Bieschke, Herbert, and Eberz (2000), who found that providing students with explicit research opportunities within a training environment could influence their beliefs and attitudes toward research.

7.2 Controlling for the Effects of Preknowledge and Attitudes

In order to be able to control for the effects of both preknowledge and attitudes, researchers can utilize a number of methods. First, it is important for study participants to have an awareness of the significance of the study that they are involved in. Unless study participants can appreciate the importance and significance of a study, they might not necessarily be motivated to make an effort to respond truthfully, based on their true skills, knowledge, and beliefs. Therefore, in order to increase the likelihood that the resulting data will have a high degree of validity, researchers should make a great effort to convey the importance of their studies to each participant to try to obtain their full compliance.

Moreover, researchers themselves, in addition to professors of educational research, should make an effort to convey to study participants the importance of ethics in regard to research participation (see Ferrero & Pinto, Chapter 15). Research participants should be made aware of their rights when participating in a research study (see Festinger et al., Chapter 6), along with their responsibilities. Just as it is important for researchers to keep the data obtained confidential, it is equally important for the study participants to keep the details of a study confidential by not engaging in crosstalk (Edlund et al., 2017; Edlund et al., 2009). According to Edlund et al. (2009; 2014; 2017), crosstalk can be significantly reduced by simply debriefing them about the problems associated with it. Therefore, both researchers and professors of educational research should make an effort to inform their students of these confidentiality issues, as well as of the pitfalls that can occur when this confidentiality is not upheld from either side.

The research design that is selected for any study can also help control for participant preknowledge and pre-existing attitudes. For example, by randomly selecting the individuals who participate in a study, or by randomly placing participants in control and experimental groups, one can minimize the possibility that any pre-existing differences in preknowledge exist among the two groups or among the sample and the population (see also Nichols & Edlund, Chapter 3).

Collecting information on the participants prior to the start of the study, as a type of pre-test, could also be an essential tool for researchers. This information could be collected in the form of a questionnaire with four types of questions. The first type of questions could include information on participants' overall attitudes in relation to research methods. The administration of a short questionnaire, such as the revised version of the Attitudes Toward Research Scale, could assist researchers in this effort (Papanastasiou, 2014). The second type of questions that could be included are attitudinal questions related to the participants' attitudes toward the topic that is examined by the specific research study. The third type of questions that could be included are questions regarding crosstalk. These questions could ask the participants whether crosstalk had occurred, whether they were already aware of some details of the study, as well as whether or not they regarded the act of crosstalk as an ethical issue for research studies. Finally, the "pre-test" could include any other items that might be useful to obtain important background information on participants (e.g., whether they ever took a research methods course and their grades in this course, see Papanastasiou, 2007). This information can enable researchers to examine the characteristics of participants that are prone to engage in crosstalk or to have negative attitudes toward the research study, and thus act in unethical or untruthful ways.

None of the methods mentioned above, however, can guarantee that study participants will always act in an ethical manner by avoiding crosstalk or by always responding in a truthful manner, regardless of their pre-existing knowledge and attitudes (Blackhart, Brown, Clark, Pierce, & Shell, 2012).

Therefore, researchers should incorporate methods to detect aberrant responses in their data as part of their regular data cleaning routines (please see Blackhart & Clark, Chapter 9 for a detailed exposition on suspicion probes and debriefing). This should be done in all research studies but especially for a) those who lack significant incentives, resulting in participants not being motivated to perform at their peak and b) those whose results will form the basis of important policy and other decisions (Osborne & Blanchard, 2011). Collecting information, such as response times (Wise, 2006; Wise & Kong, 2005), could also be used by researchers in an attempt to measure the effort made by the participants while responding to tests or questionnaires that are computer based. Participants who have spent minimal time in responding to individual questions are least likely to have made an effort to respond truthfully to the measurement instruments used in the study.

Finally, the list of suggestions mentioned above that could be used for controlling for the effects of crosstalk and attitudes points to the need for universities to offer research methods courses as a requirement for all students. These courses should be offered at both the graduate as well as the undergraduate levels. Currently, university programs in certain countries (e.g., in Greece and Cyprus) require all of their undergraduate students in the field of education to take at least one research methods course. Moreover, according to the American Psychological Association's principles for quality in undergraduate education in psychology, undergraduate students should have a basic understanding of research methods and statistics early in their careers (APA, 2011). Thus, psychology programs typically include a research methods course as one of those required for undergraduate students. Teacher education departments in Finland also require all of their students to take a research methods course as part of their five-year program that covers the Bachelor's and Master's program in Education. However, this is not a common practice in many other social and behavioral science fields, and the offering of such courses varies significantly from country to country and from program to program. The need for the offering of such courses, though, is especially important in order to increase the quality of data and of the results that are obtained from research studies. Within these courses, students can be taught about the scientific method while concurrently learning about how to perform studies with high degrees of internal validity. Learning about the variables that can interfere with the validity of research results (e.g., preknowledge issues) should be an important component of such courses. At the same time, these courses can help students become critical consumers of research so that they can critically evaluate any ideas, studies, and information that they are confronted with (Papanastasiou, 2007).

7.3 Conclusion

Any research study involving human subjects entails a number of threats in regard to the degree of the validity of the data that are obtained. Although some threats to the internal validity of studies are mentioned frequently in textbooks or in some research methods courses, not all of them can always be covered. This is the case with crosstalk, although to some extent it overlaps with the internal validity threat of testing. Where like past examinees past participants in research studies might discuss study details with future participants. Comments about a research study can then interfere with the effort that the participants make in their responses. In addition, the attitudes that people hold in regard to the importance of research can influence their willingness to participate in a study and the truthfulness of their responses. However, many novice researchers appear to be unaware of these threats to validity. Moreover, it is likely that study participants themselves might be unaware of the adverse effects their actions have on the quality of the research results. These issues strengthen the need for all students to be exposed to research training, starting from an early age, in order to minimize the existence of such threats to the validity of studies.

First of all, starting from a young age, students should be exposed to research studies, as well as to the value of research. They should learn to appreciate research as well as the ways in which our lives have improved because of the results obtained from research. For example, without medical research, people would still perish from diseases that used to be incurable in the past. Without psychological research, many more individuals would suffer from debilitating mental illnesses that can now be cured with clinical interventions. Without consumer research studies, we would not be able to make informed decisions on what purchases to make. Being able to appreciate the importance of research is the first step that needs to be made in people's research training.

Once people learn to value research and appreciate the ways in which it has improved our lives, they should be made aware of the various threats to the validity of the data so they can avoid them. Learning about the variables that can interfere with the validity of research results should be an important component in all students training. Participant preknowledge and crosstalk should also be covered as an important threat to the validity of a study's results. In addition, research courses should also cover the topic of ethics, interrelated with its effects on validity.

Professors and teachers could also utilize the Attitudes Toward Research Scale (ATR) in an attempt to identify the attitudes that students hold in regard to research at the beginning of a course (Papanastasiou, 2005). The Attitudes Toward Research Scale is a self-report questionnaire on a seven-point Likert scale that measures students' attitudes toward the field of research. The revised version of the questionnaire, which was originally developed in English as well as in Greek, includes thirteen items which measure Research Usefulness, Research Anxiety, and Positive Predisposition Toward Research (Papanastasiou, 2014). The use of this scale can be valuable for both researchers as well as professors of educational research. By administering the ATR scale to individuals, one can have an indicator of the degree to which they value research based on how useful they consider it to be, as well as identify individuals with extreme levels of attitudes toward research. Thus, according to the ways in which participants perceive research, researchers or professors of educational research could make more effort in trying to "correct" the problematic situations in an attempt to increase the quality of the data that will be obtained.

From a researcher's perspective, the ways in which research studies with human participants are set up can also influence the quality of the data that are obtained. For example, crosstalk is less likely to occur when the researchers utilize whole classes who concurrently take part in the research study at a single time point. Informing participants of the significance of the research study and about the reasons why crosstalk must be avoided should also be established as a common practice by researchers. When participants are made aware of both of these issues, they are more likely to avoid the pitfalls associated with participant preknowledge and crosstalk. Finally, researchers should also thoroughly examine their data in order to detect aberrant responses and delete them once they are detected.

KEY TAKEAWAYS

- Participant preknowledge about a study, stemming from crosstalk, has the potential to threaten the validity of a study's results.
- Students' pre-existing attitudes toward research in general, toward the specific study, or the researcher also have the potential to affect the quality of the data that are obtained.
- Researchers should make great effort to convey a) the importance of their studies to study participants, and b) the importance of refraining from crosstalk. These efforts might help decrease the possibility that crosstalk will take place as well as minimize the effects of aberrant responses due to negative pre-existing attitudes.
- Research training, starting from an early age, can help individuals gain the proper tools to critically evaluate research data and research results, while being aware of the importance of ethical participation in research studies so as to act ethically and appropriately in any study.
- Researchers could also utilize specific research designs to minimize the effects of pre-existing knowledge and attitudes on data validity.

IDEAS FOR FUTURE RESEARCH

- To what extent is research anxiety related to aberrant response sets?
- Are individuals who perceive research as useful less likely to respond in untruthful ways or engage in crosstalk?
- Is the occurrence of crosstalk comparable in experimental studies compared to descriptive, questionnaire-based studies?
- Does the occurrence of crosstalk differ from country to country and from subgroup to subgroup?
- Are individuals who are aware of the ethics related to research studies least likely to engage in crosstalk?

SUGGESTED READINGS

American Psychological Association (APA). (2011). *APA Principles for Quality Undergraduate Education in Psychology.* Washington, DC: Author.

Edlund, J. E., Sagarin, B. J., Skowronski, J. J., Johnson, S., & Kutter, J. (2009). Whatever happens in the laboratory stays in the laboratory: The prevalence and prevention of participant crosstalk. *Personality and Social Psychology Bulletin, 35,* 635–642.

Nichols, A. L., & Maner, J. K. (2008). The good-subject effect: Investigating participant demand characteristics. *The Journal of General Psychology*, *135*, 151–165.

Osborne, J. W., & Blanchard, M. R. (2011). Random responding from participants is a threat to the validity of social science research results. *Frontiers in Psychology*, *1*, 220. http://doi.org/10.3389/fpsyg.2010.00220

Papanastasiou, E. C. (2014). Revised-Attitudes Toward Research Scale (R-ATR): A first look at its psychometric properties. *Journal of Research in Education*, *2*, 146–159.

Papanastasiou, E. C., & Zembylas, M. (2008). Anxiety in undergraduate research methods courses: Its nature and implications. *International Journal of Research and Method in Education*, *31*, 155–167.

REFERENCES

American Psychological Association (APA). (2011). *APA Principles for Quality Undergraduate Education in Psychology*. Washington, DC: Author.

Bard, C. C., Bieschke, K. J., Herbert, J. T., & Eberz, A. B. (2000). Predicting research interest among rehabilitation counseling students and faculty. *Rehabilitation Counseling Bulletin*, *44*(1), 48–55. http://dx.doi.org/10.1177/003435520004400107

Better Business Bureau. (2016, January 21). BBC warns consumers of Facebook IQ test traps. Retrieved from www.bbb.org/stlouis/news-events/news-releases/2016/01/bbb-advises-caution-with-facebook-iq-tests/

Blackhart, G. C., Brown, K. E., Clark, T., Pierce, D. L., & Shell, K. (2012). Assessing the adequacy of post-experimental inquiries in deception research and factors that promote participant honesty. *Behavior Research Methods*, *44*, 24–40.

Crocker, L., & Algina, J. (2006). *Introduction toClassical and Modern Test Theory* (3rd ed.). Belmont, CA: Wadsworth Publishing Company.

Edlund, J. E., Lange, K. M., Sevene, A. M, Umansky, J., Beck, S. A., & Bell, D. J. (2017). Participant crosstalk: Issues when using the Mechanical Turk. *The Quantitative Methods for Psychology*, *13*, 174–182.

Edlund, J. E., Nichols, A. L., Okdie, B. M., Guadagno, R. E., Eno, C. A., Heider, J. D., … Wilcox, K. T. (2014). The prevalence and prevention of crosstalk: A multi-institutional study. *The Journal of Social Psychology*, *154*, 181–185.

Edlund, J. E., Sagarin, B. J., Skowronski, J. J., Johnson, S. J., & Kutter, J. (2009). Whatever happens in the laboratory stays in the laboratory: The prevalence and prevention of participant crosstalk. *Personality and Social Psychology Bulletin*, *35*, 635– 642. doi:10.1177/0146167208331255

Lichtenstein, E. (1970). "Please don't talk to anyone about this experiment": Disclosure of deception by debriefed subjects. *Psychological Reports*, *26*(2), 485–486. doi:10.2466/pr0.1970.26.2.485

Lodico, M. G., Spaulding, D., & Voegtlee, K. H. (2004, April). Promising practices in the teaching of educational research. Paper presented at the American Educational Research Association Conference, San Diego, CA.

McGuire, W. J. (1969). The nature of attitudes and attitude change. In G. Lindzey & E. Aronson (Eds.), *The Handbook of Social Psychology* (2nd ed., Vol. 3, pp. 136–314). Reading, MA: Addison-Wesley.

Murtonen, M. (2005). University students' research orientations: Do negative attitudes exist toward quantitative methods? *Scandinavian Journal of Educational Research*, *49*, 263–280.

Murtonen, M., & Lehtinen, E. (2003). Difficulties experienced by education and sociology students in quantitative methods courses. *Studies in Higher Education*, *28*, 171–185.

Murtonen, M., Olkinuora, E., Tynjala, P., & Lehtinen, E. (2008). "Do I need research skills in working life?": University students' motivation and difficulties in quantitative methods courses. *Higher Education*, *56*, 599–612.

Nichols, A. L., & Maner, J. K. (2008). The good-subject effect: Investigating participant demand characteristics. *The Journal of General Psychology, 135*, 151–165.

Onwuegbuzie, A. J., Slate, J. R., Paterson, F. R. A., Watson, M. H., & Schwartz, R. A. (2000). Factors associated with achievement in educational research courses. *Research in the Schools, 7*, 53–65.

Osborne, J. W., & Blanchard, M. R. (2011). Random responding from participants is a threat to the validity of social science research results. *Frontiers in Psychology, 1*, 220. http://doi.org/10.3389/fpsyg.2010.00220

Papanastasiou, E. C. (2005). Factor structure of the "Attitudes Towards Research" Scale. *Statistics Education Research Journal, 4*, 16–26.

Papanastasiou, E. C. (2007). Research methods attitude, anxiety and achievement: Making predictions. In M. Murtonen, J. Rautopuro, & P. Väisänen (Eds.), *Learning and Teaching of Research Methods at University* (pp. 39–50). Turku, Finland: Finnish Education Research Association.

Papanastasiou, E. C. (2014). Revised-Attitudes Toward Research Scale (R-ATR): A first look at its psychometric properties. *Journal of Research in Education, 2*, 146–159.

Papanastasiou, E. C., & Karagiorgi, Y. (2012, April). The research-oriented professional teacher: An illusion or a possibility? – the case of secondary schoolteachers in Cyprus. Paper presented at the annual meeting of the American Educational Research Association, Vancouver, British Columbia, Canada.

Papanastasiou, E. C., & Zembylas, M. (2008). Anxiety in undergraduate research methods courses: Its nature and implications. *International Journal of Research and Method in Education, 31*, 155–167.

Plous, S. (1993). *The Psychology of Judgment and Decision Making*. New York: McGraw-Hill.

Schmitt, N., Cortina, J. M., & Whitney, D. J. (1993). Appropriateness fit and criterion-related validity. *Applied Psychological Measurement, 17*, 143–150.

Szymanski, E. M., Whitney-Thomas, J., Marshall, L., & Sayger, T. V. (1994). The effect of graduate instruction in research methodology on research self-efficacy and perceived research utility. *Rehabilitation Education, 8*, 319–331.

Walkington, H. (2015). *Students as Researchers: Supporting Undergraduate Research in the Disciplines of Higher Education*. York, UK: The Higher Education Academy.

Weingart, S. D., & Faust, J. S. (2014). Future evolution of traditional journals and social media medical education. *Emergency Medicine Australasia, 26*, 62–66.

Wise, S. L. (2006). An investigation of the differential effort received by items on a low-stakes computer-based test. *Applied Measurement in Education, 19*, 95–114.

Wise, S. L., & Kong, X. (2005). Response time effort: A new measure of examinee motivation in computer-based tests. *Applied Measurement in Education, 18*, 163–183.

8 Experimenter Effects

David B. Strohmetz

You designed what you think is the perfect study. You thoroughly reviewed the literature, developed sound hypotheses based on theory, and carefully planned how you will manipulate and measure your variables. You are confident that the data will provide the answers that you are seeking. These thoughts must have been going through Bob Rosenthal's mind when he conducted his dissertation research. But, much to his dismay, he didn't realize he forgot to consider one important factor that could impact his study – namely himself, the experimenter (Rosenthal, 1994a)! Only after he started trying to understand his findings did Rosenthal realize that his a priori expectations may have influenced the study's outcomes. More specifically, despite ostensibly maintaining scientific rigor and control, Rosenthal found that his three randomly assigned groups differed on his pretreatment measure. How was this possible? The groups had not yet been exposed to the experimental treatment and the instructions to the groups had been verbatim. He didn't try to influence his participants' behavior, but, nevertheless, he had. Knowing the participants' assigned condition from the onset of the experiment, he unconsciously treated the groups differently, increasing the likelihood that his research hypotheses would be supported.

Social and behavioral science research does not happen in a vacuum. It is inherently a social enterprise. In the typical experiment, a researcher creates conditions and then observes how the targets of interest, namely the participants, behave when encountering those conditions. However, the participants are not "zombies" (see Klein et al., 2012). They are cognizant of the situation, including the actual or implied presence of the experimenter. This awareness may consciously or unconsciously influence their subsequent behaviors. The experimenter is also not an automaton. Their attributes, expectations, and motivations may influence the study. For example, the experimenter may exhibit a confirmation bias, focusing on participant responses or behaviors that tend to favor the research hypotheses while ignoring or overlooking potentially contradictory evidence (Nickerson, 1998). The experimenter's personal qualities and other characteristics could also influence the results of a study, especially if these attributes differentially influence participants depending on their treatment condition. This chapter will explore these possibilities, focusing on the impact that the experimenter may have on a study independent of the variables being investigated.

The underlying assumption of this chapter is that the experimenter is an active participant in the study and therefore can influence the results of the study. This is not a new idea. Rosenzweig (1933) viewed the researcher as an integral part of the experimental situation, whose role in influencing the outcome of the study should not be overlooked. Rosenzweig warned that one's observational and motivational attitudes, as well as one's personality, could influence the results of a study independent of the variables under investigation and, therefore, are worthy of consideration in their own right.

How might the experimenter influence the results of a study? One possibility is through the types of decisions and observations the researcher makes during the course of the study. For example, the researcher may have a systematic bias in the nature or types of observations made during the study (Rosenzweig, 1933). The experimenter could exhibit a bias in how the data are interpreted or what conclusions are drawn based on the data. More nefariously, the experimenter could engage in questionable data-analytic strategies, such as p-hacking (reanalyzing the data to achieve a desired result), that increases the likelihood of finding support for one's hypotheses (Simmons, Nelson, & Simonsohn, 2011). We can describe these types of experimenter bias as reflecting noninteractional experimenter effects as the researcher is not actually influencing the participants' behavior or responses during the study (Rosnow & Rosenthal, 1997). Instead, the bias is being introduced through the nature of the experimenter's observations and interpretation of the data. The first part of this chapter will explore potential noninteractional experimenter effects in research.

As Rosenzweig (1933) suggested, the physical and psychological characteristics of the experimenter might also impact participants' behaviors during the course of the experiment. Imagine that an experimenter is interested in studying a treatment for reducing anxiety, but the experimenter is a large, intimidating individual with a menacing grin. If there is no reduction, or possibly an increase, in participants' anxiety level, could this be due to the ineffectiveness of the treatment or simply the presence of the experimenter? We will refer to this potential problem as an example of an interactional experimenter effect in that the researcher is directly influencing the participants' responses and behaviors during the course of the study (Rosnow & Rosenthal, 1997). In the second part of this chapter we will discuss how potential interactional experimenter effects, such as the researcher's physical and psychological attributes, expectations, and experiences, may influence the outcomes of a study. We will conclude this chapter with suggestions for minimizing potential experimenter effects in your study.

8.1 Noninteractional Experimenter Effects

Experimenters have the same biases and flaws in their thinking as everyone else (Kahneman, 2011). These biases can be subtle or overt, influencing not only the experimenter's observations but also his or her conclusions. Duarte, Crawford, Stern, Haidt, Jussim, and Tetlock (2015) argued that there is a liberal bias in the social and behavioral sciences, influencing the types of research pursued, the conclusions drawn, and the likelihood of publication. Because these biases are independent of the interactions between the researcher and the participant, we refer to them as noninteractional experimenter effects. For example, a researcher might inadvertently bias a study simply through the choice of experimental stimuli. Forster (2000) gave experienced and inexperienced researchers word pairs and asked them to identify which words would elicit the faster reaction time in a lexical decision-making task. He found that experience didn't matter: The researchers picked with reasonable accuracy the faster reaction time words, increasing the likelihood that the research hypotheses would be supported.

Strickland and Suben (2012) examined this potential bias in experimental philosophy. They had undergraduate research assistants develop sentences to test a hypothesis concerning people's intuitions about another person's mental state. While they were given the identical instructions for developing these sentences, the assistants were given different information as to the specific hypotheses being tested in the study. Strickland and Suben found that the sentence stimuli developed by the assistants for use in their experiment were influenced by their study expectations. The sentences they developed, which were reflective of the expected study outcomes, were rated by others as more natural sounding than the sentence stimuli generated for the comparison group. A subsequent online experiment showed that this bias in stimuli led to different outcomes, depending on the researchers' initial hypotheses.

In a meta-analysis of apparent sex differences in influenceability, Eagly and Carli (1981) noted that gender differences were more likely to be found in studies designed by male experimenters. Eagly and Carli suggested that researchers may be prone to selecting stimuli and designing experimental situations that favor their own sex. That is, the male experimenters tended to use message topics in their research designs that favored masculine interests and expertise of their own sex, increasing the likelihood of finding that females are more persuadable and conforming in influence situations.

What does this mean? Your choices of experimental and control stimuli may be implicitly biased, increasing the likelihood of you finding support for your research hypotheses. At best, your results may overstate the magnitude of the experimental effect you are investigating. At worst, they may simply reflect spurious findings. In either case, we are limited in estimating the true magnitude of the effect.

The potential for rater bias is always an issue in studies that require researchers to make judgements about participants' behaviors or characteristics (Hoyt, 2000). This bias is not limited to social and behavioral science research. Researchers had veterinary students observe either the positive and negative social interactions of pigs, the panting behaviors of cattle, or the expressive behaviors of hens in two different videos (Tuyttens et al., 2014). Although the students were led to believe that the videos were of animals being raised in two different situations, they were, in fact, scoring the same videos. The researchers found that the students' ratings were biased by the contextual information they were provided. Even though the students were watching the same animals, in their mind, they were judging entirely different animals, and this was reflected in their judgements.

Observer bias can lead to overestimation of potential experimental effects, particularly when subjective measurements are being made. Researchers compared studies in the life sciences that either did or did not employ the use of blind experimenter protocols (Holman, Head, Lanfear, & Jennions, 2015). Blind studies involve participants not knowing the hypotheses/manipulations (this is very common in the social and behavioral sciences); a related concept is double-blind studies where both the participants and the research assistants do not know the hypothesis/manipulation being employed for a particular participant (this is somewhat less common). They found that the nonblind studies reported both more statistically significant findings and larger effect sizes than comparable studies using blind protocols where the experimenters were unaware of the participants' treatment conditions. Hróbjartsson et al. (2013) reached similar conclusions in their meta-analysis of randomized clinical trials comparing blind and nonblind experimenters. Treatment effects tended be more exaggerated and stronger in studies where the researchers were not blind to treatment conditions.

Noninteractional experimenter effects can also be introduced into a study after the data are collected. Data analysis is ultimately a subjective enterprise in that it is the researcher who decides how to analyze the data to test one's hypotheses. The statistical choices that experimenters make may be biased in favor of their hypotheses, increasing the likelihood of concluding the presence of an effect that does not exist (Simmons et al., 2011). This bias toward false positive results has important implications for the advancement of science, especially given the tendency for journals to publish studies with statistically significant findings rather than studies with null results. Failed replication studies are less likely to enter the literature and are therefore unable to challenge the veracity of previously published findings (please see Soderberg & Errington, Chapter 14 for more details on replication). Researchers may pursue lines of research that are ultimately unfruitful given that the premise of these lines is based on questionable statistical findings. The replication crisis being experienced in the social and behavioral sciences is, in part, due to the realization that false positive results are a greater problem than previously thought (e.g., Baker & Dolgin, 2017; Chang & Li, 2015; Hales, 2016; Pashler & Harris, 2012).

It is important to note that this bias is not necessarily intentional or malicious. Rather, it can emerge as a by-product of the decisions that researchers make as they collect and analyze their data (Simmons et al., 2011). We are confronted with many decisions as we conduct our study. How many participants should we include? What if we are having trouble obtaining participants? Should we periodically analyze the data we have collected to see if we should continue the study? What should we do with data that appear to be outliers? Do we include them in our dataset or do we omit them? Which variables in our dataset should we combine, or which should we transform? What statistical tests should we use to test our hypotheses? Should we first control for certain factors before we test our hypotheses, or should we compare certain conditions and not others? Many of these are decisions we tend to make only after we have collected our data. We will naturally have a justification for every such decision we make. However, given our self-serving bias, there is an increased likelihood that the decisions will be biased in favor of our hypotheses (Kahneman, 2011; Nickerson, 1998). Buchanan and Lohse (2016) observed this bias among researchers who, when presented with different statistical scenarios, indicated that they were more likely to collect additional data or check for outliers in the data when the results of their statistical tests "approached" statistical significance as opposed to just "achieving" statistical significance (i.e., p = .06 vs. p = .04). They also said they would be more likely to write up the results regarding the central hypothesis when p = .04 than when p = .06, which would contribute to the publication bias and file-drawer effect in science where there is a tendency to only publish statistically significant findings (Fanelli, 2012; Rosenthal, 1979; 1993).

8.2 Minimizing Potential Noninteractional Experimenter Effects

It is clear that the experimenter can influence the results of a study, regardless of the participants' actual behaviors during the course of it. The question is how to minimize potential noninteractional experimenter effects. First, we should only employ the use of observers, raters, and research assistants who are unaware or blind to our study's purpose and hypotheses. Second, we must place a priority on training our observers in whom and what they should be observing when collecting the data. This will help to minimize potential intrarater and interrater errors, increasing our confidence in the quality of the data being used to address our research question.

With respect to potential noninteractional experimenter effects, once the data have been collected, it is important for researchers to have developed an a priori analysis plan that will inform the statistical analysis decisions they will need to make. This plan specifies how data-related decisions will be made with respect to issues such as the number of participants to be included in the study, how outliers in the data will be treated, and what comparisons will be made using which statistical tests. To promote these types of considerations, the Center for Open Science (https://cos.io/) enables researchers to openly share their research with the scientific community. This involves not only pre-registering one's study and including one's data analysis plan but also uploading all relevant materials and datasets (Nosek et al., 2015). Increasingly, research journals are offering "badges" to studies which were pre-registered on websites such as the Open Science Framework (https://osf.io/).

These open science initiatives also serve as a bulwark against more malicious noninteractional experimenter effects, namely intentional experimenter error or deceit (see Rosenthal, 2009/1975). While although extreme cases of fraudulent research have gained public notoriety (e.g., Bhattacharjee, 2012; Johnson, 2014), the flexibility that experimenters have in their data-analysis strategies can lead to ethically questionable practices (Rosenthal, 1994b). Such practices include p-hacking and data mining, where the researcher manipulates and analyzes the data until significant findings are identified (Head, Holman, Lanfear, Kahn, & Jennions, 2015). The researcher could also engage in "HARKing," or "Hypothesizing After the Results are Known" (Kerr, 1998), where significant findings are reported as if they were a priori hypotheses rather than the result of exploratory analyses. Not surprisingly, these practices improve the chances of publication despite the increased likelihood of generating false positive results (Simmons et al., 2011). By pre-registering one's study, researchers are publicly stating their intentions in terms of how they will gather and analyze their data, increasing the scientific community's trust that the results are not due, in part, to noninteractional experimenter effects. Such initiatives, however, may not be effective in minimizing interactional experimenter effects – we will discuss these next.

8.3 Interactional Experimenter Effects

Any social or behavioral science experiment is inherently a social interaction. It involves at least two individuals: The experimenter and the participant. Even online studies involve the implied presence of the researcher. The experimenter creates the circumstances for the participant to encounter and then observes how the participant responds. However, the participant is not responding "purely" to these circumstances. Rather, the participant is also responding to the experimenter. That is, the participant enters the experimental situation and immediately begins assessing both the situation and the experimenter (Strohmetz, 2006). This has implications for the study if the participant's observations of the experimenter and subsequent impressions influence the participant's behavior independent of the variables under investigation. We will call this potential influence "interactional experimenter effects" in that they emerge as the result of the direct interactions between the experimenter and the participant.

How might the experimenter influence the participants' behaviors during the study itself? One way is through the experimenter's physical or biosocial attributes such as gender or race. Whether or not this influence is of concern depends on whether these attributes are relevant to the variables being measured (Barnes & Rosenthal, 1985) or if they influence the participants' attitudes toward the experimenter in ways that affect their behaviors (Nichols & Maner, 2008). If this is not the case, then there is little concern that the results of one's study are due, in part, to the influence of these attributes. There is also little concern if the attributes affect all participants similarly, regardless of the experimental conditions. Things are more problematic if biosocial attributes differentially affect participants depending on one's experimental condition, as now the internal validity of the experiment may have been compromised.

Concerns about the differential effect that the experimenter's gender may have on the outcome of study are not a new idea (see Rumenik, Capasso, & Hendrick, 1977). For example, how male and female participants respond in pain tolerance studies may be due, in part, to the gender of the experimenter they encounter. Levine and De Simone (1991) had male and female participants take a cold pressor test where they had to submerge their hand in a bucket of ice cold water and report every fifteen seconds on the pain they were experiencing. Overall, female participants reported higher levels of pain. More importantly, there was an interaction effect whereby the average intensity of pain ratings self-reported by male participants was lower when the experimenter was female than male. This experimenter difference did not emerge for the female participants.

Using electrodes, Aslaksen, Myrbakk, Høifødt, and Flaten (2007) induced heat pain in the forearms of male and female participants. While there were no physiological or autonomic differences in response to the pain between the male and female participants depending on whether the experimenter was male or female, there were differences in their subjective reports. Male participants described the pain as being less intense when they were being tested by a female rather than a male experimenter. Furthermore, Lundström and Olsson (2005) observed that female participants who were unknowingly exposed to pheromones reported a stronger positive mood when they interacted with a male experimenter. These findings suggest the experimenter's gender could be a potential moderator in studies involving subjective measures of arousal or discomfort.

The gender of the experimenter may also influence how participants respond to self-report measures. One study found that female participants had a more positive attitude toward casual sex when they completed a survey in the presence of a female experimenter than when they completed the same survey alone (McCallum & Peterson, 2015). In a study of heterosexual undergraduates, males led to believe that women are becoming increasingly sexually permissive provided inflated estimates of their past sexual partners when the experimenter was a female (Fisher, 2007). Barnes and Rosenthal (1985) observed that the experimenter's sex and attractiveness also influenced participants' ratings of photos.

Other physical attributes of the experimenter may have an impact on participants' behavior. Marx and Goff (2005) had black and white participants take a challenging verbal test administered by either a black or white experimenter. They found that the black participants did worse on the task than white participants when the experimenter was white. Huang (2009) observed a similar difference when comparing the performance of black and white respondents who were given a vocabulary test by either a white or black interviewer. In both of these studies, it was argued that observed differences were due to a potential stereotype threat (the adverse reaction caused by the desire to not confirm negative stereotypes; Steele, 1997; Steele & Aronson, 1995). While this explanation has been debated (e.g., Spencer, Logel, & Davies, 2016), this research does suggest that the physical attributes of the experimenter may introduce a potential confound into the study if they influence the dependent variable outside of the variables under investigation.

Psychosocial, or personal and social attributes of the experimenter may also influence the results of a study (Rosenthal, 2009/1975; Rosnow & Rosenthal, 1997). Attributes such as the researcher's anxiety level or expressed warmth can potentially influence participants' behavior during the experiment (e.g., Winkel & Sarason, 1964), even if those participating are non-human (e.g., Elkins, 1987). As Rosenzweig (1933) had suggested, an experimenter's attitude alone can influence participants' behaviors. In one study, experimenters were instructed to respond in either an indifferent or caring manner when testing five- to six-year-old preschoolers engaged in a cognitive classification task (Fraysse & Desprels-Fraysse,

1990). Children who classified objects mainly on a common property basis were most successful when the experimenter appeared to be caring during the testing. While there was no difference in performance for those who were more cognitively flexible in their classification schemes, these children were more likely to seek advice from the caring rather than indifferent experimenter to help guide their behavior.

In addition to one's research attitudes, one's experimental expectations can play a powerful role in determining the outcome of one's study. This is what Rosenthal (1994a) realized when reflecting back on his dissertation research. His awareness of his hypotheses may have inadvertently influenced his behavior right from the start of his experiment, creating a self-fulfilling prophecy (Merton, 2016/1948). Rosenthal called this phenomenon an experimenter expectancy effect (Rosenthal, 2009/1975). The underlying assumption is that awareness of one's hypotheses can lead one to differentially interact with participants in ways independent of the treatment under investigation. These interactions, although often unintentional, can influence the outcome of the study in ways that increase the likelihood the research hypotheses are supported.

Rosenthal's dissertation experience led him to systematically study potential experimenter expectancy effects involving human and non-human research subjects (Rosenthal, 1994a). In an early experimental investigation, Rosenthal and Fode (1963) recruited students to participate in the supposed development of a test of empathy. As part of this project, the students were first to gain practice as a researcher by replicating "well established findings" involving people's ability to discern how much a person is experiencing success or failure simply based on a photo. Rosenthal and Fode told half of the researchers that the photos they were using had previously been rated, on average, as people who were successful. The other half were told their photos had been previously rated as people who were failures. The experimenters were also told that they would be paid more for their efforts if they "did a good job" – that their results replicated previous findings. All experimenters were instructed to limit their interactions with the participants to only reading the standardized instructions to them. Nevertheless, differences emerged between the two expectation groups, despite all of the participants rating the exact same photos. The ratings the experimenters obtained were consistent with their expectations of whether they had photos of people previously rated by others as being successful or failures. That is, the experimenters obtained results consistent with the type of results they were motivated to obtain, despite supposedly having structured and limited interactions with the participants.

More recently, Gilder and Heerey (2018) demonstrated how experimenter beliefs may account for changes in participants' behaviors in a priming study. In a series of studies, Gilder and Heerey either informed or misinformed experimenters as to the assigned power priming condition for each of their participants. Although Gilder and Heerey overall failed to find any priming effects, they did find that participants responded differently based on the experimenters' beliefs about their assigned priming condition. In addition, participants who the experimenter believed were in the high power condition rated their experimenter more positively on traits such as trustworthiness and friendliness. This suggests that an experimenter's study expectations may subsequently alter how they interact with participants independent of the actual treatment manipulations.

The impact of experimenter expectations on subsequent research findings is not limited to research involving humans (as noted above). In the famous case of "Clever Hans," Oskar Pfungst (1965/1911) demonstrated that the world-renowned "intellectual feats" of a horse were due to the unwitting cues provided by the horse's owner not by the horse's innate cognitive abilities. In an experimental study involving the operant conditioning of rats, Rosenthal and Lawson (1964) led experimenters to believe that the rats they were working with were either "Skinner box bright" or "Skinner box dull." Even though these labels were randomly assigned, the "bright" rats outperformed the "dull" rats on the same learning tasks, suggesting that it was the experimenter's expectations that influenced the rats' performance.

Rosenthal and Jacobson (1968) conducted an analogous investigation into how teacher expectations may influence children's learning. Teachers in an elementary school were led to believe that arbitrarily selected children would experience more intellectual gains over that school year. This belief became a self-fulfilling prophecy as these children did experience greater gains compared to the other students. Despite criticisms, it is clear that teacher expectations can influence subsequent student behaviors similarly to how experimenters can influence research participants' behaviors (e.g., Rosenthal, 1995; Rosenthal & Rubin, 1978).

8.4 Minimizing Interactional Experimenter Effects in Research

Accounting for potential experimenter effects is simply part of a good research design (see Wagner & Skowronski, Chapter 2). You should consider experimenter effects as a factor to either control or to investigate as you would any other experimental or confounding variable. For example, how might you control for potential experimenter expectancy effects? One strategy is to have participants interact only with experimenters who are unaware, or blind, to the study's purpose or hypotheses. If a person is unaware of the study's intended outcomes, it is less likely that he or she will be able to bias the study in favor of the desired results. A better suggestion would be to use multiple blind experimenters when conducting the actual study so that a priori beliefs about the desired study outcomes become simply a nuisance or random variable within the experiment.

Ensuring that one is not implicitly biased when developing the study materials and stimuli may be more difficult as it is typically the lead researcher who designs the actual study. The open science movement can help with identifying potential biases in one's study materials (Nosek et al., 2015). When all study materials and stimuli are made available on an open science website rather than simply summarized in a research article, others have the ability to evaluate the potential for hidden prejudices that may also account for the results (e.g., Eagly & Carli, 1981). Preregistering one's data-analysis plan and posting the final dataset also create a level of accountability to minimize potential questionable data-analytical strategies which increase the likelihood of finding support for one's hypotheses. These practices will help others evaluate whether your results are due, in part, to noninteractional experimenter effects.

To control for potential variations in how an experimenter interacts with the different treatment groups, you should standardize his or her interactions across all participants. Ideally, you would limit the interactions between the experimenter and participant during the course of the study to reduce potential opportunities for unintentional experimenter influences. This could be done with the use of computers and other technologies to manipulate the independent variables and record the participants' subsequent responses. In situations where this is not appropriate or practical, effort should be made to structure and standardize the experimenter–participant interactions as much as possible and to monitor these interactions for potential deviations which may bias the results of the study.

Rather than controlling for potential experimenter effects, one could incorporate them as another variable under investigation within the study. For example, suppose you were concerned that the sex of the experimenter could differentially affect participants' behaviors depending on their treatment condition. In this instance, you could conduct the study with both a female and male experimenter and use sex of experimenter as another study variable. This will allow you to statistically test whether participants' study behaviors or responses were due, in part, to an interaction between the experimenter's sex and the treatment conditions.

To identify potential experimenter expectancy effects, you could also use an expectancy control design (Rosenthal & Rosnow, 2008). In this design, you would treat experimenter expectancies as an additional independent variable in your study. Though costly in terms of resources, you could randomly assign experimenters to different experimental expectations groups and evaluate the impact that such expectancies might have on the outcome variable. Burnham (as cited in Rosenthal & Rosnow, 2008) used this design to examine the influence of experimenter expectancies on a learning discrimination study involving rats. In this study, the rats had one of two types of surgery: Lesioning which involved removing part of the rat's brain or sham surgery where the rat's brain was not actually damaged. In addition, the rats were randomly labeled as "lesioned rats" or "non-lesioned rats" regardless of the actual surgery the rat received. Not surprisingly, the lesioned rats performed worse on the discrimination task than the sham surgery rats. However, rats who were labeled as "lesioned rats" did worse on this discrimination task than the "non-lesioned" rats, regardless of their actual surgery status. What is interesting in this study is that the effect of the experimenter expectancies on the rat's performance was not that different than the effect of the actual surgery (Rosenthal & Rosnow, 2008). If you are concerned with potential experimenter expectancy effects in your study, it might be wise to consider such effects as simply another variable to be manipulated and investigated as part of your study. Doing so will increase your confidence in being able to rule out potential threats to the internal validity of your study.

8.5 Conclusion

Conducting social and behavioral scientific research is unlike other types of scientific research in that we are studying sentient organisms who are not only reacting to the conditions we have established for them, but also to the experimenters themselves. We may unwittingly influence the results of our study, either through our observational judgements, our data-related decisions, our biosocial or psychosocial attributes, or even by our experimental expectations. Awareness of these potential influences not only forces us to examine our own role within the psychological experiment but also leads us to strive to design and conduct better studies. This is a lesson that Rosenthal (1994a) learned after completing his dissertation research and one that will help you to become a better scientist.

KEY TAKEAWAYS

- Recognize that you are also part of the experimental situation and therefore can influence participants' responses and behaviors independent of the variables you are investigating.
- Increase the transparency in how you conducted your study and the basis for your conclusions by pre-registering your study on a website such as the Center for Open Science (https://cos.io).
- Employ the use of observers, raters, and/or research assistants who are blind to the study's purposes and hypotheses.
- Standardize and/or minimize the interactions between the experimenter and the participant during the course of study to reduce potential expectancy effects. Employ the use of technology to manipulate variables and to record the participants' responses and behaviors during the study.
- Consider who interacted with participants as another variable in your study so that you can assess to what extent the study outcomes may be due to experimenter effects.

IDEAS FOR FUTURE RESEARCH

- To what extent is the replication crisis driven by experimenter effects (and which effects specifically)?
- What effect will the open science movement have on the confidence the social and behavioral sciences have in our results?
- What paradigms are most susceptible to experimenter effects and what structural steps can be taken to minimize the effects?

SUGGESTED READINGS

Nichols, A. L., & Edlund, J. E. (2015). Practicing what we preach (and sometimes study): Methodological issues in experimental laboratory research. *Review of General Psychology, 19*, 191–202. doi:10.1037/gpr0000027

Nosek, B. A., Alter, G., Banks, G. C., Borsboom, D., Bowman, S. D., Breckler, S. J., … Yarkoni, T. (2015). Promoting an open research culture: Author guidelines for journals could help to promote transparency, openness, and reproducibility. *Science, 348*(6242), 1422–1425. doi:10.1126/science.aab2374

Rosenthal, R. (1994). Interpersonal expectancy effects: A 30-year perspective. *Current Directions in Psychological Science, 3*, 176–179. doi:10.1111/1467-8721.ep10770698

Rosnow, R. L., & Rosenthal, R. (1997). *People Studying People: Artifact and Ethics in Behavioral Research*. New York: Freeman.

Simmons, J. P., Nelson, L. D., & Simonsohn, U. (2011). False-positive psychology: Undisclosed flexibility in data collection and analysis allows presenting anything as significant. *Psychological Science, 22*, 1359–1366. doi:10.1177/0956797611417632

REFERENCES

Aslaksen, P. M., Myrbakk, I. N., Høifødt, R. S., & Flaten, M. A. (2007). The effect of experimenter gender on autonomic and subjective responses to pain stimuli. *Pain, 129*, 260–268. doi:10.1016/j.pain.2006.10.011

Baker, M., & Dolgin, E. (2017, January 19). Reproducibility project yields muddy results. *Nature, 541*, 269–270. doi:10.1038/541269a

Barnes, M. L., & Rosenthal, R. (1985). Interpersonal effects of experimenter attractiveness, attire, and gender. *Journal of Personality and Social Psychology, 48*, 435–446. doi:10.1037/0022-3514.48.2.435

Bhattacharjee, Y. (2013, April 26). The mind of a con man. *The New York Times*. Retrieved from www.nytimes.com

Buchanan, T. L., & Lohse, K. R. (2016). Researchers' perceptions of statistical significance contribute to bias in health and exercise science. *Measurement in Physical Education and Exercise Science, 20*, 131–139. doi:10.1080/1091367X.2016.1166112

Chang, A. C., & Li, P. (2015). Is economics research replicable? Sixty published papers from thirteen journals say "usually not." *Finance and Economics Discussion Series* 2015–083. Washington, DC: Board of Governors of the Federal Reserve System. doi:10.17016/FEDS.2015.083

Duarte, J. L., Crawford, J. T., Stern, C., Haidt, J., Jussim, L., & Tetlock, P. E. (2015). Political diversity will improve social psychological science. *Behavioral and Brain Sciences, 38*. doi:10.1017/S0140525X14000430,e130

Eagly, A. H., & Carli, L. L. (1981). Sex of researchers and sex-typed communications as determinants of sex differences in influenceability: A meta-analysis of social influence studies. *Psychological Bulletin, 90*, 1–20. doi:10.1037/0033-2909.90.1.1

Elkins, R. L. (1987). An experimenter effect on place avoidance learning of selectively-bred taste-aversion prone and resistant rats. *Medical Science Research, 15*, 1181–1182.

Fanelli, D. (2012). Negative results are disappearing from most disciplines and countries. *Scientometrics, 90*, 891–904. doi:10.1007/s11192-011-0494-7

Fisher, T. D. (2007). Sex of experimenter and social norm effects on reports of sexual behavior in young men and women. *Archives of Sexual Behavior, 36*, 89–100. doi:10.1007/s10508-006-9094-7

Forster, K. L. (2008). The potential for experimenter bias effects in word recognition experiments. *Memory and Cognition, 28*, 1109–1115.

Fraysse, J. C., & Desprels-Fraysse, A. (1990). The influence of experimenter attitude on performance of children of different cognitive ability levels. *The Journal of Genetic Psychology, 151*, 169–179.

Gilder, T. S. E., & Heerey, E. A. (2018). The role of experimenter belief in social priming. *Psychological Science*. https://doi.org/10.1177/0956797617737128 [Prepublished January 29.]

Hales, A. H. (2016). Does the conclusion follow from the evidence? Recommendations for improving research. *Journal of Experimental Social Psychology, 66*, 39–46. doi:10.1016/j.jesp.2015.09.011

Head, M. L., Holman, L., Lanfear, R., Kahn, A. T., & Jennions, M. D. (2015). The extent and consequences of p-hacking in science. *PLoS Biology, 13*, e1002106. doi:10.1371/journal.pbio.1002106

Holman, L., Head, M. L., Lanfear, R., & Jennions, M. D. (2015). Evidence of experimental bias in the life sciences: Why we need blind data recording. *PLoS: Biology*, *13*(7), e1002190. doi:10.1371/journal. pbio.1002190

Hoyt, W. T. (2000). Rater bias in psychological research: When is it a problem and what can we do about it? *Psychological Methods*, *5*, 64–86. doi:10.1037//1082-9S9X.5.1.64

Hróbjartsson, A., Skou Thomsen, A. S., Emanuelsson, F., Tendal, B., Hilden, J., Boutron, I., ... & Brorson, S. (2013). Observer bias in randomized clinical trials with measurement scale outcomes: A systematic review of trials with both blinded and nonblinded assessors. *Canadian Medical Association Journal*, *185*, E201–E211. doi:10.1503 /cmaj.120744

Huang, M. (2009). Race of the interviewer and the black–white test score gap. *Social Science Research*, *38*, 29–38. doi:10.1016/j. ssresearch.2008.07.004

Johnson, C. Y. (2014, May 30). Harvard report shines light on ex-researcher's misconduct. *Boston Globe*. Retrieved from www.bostonglobe.com

Kahneman, D. (2011). *Thinking, Fast and Slow*. New York: Farrar, Straus, and Giroux.

Kerr, N. L. (1998). HARKing: Hypothesizing after the results are known. *Personality and Social Psychology Review*, *2*, 196–217. doi:10.1207/ s15327957pspr0203_4

Klein, O., Doyen, S., Leys, C., de Saldanha da Gama, P. M., Miller, S., Questienne, L., & Cleeremans, A. (2012). Low hopes, high expectations: Expectancy effects and the replicability of behavioral experiments. *Perspectives on Psychological Science*, *7*, 572–584. doi:10.1177/ 1745691612463704

Levine, F. M., & De Simone, L. L. (1991). The effects of experimenter gender on pain report in male and female subjects. *Pain*, *44*, 69–72. https://doi.org/10.1016/0304-3959(91)90149-R

Lundström, J. N., & Olsson, M. J. (2005). Subthreshold amounts of social odorant affect mood, but not behavior, in heterosexual women when tested by a male, but not a female, experimenter. *Biological Psychology*, *70*, 197–204. doi:10.1016/j.biopsycho.2005.01.008

Marx, D. M., & Goff, P. A. (2005). Clearing the air: The effect of experimenter race on target's test performance and subjective experience. *British Journal of Social Psychology*, *44*, 645–657.

McCallum, E. B., & Peterson, Z. D. (2015). Effects of experimenter contact, setting, inquiry mode, and race on women's self-report of sexual attitudes and behaviors: An experimental study. *Archives of Sexual Behavior*, *44*, 2287–2297. doi:10.1007/s10508-015-0590-5

Merton, R. K. (2016). The self-fulfilling prophecy. *Antioch Review*, *74*, 504–521. [Originally published 1948.]

Nichols, A. L., & Maner, J. K. (2008). The good-subject effect: Investigating participant demand characteristics. *The Journal of General Psychology*, *135*, 151–165. doi:10.3200/ GENP.135.2.151-166

Nickerson, R. S. (1998). Confirmation bias: A ubiquitous phenomenon in many guises. *Review of General Psychology*, *2*, 175–220. doi:10.1037/1089-2680.2.2.175

Nosek, B. A., Alter, G., Banks, G. C., Borsboom, D., Bowman, S. D., Breckler, S. J., ... Yarkoni, T. (2015). Promoting an open research culture: Author guidelines for journals could help to promote transparency, openness, and reproducibility. *Science*, *348*(6242), 1422–1425. doi:10.1126/science.aab2374

Pashler, H., & Harris, C. R. (2012). Is the replicability crisis overblown? Three arguments examined. *Perspectives on Psychological Science*, *7*, 531–536. doi:10.1177/ 1745691612463401

Pfungst, O. (1965/1911). *Clever Hans (The Horse of Mr. on Osten)*. New York: H. Holt. [Reissued 1965 by Holt, Rinehart & Winston, New York.]

Rosenthal, R. (1979). The file drawer problem and tolerance for null results. *Psychological Bulletin*, *86*, 638–641. doi:10.1037/0033-2909.86.3.638

Rosenthal, R. (1993). Cumulating evidence. In G. Keren & C. Lewis (Eds.), *A Handbook for Data Analysis in the Behavioral Sciences:*

Methodological Issues (pp. 519–559). Hillsdale, NJ: Lawrence Erlbaum Associates.

Rosenthal, R. (1994a). Interpersonal expectancy effects: A 30-year perspective. *Current Directions in Psychological Science*, *3*, 176–179. doi:10.1111/1467-8721.ep10770698

Rosenthal, R. (1994b). Science and ethics in conducting, analyzing, and reporting psychological research. *Psychological Science*, *5*, 127–134. doi:10.1111/j.1467-9280.1994.tb00646.x

Rosenthal, R. (1995). Critiquing Pygmalion: A 25-year perspective. *Current Directions in Psychological Science*, *4*, 171–172. doi:10.1111/1467-8721.ep10772607

Rosenthal, R. (2009). Experimenter effects in behavioral research. In **R. Rosenthal** & **R. L. Rosnow** (Eds.), *Artifacts in Behavioral Research* (2nd ed., pp. 287–666). Oxford: Oxford University Press. [Originally published 1975.]

Rosenthal, R., & Fode, K. (1963). Psychology of the scientist: V. Three experiments in experimenter bias. *Psychological Reports*, *12*, 491–511.

Rosenthal, R., & Jacobson, L. (1968). *Pygmalion in the Classroom: Teacher Expectation and Pupils' Intellectual Development*. New York: Holt, Rinehart & Winston.

Rosenthal, R., & Larson, R. (1964). A longitudinal study of the effects of experimenter bias on the operant learning of laboratory rats. *Journal of Psychiatric Research*, *2*, 61–72.

Rosenthal, R., & Rosnow, R. L. (2008). *Essentials of Behavioral Research: Methods and Data Analysis* (3rd ed.). Boston, MA: McGraw-Hill.

Rosenthal, R., & Rubin, D. B. (1978). Interpersonal expectancy effects: The first 345 studies. *Behavioral and Brain Sciences*, *1*, 377–415. doi:10.1017/S0140525X00075506

Rosenzweig, S. (1933). The experimental situation as a psychological problem. *Psychological Review*, *40*(4), 337–354. doi:10.1037/h0074916

Rosnow, R. L., & Rosenthal, R. (1997). *People Studying People: Artifact and Ethics in Behavioral Research*. New York: Freeman.

Rumenik, D. K., Capasso, D. R., & Hendrick, C. (1977). Experimenter sex effects in behavioral research. *Psychological Bulletin*, *84*, 852–877. doi:10.1037/0033-2909.84.5.852

Simmons, J. P., Nelson, L. D., & Simonsohn, U. (2011). False-positive psychology: Undisclosed flexibility in data collection and analysis allows presenting anything as significant. *Psychological Science*, *22*, 1359–1366. doi:10.1177/0956797611417632

Spencer, S. J., Logel, C., & Davies, P. G. (2016). Stereotype threat. *Annual Review of Psychology*, *67*, 415–437. doi:10.1146/annurev-psych-073115-103235

Steele, C. M. (1997). A threat in the air: How stereotypes shape intellectual identity and performance. *American Psychologist*, *52*, 613–629. doi:10.1037/0003-066X.52.6.613

Steele, C. M., & Aronson, J. (1995). Stereotype threat and the intellectual test performance of African Americans. *Journal of Personality and Social Psychology*, *69*, 797–811. doi:10.1037/0022-3514.69.5.797

Strickland, B., & Suben, A. (2012). Experimenter philosophy: The problem of experimenter bias in experimental philosophy. *Review of Philosophical Psychology*, *3*, 457–467. doi:10.1007/s13164-012-0100-9

Strohmetz, D. B. (2006). Rebuilding the ship at sea: Coping with artifacts in behavioral research. In **D. A. Hantula** (Ed.), *Advances in Social and Organizational Psychology: A Tribute to Ralph Rosnow* (pp. 93–112). Mahwah, NJ: Lawrence Erlbaum Associates.

Tyttens, F. A. M., de Graaf, S., Heerkens, J. L. T., Jacobs, L., Nalon, E., Ott, S., … & Ampe, B. (2014). Observer bias in animal behavior research: Can we believe what we score, if we score what we believe? *Animal Behaviour*, *90*, 273–280.

Winkel, G. H., & Sarason, I. G. (1964). Subject, experimenter, and situational variables in research on anxiety. *Journal of Abnormal and Social Psychology*, *68*, 601–608.

9 Suspicion Probes and Debriefing in the Social and Behavioral Sciences

Ginette C. Blackhart

Travis D. Clark

In research conducted within the social and behavioral sciences, the quality of data often hinges on participants' naïveté to study details. Orne (1962) suggested that the most obvious methodological technique to assess whether participants perceive demand characteristics (demand characteristics are subtle cues alerting participants to what is expected of them within a study) present within a study is a postexperimental inquiry (PEI, also referred to as an awareness check, awareness probe, or suspicion probe). As a result, many researchers in the social and behavioral sciences use some form of postexperimental inquiry prior to the debriefing process to assess participant knowledge and/or suspicion of the study purpose/hypothesis. The purpose of this chapter is to discuss issues related to probing for suspicion and debriefing participants. We begin by discussing issues related to research employing deception and to participant suspicion. We then discuss the debriefing process and probing participants for suspicion, including the adequacy of postexperimental inquiries and how to handle data collected during the postexperimental inquiry process. Finally, we provide readers

with recommendations concerning probing for participant suspicion and debriefing.

9.1 Deception in Research

Although suspicion probes and debriefing should be used in most research conducted within the social and behavioral sciences, these should especially be included in research employing deception. Deception is a methodological technique where participants are intentionally misinformed or not fully aware of the specific purpose of a study. Hertwig and Ortmann (2008) state that, although deception in research is not easily defined, a consensus has emerged across disciplines defining it as "intentional and explicit provision of erroneous information … whereas withholding information about research hypotheses, the range of experimental manipulations, or the like ought not to count as deception" (p. 222). The use of deception in psychological research has a long history, nearly as old as the discipline itself (Nicks, Korn, & Mainieri, 1997). Furthermore, although often associated with psychological research, deception is commonly employed within other social and behavioral science disciplines, such as socio-logical and marketing/advertising/consumer research (e.g., Barrera & Simpson, 2012; Hertwig & Ortmann, 2008). Deception is typic-ally employed when researchers assume that full awareness of the experimental protocol or hypothesis would change participants' behav-iors and therefore alter the results of the study. Consequently, deception is used when researchers believe that they cannot obtain viable or desired information through other means, based on the notion that awareness of the exact nature of the study will comprom-ise certain psychological or behavioral processes. Thus, use of deception enables researchers to collect accurate, uncompro-mised data (see Nichols and Edlund, Chapter 3 for a more detailed discussion of deception).

Golding and Lichtenstein (1970) formu-lated four factors necessary for the legitimacy of deception in research: 1) suspiciousness of the study protocol will not affect its response outcomes, 2) participants arrive with little or no knowledge of the study, 3) the study does not indicate to participants that they are being deceived, and 4) knowledge of the study gained before or during the protocol can be assessed by the experimenter. Viola-tions of these assumptions, if undetected, seriously compromise the integrity of experi-mental results. Empirical research has shown, however, that each of these assumptions is often violated in social and behavioral sci-ences research in which deception is used.

The ethical guidelines for research pub-lished by the American Psychological Associ-ation (APA, 2017), the American Sociological Association (ASA, 1999), and the National Communication Association (2017) state that participants must be debriefed as early as is feasible (preferably at the conclusion of their participation within the study) and that they are permitted to withdraw their data from a study involving deception. The debriefing process should provide participants with information about the nature, results, and conclusions of the research; must correct any misconceptions participants might have about the study; and should explain that deception is often an essential feature in the design and conduct of an experiment.

9.2 Participant Suspicion

For most experimental research conducted in the social and behavioral sciences, it is assumed, with little scrutiny, that participants arrive to a study naïve to the true nature of the research. In reality, participants can get information about a study from a variety of sources, such as from other participants, the experimenter, or other elements within the study that create demand characteristics (see Papanastasiou, Chapter 7 for a detailed exposition on these issues).

One source of information is from other participants through participant crosstalk – the tendency for individuals in a participant pool or population of interest to communicate with one another about the details of a study. Research has shown that participant crosstalk is quite common (e.g., Edlund, Lange, Sevene, Umansky, Beck, & Bell, 2017; Edlund, Sagarin, Skowronski, Johnson, & Kutter, 2009; Lichten-stein, 1970) and in undergraduate participant pools, often increases over the course of an academic semester (Edlund et al., 2009).

Additionally, participating in research involving deception may increase suspicion when those same participants volunteer for future studies, even when deception is not included in studies in which participants may subsequently participate. In fact, simply knowing that deception is used in research may increase participants' suspicion about research (Hertwig & Ortmann, 2008). Hertwig and Ortmann examined whether previous experience with deception in research (i.e., participants previously participating in a study involving deception) led to greater suspicion and whether this influenced their behaviors. Based on their review of prior research, they concluded that, whereas first-hand experience with deception increased suspicion and influenced participants' behaviors, knowledge that participants might be deceived in a study seemingly did not influence participants' behaviors. Similarly, Cook and Perrin (1971) found greater bias for participants who had previously experienced deception than for participants who knew about experimental deception but had not previously experienced it personally. Epley and Huff (1998) found that participants who had experienced deception continued to exhibit suspicion three months later. In fact, participants reported being slightly *more* suspicious three months later than they were immediately after being debriefed.

Unfortunately, prior research has shown that non-naïve participants often behave differently than naïve participants. Nichols and Maner (2008) informed participants of the ostensible study hypothesis and examined whether this would influence participants' behaviors. They found that a majority engaged in behaviors designed to confirm the hypothesis, which is consistent with Orne's (1962) supposition that most participants are motivated to conform to what they believe is expected of them. This effect was stronger for participants who reported positive attitudes toward the experiment and toward the experimenter (Nichols & Maner, 2008). Others have concluded that, rather than behaving in a manner designed to confirm the hypothesis, participants' motives may predict whether they act as "good" or "bad" subjects (Shimp, Hyatt, & Snyder, 1991). For example, examining participants' motives for their behaviors across two studies, Allen (2004) found that participants' motives might vary from one study to another depending on the demand characteristics present within a study. That is, depending on the demand characteristics present and participants' personal motives, participants may choose to engage in behaviors that confirm or disconfirm the hypothesis.

Although some researchers have concluded that the effects of suspicion on the outcomes of a research study are negligible (e.g., Kimmel, 1998), other research suggests that suspiciousness of or having information about the study protocol can change behavioral outcomes (e.g., Hertwig & Ortmann, 2008; Shimp et al., 1991). For instance, Gardner (1978) conducted a series of studies during the period in which ethics guidelines began to mandate providing participants with informed consent. He discovered that participant discomfort following a noise blast vanished in those for whom informed consent was required. He initially discovered this effect in different groups of participants before and after the guidelines and replicated this effect by experimentally manipulating the level of informed consent (see Festinger et al., Chapter 6 for more issues associated with informed consent). Additionally, Adair, Dushenko, and Lindsay (1985) showed that the level of information provided to participants moderates other established psychological phenomena. Further complicating this issue is the fact that participants in a negative mood state are more suspicious and are thus more likely to detect deception (Forgas & East, 2008). As a result, studies that employ deception *and* induce a negative mood state (e.g., studies where participants are given false failure feedback or where participants are led to believe they have been socially excluded by other participants) may be especially likely to elicit suspicion and increase the likelihood that participants will detect deception or simply act differently because they believe deception is being used within the study.

Hertwig and Ortmann (2008) reviewed research examining the effects of suspicion on experimental outcomes. They found that, within studies examining conformity, participants who were suspicious of deception were less likely to conform than participants who were not suspicious. They also found that the more explicit the information participants were given (e.g., told they would be deceived, received detailed information about the true purpose of the study, or acknowledged awareness of the experimental manipulation), the more participant suspicion altered participants' behaviors.

As a result, it can neither be assumed that participants are naïve when they participate in a study nor that any suspicion they may have of the study will be negligible. Furthermore, for at least some areas of research, it appears that suspicion and/or the knowledge that they might be deceived may influence participants' behaviors as well. It is, therefore, important to adequately probe for participant knowledge and suspicion as part of the study protocol.

9.3 Debriefing and Probing for Suspicion

Because prior knowledge or suspicion about a study's purpose may influence the outcome of the study, researchers should probe for knowledge and suspicion before participants leave the study. It is recommended that this take place during the debriefing process (Orne, 1962).

9.3.1 Debriefing

Following participation in research, researchers should provide participants with an opportunity to obtain information about the nature, results, and conclusions of the research in which they participated (regardless of whether deception is used within the research). Further, the ethical guidelines of the APA and of the ASA state that, when deception is employed in research, researchers are to provide participants with an opportunity to obtain information about the research and to explain any deception they experienced as a result of participating in the study.

There are two important components of the debriefing when deception has been employed in the design of an experiment: Dehoaxing and desensitizing (Holmes, 1976a; 1976b; Kelman, 1967). **Dehoaxing** is revealing the truth about the purpose of the study. That is, the researcher reveals to participants that they were deceived, how they were deceived, and what the study is actually about. **Desensitizing** is eliminating any negative feelings participants might have experienced as a result of participating in the study. Participants may feel anger, shame, or other negative emotions when they find out they were deceived. It is, therefore, important that experimenters sincerely apologize to participants for deceiving them and explain to participants *why* they were deceived to reduce any negative feelings they may have.

During the debriefing process, participants should also be given an opportunity to ask questions about the study and to have those questions answered to the best of the experimenter's ability. Finally, as outlined in the APA guidelines (and as required by many Institutional Review Boards), participants should be given an opportunity to withdraw their data from the study.

In addition to preventing harm to participants, as well as probing for suspicion (discussed below), it is also important to maintain experimental control during the debriefing process. One way to do this is to reduce participant crosstalk. For example, Edlund and colleagues (Edlund et al., 2009; Edlund et al., 2014) examined classroom- and laboratory-based interventions designed to reduce participant crosstalk. Edlund et al. (2009) found that, although the classroom intervention significantly reduced participant crosstalk, the combination of the classroom and laboratory interventions further reduced participant crosstalk. Edlund et al. (2014) found similar results across several higher education institutions.

Within our research labs, we have developed a standardized debriefing that we use for all of our studies, regardless of whether deception is included. We first begin with the following: "Thank you for participating in this research today. We would like to give some explanation as to what we are studying." We then include details about the

study. If deception was not used within the study, we often reiterate information that was included in the informed consent document, perhaps elaborating on that information. When deception is used, we tell participants what the study is really about. For instance, we might state, "You were initially told that the purpose of this study was to . . . This was a cover story. The research you participated in today is actually designed to examine . . ." and then give participants more details about the purpose of the study. Furthermore, when deception is used, we state,

> *We want to thank you for your participation in this research and apologize for being misleading. It was necessary for us to not be completely forthcoming about the purpose of the study until now as participants' behaviors may have been very different if they knew up front that this is what we were studying.*

This portion of the debriefing is to explain to participants why deception was necessary and to help desensitize them from any negative feelings they might have experienced upon learning they were deceived. We then give participants information about mental health resources on campus (if they are university students) and contact information for the principal investigator and the Institutional Review Board should they have questions or concerns about the study. Finally, we include the following statement in an attempt to reduce participant crosstalk:

> *Because other people you know may participate in this study in the future, we ask that you NOT share the details of this*

study with others. If other people (e.g., romantic partner, friends, roommates, other students) ask you what this study is about or what you did in this study, please only share with them the information that was posted about this study when you signed up to participate.

9.3.2 Probing for Suspicion

In order to determine whether participants have any knowledge or suspicion about the experiment, researchers should probe for knowledge/suspicion by using some form of a postexperimental inquiry (PEI) prior to debriefing (Aronson, Ellsworth, Carlsmith, & Gonzales, 1990; Brzezinski, 1983; Orne, 1962). The purpose of this inquiry is to assess whether participants are suspicious of the cover story or of any feedback or information given, whether participants thought they were being deceived, whether they had knowledge about any aspect of the study prior to participating, and whether they detected or perceived other demand characteristics.

Many researchers use a **funnel debriefing** procedure (see Blackhart, Brown, Clark, Pierce, & Shell, 2012) in which the PEI begins with very basic questions (e.g., "Do you have any questions about the study?") and ends with more specific questions about the study before participants are debriefed (e.g., "Did your experimenter do anything to cause you to be suspicious?" or "Did you expect to be deceived in this study?"; Bargh & Chartrand, 2000; Orne, 1962; Page, 1973). The goal of such a procedure is to probe participants for suspicion and knowledge about the study in a systematic way. The purpose of asking such questions *prior* to the debriefing is to gauge

whether participants truly had any suspicion or knowledge about the study. If participants are queried after being debriefed, they may engage in the hindsight bias and state they knew the purpose of the study or were suspicious of the study when they truly were not aware or suspicious before being debriefed.

There is no standard funnel debriefing procedure within the social and behavioral sciences. That is, the methods researchers use to probe for suspicion and/or awareness vary greatly. There are, however, examples provided in the literature that share common themes that may help guide researchers in selecting PEI questions. One example is given by Golding and Lichtenstein (1970), who asked thirteen graded questions designed to give participants an opportunity to "confess" throughout the interview. Page and his colleagues (Page, 1969; Page & Scheidt, 1971) also reported the questions included on their PEIs. In examining the PEIs used by these researchers, it becomes apparent that there are commonalities in the questions posed to participants, such as asking participants what they thought the study was about, what they thought the purpose of the study was and what they were supposed to do, whether they thought the study was about something else, what they thought the hypothesis was, and whether they were suspicious and/or had prior knowledge about the study. In addition, Page (1971) examined whether a single-question or multiple-question PEI resulted in better detection of participants' awareness of demand characteristics within the study. Across two studies, Page (1971) found that a multiple-question approach resulted in superior detection of awareness. Further, researchers must take care to word PEI questions in such a way as to not increase participant suspicion or create greater demand characteristics within the study (Orne, 1962; Shimp et al., 1991).

Based on the way other researchers have structured their PEIs, as well as on empirical evidence from studies that have assessed the adequacy of PEIs (discussed below), our labs frequently include the following questions on our funnel debriefing in order to determine whether participants had any information about the study prior to participating, knew or guessed the study hypothesis, knew they were being deceived, or were generally suspicious of the study.

1. In your own words, what was the present study about?
2. Did anything odd or strange happen while you were participating in this study? If yes, please explain what happened.
3. Did you believe, at any time, that the experiment dealt with anything other than what the experimenter had described to you? If yes, what do you believe the study was really about?
4. Did your experimenter do anything to cause you to be suspicious? If yes, what did the experimenter do to cause you to be suspicious?
5. Did you feel that certain behaviors or reactions were expected from you at any time? If yes, what behaviors or reactions do you think were expected of you?
6. Did any of the above affect your behavior in any way? If yes, how did they affect your behavior?
7. Sometimes people may hear something about a study before they participate in that study. Did you receive any information about this study, before

participating or during participation, from any source (e.g., from other students, another participant with whom you interacted today, your psychology instructor, psychology textbooks, previous research you have participated in)? If yes, please tell us what information you received before participating in the study or during participation in the study (we are not interested in finding out how or from whom that information was obtained).
8. Sometimes psychology studies include elements of deception. Did you expect to be deceived in this study? If yes, how did you expect to be deceived?

We may also include questions about other participants, experimental confederates, and/or the experimental tasks. Asking participants questions similar to these, prior to the debriefing, may enable researchers to determine whether participants knew about the purpose of the study, whether any aspect of the study gave away its purpose or hypothesis or alerted participants to what was expected of them, and whether they were suspicious of the study (though see further discussion of this issue below). These questions can be easily modified to reflect other types of research and research in other social and behavioral science disciplines.

Other researchers have offered additional suggestions. Allen (2004) looked at whether the PEI could examine mediators of demand bias as identified by Rosenthal and Rosnow (1991), which include receptivity, capability, and motivation (conformity, independence, or anticonformity). To assess this, after asking participants other questions about the study, he asked participants whether their behaviors within the study or responses to the task were based on any of five motives listed, asking them to select just one. They were, "My ratings reflected my desire to get done with the study as quickly as possible," "My ratings reflected my desire to conform with what was expected in this study," "My ratings reflected my desire to not conform with what was expected," "I am not aware that I had any particular motive," and "Other – if none of the above is accurate, please explain your motives here" (p. 73). These items can also be altered to the specifics of the experiment being conducted.

Rubin, Paolini, and Crisp (2010) developed an alternative method for assessing participant knowledge, the Perceived Awareness of the Research Hypothesis Scale (PARH). The PARH is a quantitative scale specifically designed to assess the influence of demand characteristics on participants' behaviors. The scale includes four questions on a seven-point Likert scale ranging from *strongly disagree* (1) to *strongly agree* (7). These questions are, "I knew what the researchers were investigating in this research," "I wasn't sure what the researchers were trying to demonstrate in this research," "I had a good idea about what the hypotheses were in this research," and "I was unclear about exactly what the researchers were aiming to prove in this research." According to Rubin (2016), there are several advantages to using the PARH over other PEIs. One advantage is that it directly assesses participant awareness of the research hypotheses instead of asking vague, open-ended questions that may not be all that informative, but also without asking questions so direct that they may alert participants to the hypotheses of the research

prior to debriefing. Another advantage of the PARH is that it includes close-ended questions rather than open-ended questions. This provides a quicker assessment of participant knowledge (rather than having participants complete multiple open-ended questions). It is also easier for researchers to quickly assess for participant knowledge using the PARH than it is for researchers to read through and interpret participants' open-ended responses as an indication of knowledge. A third advantage is that the quantitative questions allow researchers to empirically test for associations between participants' perceived knowledge of the research hypotheses and other factors assessed within the study. This approach may be especially useful when conducting online studies.

As stated by Rubin (2016), however, the PARH does not enable researchers to determine whether participants actually know the research hypotheses but whether participants *perceive* knowing the research hypotheses. This could be problematic in determining how to handle data collected from participants who perceive they know the research hypothesis. If it is determined that a participant truly knew the research hypothesis, the researcher may choose to discard that participant's data or to control for knowledge through statistical analyses. Not knowing whether the participant actually knew the hypothesis or just thought he or she knew the hypothesis, however, creates difficulty for the researcher in determining how to treat data for participants who perceived they knew the hypothesis. In addition, the PARH does not assess for general suspicion. As a result, it may be useful for researchers to include the PARH at the end of a study prior to debriefing and to also include two or three open-ended questions that will allow the researcher to determine whether participants actually know the hypothesis and to assess for participant suspicion. This method, however, should be tested empirically.

9.3.3 Adequacy of Postexperimental Inquiries

Although a PEI remains the best methodological tool for researchers to assess for participant knowledge and suspicion, the ability of PEIs to detect whether participants have knowledge about a study, believe the ostensible purpose of the study, or are suspicious of any aspect of the study is questionable. Orne (1962) suggested,

> Most subjects are cognizant that they are not supposed to know any more about an experiment than they have been told and that excessive knowledge will disqualify them from participating, or, in the case of a postexperimental inquiry, such knowledge will invalidate their performance. As we pointed out earlier, subjects have a real stake in viewing their performance as meaningful. For this reason, it is commonplace to find a pact of ignorance resulting from the intertwining motives of both experimenter and subject, neither wishing to create a situation where the particular subject's performance needs to be excluded from the study. For these reasons, inquiry procedures are required to push the subject for information without, however, providing in themselves cues as to what is expected. The general question, which needs to be explored, is the subject's perception of the experimental purpose and the specific hypotheses of the experimenter. This can best be done by an open-ended procedure starting with the very general question of, "What do you think that the experiment is about?" and only much later asking specific questions. (pp. 780–781)

This so-called "pact of ignorance" to which Orne (1962) referred may explain why PEIs are not always accurate. Although Orne provides some suggestions for the types of questions to include in PEIs to best assess for participant knowledge and suspicion, and some researchers report highly accurate PEIs (see, for example, Aronson, 1966), the majority of studies experimentally assessing the accuracy of PEIs have shown that these methods may not be all that effective in accomplishing this aim. For instance, several studies have employed a technique in which a confederate gave participants information about the study prior to participation. Two studies (McMillen & Austin, 1971; Nichols & Maner, 2008) found that *none* of the participants admitted to having received information about the study during the PEI. Likewise, Taylor and Shepperd (1996) found that, although participants discussed feedback with one another that they had received before interacting as a group, uncovering the deception in the study, none of the participants admitted to knowing the feedback they received was false during the PEI. Sagarin, Rhoads, and Cialdini (1998) had similar results, finding that only one out of eighty-one participants admitted to having received the answer for the task from a confederate. Additionally, Levy (1967) found one of sixteen participants admitted to having received information about the study before participating.

Golding and Lichtenstein (1970) designed a study to determine how likely participants were to admit to having prior information about the study and to possessing suspicion and/or awareness of the study purpose. They also examined whether certain factors increased participants' admission rates. They found that, although awareness and admission rates were low, participants told that their answers were important for the sake of the scientific integrity of the study had higher admission rates. Newberry (1973) further examined factors that increased admission to having received information about the study before participating. Overall, he found that 30–80% of participants were not forthcoming about receiving information from a confederate. Admission rates were lower when participants were questioned by an experimenter in person than by a questionnaire.

Blackhart et al. (2012) used an experimental confederate to directly supply half of the participants with ostensible information about the study. Although the informed participants reported significantly more awareness of the study hypothesis and admission of receiving information than the naïve participants, the self-reported levels of awareness and receipt of information were abysmally lower than the true, experimentally introduced levels of information. This was true even when mood was manipulated, when participants were offered a reward for correctly guessing the true purpose of the study, and when the interview method was manipulated. Clark (2013) replicated the Blackhart et al. design with the addition of several conditions meant to increase participants' response rates. None of

these changes was successful. Thus, the trend seen in the majority of studies is that PEI accuracy is generally low.

As a result of these disappointing results, Blackhart et al. (2012) attempted to discover why participants report such low rates of awareness. The most commonly cited reasons were concerns about ruining the study (44%), concerns about not receiving research credit or payment (39%), concerns about getting someone in trouble (31% expressed concern about getting themselves into trouble and 35% expressed concern about getting someone else into trouble), and concerns about appearing foolish (26%).

Despite evidence that PEIs are inaccurate, steps can be taken to improve their accuracy. Golding and Lichtenstein (1970) introduced a scientific integrity prompt to some participants that significantly increased PEI accuracy. Based on Golding's and Lichentenstein's (1970) findings, as well as our own empirical findings (Blackhart et al., 2012), we include a similar prompt prior to asking participants to complete the PEI. Our scientific integrity prompt is designed to both promote scientific integrity (see Golding & Lichtenstein, 1970) and enlist the aid of participants (see Aronson et al., 1990). Specifically, the experimenter reads the following to participants before they complete the PEI:

We would like your feedback about the design of the study. We want to be sure that our experimental design is sound and we need your feedback to help us improve this study. In addition, we want to know whether anything odd or irregular happened as you participated in the study today. These things sometimes happen and, as long as we know about them, we can correct for them and make sure that our findings are valid and reliable. It is therefore extremely important for the scientific validity of the study that you tell us whether anything like this happened today. Please be as honest as possible in your answers; no feedback we receive, including negative feedback, will result in a loss of research credit, nor will it affect how we use your data. In fact, negative feedback is an important way for us to improve upon our design for future studies. Be as detailed as you feel is necessary to fully answer each question. You may spend as much time on these questions as you want, but we ask that you spend a minimum of 5 minutes answering these questions.

In addition to giving participants a scientific integrity prompt prior to the funnel debriefing procedure, Blackhart et al. (2012) experimentally manipulated whether the PEI took place as an in-person interview with the experimenter or on a computer questionnaire participants completed alone. We found that participants responding anonymously on a computer admitted to more suspicion and awareness than participants completing the in-person interview with the experimenter, which is similar to results reported by Newberry (1973). We also examined whether offering participants a reward for correctly stating the purpose of the study would increase admission of suspicion and/or awareness, as Cryder, Lerner, Gross, and Dahl (2008) and Taylor and Shepperd (1996) suggested that doing so might increase truthfulness on the PEI. We found, however, that offering

participants an extra credit or cash reward for correctly stating the purpose of the study had mixed efficacy (Blackhart et al., 2012). As a result, we caution researchers against offering such rewards. Clark (2013) included a descriptive norm prompt indicating to participants that previous participants had been very open with feedback (particularly negative feedback) about study procedures. Although there was some evidence that this increased PEI accuracy, the low base rates of PEI admission make it difficult to conclude whether this technique was truly effective.

9.3.4 Handling Data Collected from Postexperimental Inquiries

Despite the fact that the validity of PEIs is questionable in regard to detecting participant knowledge and suspicion, it currently remains our best tool for assessing participant knowledge and suspicion. But what should researchers do with the information collected from the PEI?

Bargh and Chartrand (2000) suggest that any participants exhibiting "genuine awareness" should have their data excluded from statistical analyses, a practice followed by many social and behavioral science researchers using a PEI to assess for participant knowledge and suspicion (Shimp et al., 1991). Shimp et al. (1991), however, point out that participants who guess the hypothesis may differ in systematic ways from participants who do not. For instance, they may be more alert and engaged, more intelligent, more sophisticated when it comes to their knowledge about research, have a higher need for cognition, or differ in other ways (Shimp et al., 1991). As a result, this could introduce bias that threatens not only the internal validity of the study but that also threatens the study's external and construct validity. Shimp et al. (1991) thus recommend that researchers conduct a sensitivity analysis in which they run statistical analyses including and excluding participants aware of the hypothesis or true purpose of the study. If the results are incongruous, researchers should note the differences and allow readers to draw their own conclusions for inconsistencies.

How researchers treat data from suspicious, but not knowledgeable, participants varies. Some choose to discard data from suspicious participants, whereas others choose to statistically control for suspicion in order to determine whether it influences study outcomes. As suspicion may influence some study outcomes and not others, it is wise for researchers to at least assess whether suspicion influences and/or is correlated with study variables. Researchers may, therefore, conduct a sensitivity analysis, running statistical analyses with and without suspicious participants included in the analyses to examine whether differences exist.

9.4 Conclusion

The purpose of this chapter is to provide researchers with recommendations for debriefing participants and for probing for participant suspicion and knowledge about the true purpose of the study. First and foremost, whether or not participants are deceived as part of an experimental protocol, participants must be debriefed once their participation in the study has concluded. If deception is included in the design of the study, participants must be dehoaxed (told the truth about the study) and desensitized (reducing any negative feelings that may have resulted from participating

in the study) as part of the debriefing procedure. In addition, participants should be given an opportunity to ask questions and to withdraw their data from the study should they choose to do so. Finally, researchers should include the intervention discussed by Edlund et al. (2009) in order to attempt to reduce participant crosstalk.

Although prior research indicates that PEIs may not detect all instances of participant suspicion and/or knowledge about an experiment, no other tools currently exist by which to assess for participant suspicion/awareness. As a result, it is recommended that researchers use some form of a PEI to assess for participant suspicion and knowledge about the study. The most common PEI method is the funnel debriefing procedure (Blackhart et al., 2012), whereby participants are first asked general questions and then answer more specific questions about the study prior to being debriefed. Although there is no standard funnel debriefing procedure established in the social and behavioral sciences, this chapter offers some recommendations based on prior research. First, researchers should use a scientific integrity prompt prior to the funnel debriefing to inform participants of the importance of being honest and forthcoming in their answers on the PEI (Blackhart et al., 2012; Golding & Lichtenstein, 1970). Second, participants should be asked multiple questions on the PEI rather than a single question in order to better assess for participant suspicion and/or knowledge (Page, 1971). Third, researchers must do their best to ask participants questions without raising suspicion, without increasing demand characteristics, and in a manner that will hopefully prompt participants to be truthful in their responses (Aronson et al., 1990; Orne, 1962; Shimp et al., 1991). Fourth, researchers may want to use the PARH (Rubin, 2016; Rubin et al., 2010) along with a few open-ended questions to assess for participant knowledge and suspicion. Researchers may also want to assess participants' motives for their behavior(s) (see Allen, 2004). Fifth, it may be beneficial for participants to complete the PEI alone on a computer rather than in an in-person interview with an experimenter (Blackhart et al., 2012; Newberry, 1973). Finally, although many researchers simply remove knowledgeable participants from statistical analyses, it may be wise to instead conduct sensitivity analyses in which researchers conduct statistical analyses with and without knowledgeable and/or suspicious participants to determine whether participants' knowledge or suspicion influence the outcome(s) of the study. PEI efficacy may be low, in part, because PEI use is infrequent across the social and behavioral sciences and because PEIs are unstandardized. As a result, researchers should increase the frequency with which they include PEIs within their research studies and use the suggestions included in this chapter when creating a PEI.

Additional research is needed in order to identify other ways to improve PEIs in their ability to detect participant knowledge and suspicion and/or to identify other viable alternatives to PEIs. This is suggested because, although deception should be used as a last resort, there remain instances in which deception is a necessary research tool and alternatives to deception may not always produce comparable results. There are, therefore, times when employing deception in research may be necessary in order for researchers to collect accurate, uncompromised data. If deception is used, however, researchers must ensure that participants are not aware of the true purpose of the study. It should also be acknowledged that, although PEIs are more commonly used in research involving deception, PEI data has utility beyond its use as an awareness check or a suspicion probe. Further, although speculative, it is possible that the replication crisis in the social and behavioral sciences is due, in small part, to infrequent use of PEIs or to poorly designed PEIs. By forgoing proper standardized awareness and suspicion probes, researchers may be masking the effects of participant suspicion or knowledge on results. This may be in addition to other questionable research practices, all combining to contribute to findings that cannot be replicated. Research is needed to determine whether this is the case.

Although imperfect, the postexperimental inquiry remains our best tool to assess for participant suspicion and knowledge about the purpose of a study.

KEY TAKEAWAYS

- All participants should be debriefed, preferably at the conclusion of their participation in a study, regardless of whether deception is used within the study design. If deception is used, participants must be dehoaxed and desensitized.
- PEIs should be used in all experiments where the results may be influenced by participant knowledge of the study design,

even if the design does not include deception. This is true for cases in which the researcher may not truly expect there to be differences – the PEI data serves as a check to this assumption and may turn up other interesting findings.

- PEI efficacy may be low, in part, because PEI use is 1) infrequent and 2) unstandardized within the social and behavioral sciences. Researchers may use the suggestions provided in this chapter to create their own standardized PEI process.
- Researchers should report whether they use a PEI within their research, and their PEI methods, so that PEI results can be compared across diverse methodologies.

IDEAS FOR FUTURE RESEARCH

- What other ways can we improve PEIs in their ability to detect participant knowledge and suspicion and/or to identify other viable alternatives to PEIs?
- What is the viability of the PARH, in addition to two or three open-ended questions, as an alternative PEI method?
- Is the lack of standardized and widespread use of PEIs in research contributing to the replicability crisis in the social and behavioral sciences?

SUGGESTED READINGS

Barrera, D., & Simpson, B. (2012). Much ado about deception: Consequences of deceiving research participants in the social sciences. *Sociological Methods & Research*, *41*, 383–413. doi:10.1177/0049124112452526

Blackhart, G. C., Brown, K. E., Clark, T., Pierce, D. L., & Shell, K. (2012). Assessing the adequacy of postexperimental inquiries in deception research and the factors that promote participant honesty. *Behavior Research Methods*, *44*, 24–40. doi:10.3758/s13428-011-0132-6

Hertwig, R., & Ortmann, A. (2008). Deception in experiments: Revisiting the arguments in its defense. *Ethics & Behavior*, *18*, 59–92. doi:10.1080/10508450701712990

Rubin, M. (2016). The Perceived Awareness of the Research Hypothesis Scale: Assessing the influence of demand characteristics. *Figshare*. doi:10.6084/m9.figshare.4315778

Shimp, T. A., Hyatt, E. M., & Snyder, D. J. (1991). A critical appraisal of demand artifacts in consumer research. *Journal of Consumer Research*, *18*, 273–282. doi:10.1086/209259

REFERENCES

Adair, J. G., Dushenko, T. W., & Lindsay, R. C. L. (1985). Ethical regulations and their impact on research practice. *American Psychologist*, *40*, 59–72. doi:10.1037/0003-066X.40.1.59

Allen, C. T. (2004). A theory-based approach for improving demand artifact assessment in advertising experiments. *Journal of Advertising*, *33*, 63–73. doi:10.1080/00913367.2004.10639157

American Psychological Association (APA). (2017). Ethical Principles of Psychologists and Code of Conduct (2002, amended June 1, 2010 and January 1, 2017). Retrieved from www.apa.org/ethics/code/

American Sociological Association (ASA). (1999). Code of Ethics of the ASA Committee on Professional Ethics. Retrieved from www.asanet.org/membership/code-ethics

Aronson, E. (1966). Avoidance of inter-subject communication. *Psychological Reports*, *19*, 238. doi:10.2466/pr0.1966.19.1.238

Aronson, E., Ellsworth, P. C., Carlsmith, J. M., & Gonzales, M. H. (1990). *Methods of Research in Social Psychology*. New York: McGraw-Hill.

Bargh, J. A., & Chartrand, T. L. (2000). The mind in the middle: A practical guide to priming and automaticity research. In H. T. Reis &

C. M. Judd (Eds.), *Handbook of Research Methods in Social and Personality Psychology* (pp. 253–285). New York: Cambridge University Press.

Barrera, D., & Simpson, B. (2012). Much ado about deception: Consequences of deceiving research participants in the social sciences. *Sociological Methods & Research, 41*, 383–413. doi:10.1177/0049124112452526

Blackhart, G. C., Brown, K. E., Clark, T., Pierce, D. L., & Shell, K. (2012). Assessing the adequacy of postexperimental inquiries in deception research and the factors that promote participant honesty. *Behavior Research Methods, 44*, 24–40. doi:10.3758/s13428-011-0132-6

Brzezinski, J. (1983). Empirical control of the demand characteristics of an experimental situation in psychology. *Polish Psychological Bulletin, 14*, 149–158.

Clark, T. D. (2013). Using social influence to enhance post-experimental inquiry success. (Unpublished Master's thesis). University of North Dakota, Grand Forks, ND.

Cook, T. D., & Perrin, B. F. (1971). The effects of suspiciousness of deception and the perceived legitimacy of deception on task performance in an attitude change experiment. *Journal of Personality, 39*, 204–224. doi:10.1111/j.1467-6494.1971.tb00037.x

Cryder, C. E., Lerner, J. S., Gross, J. J., & Dahl, R. E. (2008). Misery is not miserly: Sad and self-focused individuals spend more. *Psychological Science, 19*, 525–530. doi:10.1111/j.1467-9280.2008.02118.x

Edlund, J. E., Lange, K. M., Sevene, A. M., Umansky, J., Beck, C. D., & Bell, D. J. (2017). Participant crosstalk: Issues when using the Mechanical Turk. *The Quantitative Methods for Psychology, 13*, 174–182. doi:10.20982/tqmp.13.3.p174

Edlund, J. E., Nichols, A. L., Okdie, B. M., Guadagno, R. E., Eno, C. A., Heider, J. D., … Wilcox, K. T. (2014). The prevalence and prevention of crosstalk: A multi-institutional study. *The Journal of Social Psychology, 154*, 181–185. doi:10.1080/00224545.2013.872596

Edlund, J. E., Sagarin, B. J., Skowronski, J. J., Johnson, S., & Kutter, J. (2009). Whatever happens in the laboratory stays in the laboratory: The prevalence and prevention of participant crosstalk. *Personality and Social Psychology Bulletin, 35*, 635–642. doi:10.1177/0146167208331255

Epley, N., & Huff, C. (1998). Suspicion, affective response, and educational benefit as a result of deception in psychology research. *Personality and Social Psychology Bulletin, 24*, 759–768. doi:10.1177/0146167298247008

Forgas, J. P., & East, R. (2008). On being happy and gullible: Mood effects on skepticism and the detection of deception. *Journal of Experimental Social Psychology, 44*, 1362–1367. doi:10.1016/j.jesp.2008.04.010

Gardner, G. T. (1978). Effects of federal human subjects regulations on data obtained in environmental stressor research. *Journal of Personality and Social Psychology, 36*, 628–634. doi:10.1037/0022-3514.36.6.628

Golding, S. L., & Lichtenstein, E. (1970). Confession of awareness and prior knowledge of deception as a function of interview set and approval motivation. *Journal of Personality and Social Psychology, 14*, 213–223. doi:10.1037/h0028853

Hertwig, R., & Ortmann, A. (2008). Deception in social psychological experiments: Two misconceptions and a research agenda. *Social Psychology Quarterly, 71*(3), 222–227. doi:10.1080/10508450701712990

Holmes, D. S. (1976a). Debriefing after psychology experiments I: Effectiveness of postdeception dehoaxing. *American Psychologist, 31*, 858–867. doi:10.1037/0003-066X.31.12.858

Holmes, D. S. (1976b). Debriefing after psychology experiments II: Effectiveness of postexperimental desensitizing. *American Psychologist, 31*, 868–875. doi:10.1037/0003-066X.31.12.868

Kelman, H. C. (1967). Human use of human subjects: The problem of deception in social psychological experiments. *Psychological Bulletin, 67*, 1–11. doi:10.1037/h0024072

Kimmel, A. J. (1998). In defense of deception. *American Psychologist*, *53*, 803–805. doi:10.1037/0003-066X.53.7.803

Levy, L. (1967). Awareness, learning, and the beneficent subject as expert witness. *Journal of Personality and Social Psychology*, *6*, 365–370. doi:10.1037/h0024716

Lichtenstein, E. (1970). "Please don't talk to anyone about this experiment": Disclosure of deception by debriefed subjects. *Psychological Reports*, *26*, 485–486. doi:10.2466/pr0.1970.26.2.485

McMillen, D., & Austin, J. (1971). Effect of positive feedback on compliance following transgression. *Psychonomic Science*, *24*, 59–61. doi:10.3758/BF03337892

National Communication Association. (2017). A code of professional ethics for the communications scholar/teacher. Retrieved from www.natcom.org/advocacy-public-engagement/public-policy/public-statements

Newberry, B. (1973). Truth telling in subjects with information about experiments: Who is being deceived? *Journal of Personality and Social Psychology*, *25*, 369–374. doi:10.1037/h0034229

Nichols, A. L., & Maner, J. (2008). The good-subject effect: Investigating participant demand characteristics. *The Journal of General Psychology*, *135*, 151–165. doi:10.3200/GENP.135.2.151-166

Nicks, S. D., Korn, J. H., & Mainieri, T. (1997). The rise and fall of deception in social psychology and personality research, 1921 to 1994. *Ethics & Behavior*, *7*, 69–77. doi:10.1207/s15327019eb0701

Orne, M. T. (1962). On the social psychology of the psychological experiment: With particular reference to demand characteristics and their implications. *American Psychologist*, *17*, 776–783. doi:10.1037/h0043424

Page, M. M. (1969). Social psychology of a classical conditioning of attitudes experiment. *Journal of Personality and Social Psychology*, *11*, 177–186. doi:10.1037/h0027025

Page, M. M. (1971). Postexperimental assessment of awareness in attitude conditioning. *Educational and Psychological Measurement*, *31*, 891–906. doi:10.1177/001316447103100411

Page, M. M. (1973). Effects of demand cues and evaluation apprehension in an attitude change experiment. *The Journal of Social Psychology*, *89*, 55–62. doi:10.1080/00224545.1973.9922567

Page, M. M., & Scheidt, R. J. (1971). The elusive weapons effect: Demand awareness, evaluation apprehension, and slightly sophisticated subjects. *Journal of Personality and Social Psychology*, *20*, 304–318. doi:10.1037/h0031806

Rosenthal, R., & Rosnow, R. L. (1991). *Essentials of Behavioral Research: Methods and Data Analysis* (2nd ed.). New York: McGraw-Hill.

Rubin, M. (2016). The Perceived Awareness of the Research Hypothesis Scale: Assessing the influence of demand characteristics. *Figshare*. doi:10.6084/m9.figshare.4315778

Rubin, M., Paolini, S., & Crisp, R. J. (2010). A processing fluency explanation of bias against migrants. *Journal of Experimental Social Psychology*, *46*, 21–28. doi:10.1016/j.jesp.2009.09.006

Sagarin, B., Rhoads, K., & Cialdini, R. (1998). Deceiver's distrust: Denigration as a consequence of undiscovered deception. *Personality and Social Psychology Bulletin*, *24*, 1167–1176. doi:10.1177/01464672982411004

Shimp, T. A., Hyatt, E. M., & Snyder, D. J. (1991). A critical appraisal of demand artifacts in consumer research. *Journal of Consumer Research*, *18*, 273–282. doi:10.1086/209259

Taylor, K., & Shepperd, J. (1996). Probing suspicion among participants in deception research. *American Psychologist*, *51*, 886–887. doi:10.1037/0003-066X.51.8.886

Part Three: The Social and Behavioral Scientist's Toolkit

10 Physiological Measures

Eric J. Vanman

Michael C. Philipp

Several writers of the ancient world reported the story of Antiochus, who was the son of the King of the Seleucid Empire in 294 BC (Fraser, 1969). In that year, Antiochus suddenly became quite ill, but the court physicians were unable to diagnose his disease. One physician, Erasistratus, stymied by how there seemed to be no physical cause, suspected that Antiochus might be in love. Confirming his hypothesis, Erasistratus noticed, over time, that whenever the King's new young wife, Stratonice, entered the room, Antiochus would turn red, his skin would grow warm, and his pulse would quicken. These symptoms would not occur at any other time. Erasistratus pronounced to the King that his son was love sick and that there was no cure, given the situation. When Erasistratus further revealed that Stratonice was the object of Antiochus' affection, the King selflessly gave up his wife to Antiochus so that he could save his son's life. Antiochus and Stratonice married and had five children.

This ancient tale well exemplifies how the body manifests changes that accompany many psychological and social processes (e.g., blood shunted to the periphery, sweating, heart rate increasing). Indeed, social and behavioral scientists have long used noninvasive physiological measures with humans to extend the depth and range of their research. The use of such measures began in the late nineteenth century, with scientists initially focused on recording the electrical activity of the heart or the sweating of the skin. By 1930, recordings of brain and muscle activity had been added to the armamentarium. By the end of the century, nearly any physiological system in the body could be recorded, including sexual response, digestion, and hormone production. In this chapter, we will review the major human physiological research methods in use today, as well as providing a brief introduction to neuroimaging.

10.1 Basic Principles

When a researcher chooses to use a biological measure in social and behavioral science research, he or she is acknowledging (whether explicitly or not) that an understanding of the mind, behavior, and the social context cannot be fully explained by a psychological or behavioral approach alone. Humans are affected by the people around them and the social structures in which they live, but human behavior is also fundamentally biological. Likewise, relying solely on biological reductionism to explain why someone votes for a particular political candidate will not yield satisfying explanations. Instead, using physiological measures in the social and behavioral sciences is best framed using a multilevel integrative analysis (Cacioppo, Berntson, Sheridan, & McClintock, 2000). Cacioppo and Berntson (1992) proposed three general principles that underlie such an analysis.

First, the **principle of multiple determinism** specifies that a particular behavior may have multiple antecedents across different levels of explanation. When explaining why people do what they do, the social structure of their community may play just as an important role as the nutritional quality of the local diet. In another example, violence toward another person might be influenced by perceived inequities between the two parties, as well as the effects of alcohol on frontal lobe inhibitory control. Consequently, a comprehensive explanation of the phenomenon of interest may require attention to multiple levels of analysis, such as genetics, neurochemistry, social context, and societal structure.

Second, **the principle of nonadditive determinism** states that properties of the whole (event) are not always predictable from an understanding of the parts alone. For example, when mice are infected with a respiratory virus, most will survive if left in their home cage with other familiar mice. However, when a dominant, stranger mouse is placed in the cage, the majority of mice infected with the virus will die (Johnson et al., 2006). Importantly, noninfected mice who are visited by such an intruder mouse will not show these mortality effects. Thus, an explanation of these results based on either the social (i.e., whether a stranger mouse is present) or the biological (i.e., whether the virus is present) factors alone would be insufficient – it is only an understanding of the interactive effect of the two that can explain the final outcome.

Finally, the **principle of reciprocal determinism** specifies that any level of analysis (e.g., cardiovascular functioning vs. social structure) can have mutual influences. Mutual back-and-forth influences can exist among levels of analysis. For example, viewing sexually explicit and violent material affects sympathetic nervous system arousal in human males, but that arousal can, in turn, affect the perceptions and responses to the pornographic material, such that the male might notice additional sexual cues and become even less inhibited as arousal increases (Zillman, 1984). Again, if one considers either level of analysis by itself (e.g., autonomic nervous system activity vs. mental state), such reciprocal influences would be overlooked when developing a comprehensive explanation for the behavior.

What these three principles have in common is the assertion that, when a researcher is considering using a physiological measure, such as salivary cortisol or heart rate, it is just as necessary for a scientist to have an understanding of the physiological basis of this measure as it is important to know about the other social variables that are relevant. For example, one might consider taking cortisol samples from participants in a prison population as a measure of stress in addition to administering a psychological stress questionnaire. Knowing how other variables also influence cortisol production, such as time of day, diet, or hours since waking, could prevent erroneous conclusions being drawn about the effects of prison crowding or living conditions – the main interest of the researcher. We, therefore, highly recommend that the review we provide here of physiological measures be considered a first step for the budding psychophysiologist or neuroscientist. Interest in any one measure here should be supplemented by a more in-depth analysis of that measure, which can be found in books devoted to psychophysiological measures (e.g., Blascovich, Vanman, Mendes, & Dickerson, 2011; Cacioppo, Tassinary, & Berntson, 2016).

10.2 Autonomic Measures

The human nervous system is divided into two main branches – the central, which consists of the brain and the spinal cord, and the peripheral, which consists of the nerves everywhere else. The peripheral system can be further subdivided into the somatic (largely involving the striated muscles) and the autonomic. Because the autonomic nervous system (ANS) is involved in the regulation of many organs and systems in the body, social and behavioral scientists have long been interested in its function, particularly in the way the two main branches of the ANS, the sympathetic and parasympathetic, coactivate the body in response to changes in the environment. Perhaps, the simplest way the ANS is considered by some researchers is as an index of arousal. This is intuitive, given that patterns of our ANS activity are related to how active or alert we feel (Feldman Barrett, Quigley, Bliss-Moreau, & Aronson, 2004).

However, there is not a simple one-to-one correspondence between ANS activity and our perceptions of it. Although the arousal concept has long been shown to be fraught with problems (e.g., Lacey, 1967; Picard, Fedor, & Ayzenberg, 2015), single measures of ANS activity (e.g., heart rate) are commonly used as an index of arousal by social and behavioral scientists. We review two major kinds of ANS measures here: a) electrodermal activity (EDA), which is sometimes referred to by its older name, the Galvanic Skin Reflex, and b) cardiovascular activity.

10.2.1 Electrodermal activity (EDA)

The human body is covered with sweat glands in the skin. There are two kinds: Apocrine, which are found primarily at the armpits and genital areas, and eccrine, which are located all over the body, but have their densest concentrations on the surfaces of the hands and feet. Apocrine glands are associated with body odor, and their function in humans is unclear (Wilke, Martin, Terstegen, & Biel, 2007), although, for most mammals, the sweat produced acts as a pheromone or warning signal. Eccrine glands, which form the basis of electrodermal activity (EDA) measurement, have several functions, including thermoregulation, preventing skin abrasion, and enhancing grip. Eccrine sweat glands are primarily driven by sympathetic nervous innervation that is cholinergic. Sudomotor fibers originate in the sympathetic chain and terminate at sudomotor cells of the gland. The sweat that forms in the secretory portion of the gland, as well as the sweat that is pushed up the eccrine sweat gland, reduces the electrical resistance of the skin at a particular location, such as the fingertips. A scientist recording

EDA applies a small constant voltage between two electrodes, and the sweat glands just below the surface act as a variable resistor to the current travel between the electrodes. By computing the reciprocal of that resistance, one can determine the amount of skin conductance (SC) at any moment, which is then amplified and recorded continuously during the research study (for more details, see Dawson, Schell & Filion, 2016).

Like other autonomic measures, EDA has often been considered a measure of arousal. It is involved in the orienting response, fear conditioning, anxiety, and the detection of deception (see Dawson et al., 2016 for a review). It has also been used to investigate cognitive processes in clinical populations, such as people at risk for suicide (e.g., Sarchiapone et al., 2018) or those diagnosed with schizophrenia (e.g., Cella, Okruszek, Lawrence, Zarlenga, He, & Wykes, 2018).

To record EDA, two Ag/AgCl electrodes are typically attached to the volar surfaces on the medial phalanges or the thenar and hypothenar eminences of the palms. Skin preparation involves washing the skin at those sites with soap and water, but the skin is not abraded. Fowles, Christie, Edelberg, Grings, Lykken, and Venables (1981) recommended that the gel used in the electrodes consist of unibase solution and physiological saline, which has properties consistent with skin that will remain constant over the recording session. EDA recording is unobtrusive, but movement can create artifacts, so many researchers instruct participants to keep the arm being recorded still throughout testing.

For illustrative purposes, imagine a researcher is interested in how people respond emotionally to different political candidates. The researcher might present photos for a few seconds of candidates (one at a time) standing for a local election and who come from different political parties. Observing an EDA recording during the presentations might reveal slight upward deviations from an otherwise smooth tracing that slowly drops during the course of the experiment, as indicated in Figure 10.1. A few different parameters can be extracted from that EDA recording. The SCL is the tonic skin conductance level that represents the background level of EDA. It usually starts moderately high (ca. 10 μSiemens) and slowly drops during an experiment. Superimposed on this background level, however, are small phasic changes that can be on the order of 0.05–5 μSiemens, which are called skin conductance responses (SCRs). These SCRs can be spontaneous or in response to a known stimulus. The SCR itself is comprised of several components that can also be measured, including the amplitude of the response, the latency (usually 1–3 seconds after a stimulus is presented), and the rise time (see Figure 10.1). When a longer time period is of interest (e.g., participants are listening to a political message), researchers will usually count the number of SCRs that occur during the time period and/or include the average SCL and compare this to a control period.

10.2.2 Cardiovascular System

The human heart is a busy organ. At rest, it pumps approximately 5 liters of blood per minute. During exercise, this can exceed 25 liters per minute. The heart beats about 100,000 beats per day. All this activity is due to how the cardiovascular system transports oxygen from the lungs and nutrients from the gut, transports waste products, transports

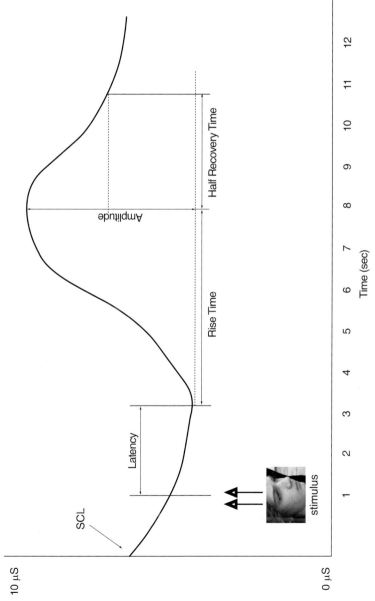

FIGURE 10.1 Components of the skin conductance response (SCR)

hormones, and is central in the thermal exchange between the core and the periphery. Social and behavioral scientists, of course, have generally been interested in how this system is affected by psychological and social variables as well. Although the heart has an intrinsic pacemaker, the rate at which the heart beats is regulated extrinsically via both parasympathetic and sympathetic influences. The parasympathetic system, via the vagus nerve, slows heart rate and is largely what causes all changes that occur below 100 bpm. The sympathetic system speeds up the heart rate, but its impact is not as strong as the parasympathetic system. Its main effect is on the contractility of the heart – how much blood gets squeezed through the heart during each cycle of contraction.

Cardiovascular measures are used in a wide range of applications in behavioral research. For example, they have been used to measure threat and challenge responses that participants experience when confronting someone from a different ethnic group (Mendes, Blascovich, Lickel, & Hunter, 2002). Cardiovascular measures have been used to index arousal to many emotional stimuli, including music (e.g., Kuan, Morris, Kueh, & Terry, 2018). Some patterns of cardiovascular activity are related to dispositional differences in emotional styles, such as hostile tendencies (Demaree & Everhart, 2004), or changes in cognitive workload, such as during a driving simulation (Rodriguez-Paras, Susindar, Lee, & Ferris, 2017).

As noted in the examples above, several measures of cardiovascular activity are commonly used today. The oldest and perhaps the most pervasive is based on the electrocardiogram (ECG), which involves recording the changes in voltage when the heart contracts. The ECG has a rather distinct waveform, with components that reflect how the four chambers of the heart (the two atria and two ventricles) depolarize and repolarize, causing the blood to be pumped through the organ. The largest component, the QRS complex, is easy to detect when amplified and displayed on a computer monitor. It occurs once for each "beat" of the heart. Detecting the R spike in this complex allows one to compute the heart rate as well as the heart period, which is the interbeat interval. Phonocardiography, which measures the "lub-dub" sound associated with a heartbeat, and photoplethysmography, which involves monitoring the changes of blood flow in the periphery associated with each beat of the heart, also provide ready measures of heart rate. In fact, modern fitness trackers, like the Fitbit or Apple Watch, use photoplethysmography to provide reliable measures of heart rate that are also highly accurate. Any of these measures will also enable the computation of heart rate variability (HRV). HRV is determined by several physiological variables, but behavioral researchers have been particularly focused on high-frequency changes in the HRV (e.g., time-domain estimates) because they index parasympathetic control.

Other measures of cardiovascular activity include blood pressure and impedance cardiography, which measure contractility and peripheral resistance as the blood travels through the body. Blood pressure typically involves some sort of auscultatory technique, which means that the blood flow in the arm is temporarily occluded by a blood pressure cuff so that the underlying pressure can be assessed. Measuring blood pressure continuously can therefore be uncomfortable for the research participant. Impedance cardiography involves the placement of spot electrodes or electrode bands at the neck and stomach regions. A low-energy, high-frequency alternating current is then transmitted through the chest. The current follows the path of least resistance – the blood-filled aorta. The blood volume and velocity of blood flow in the aorta change with each heartbeat. The corresponding changes in impedance (the resistance to current) are used with an ECG to provide hemodynamic parameters that correspond to events in the cardiac cycle, such as ventricular contraction and cardiac output (the total blood pushed out of the heart during each beat).

Which of these measures the researcher chooses depends on practical and financial concerns. The necessary equipment to record heart rate or blood pressure is fairly inexpensive (under US $100). Monitoring such activity over longer periods through a wearable device will cost several hundred dollars more and also requires that the researcher trusts that the equipment will not be lost, stolen, or broken once it leaves the laboratory. Impedance cardiography is more invasive, in that it requires electrodes or bands being attached to the torso. It is also not very ambulatory. Moreover, it is the most expensive of the methods reviewed here, with necessary start-up equipment costing US $5000 or more.

10.3 EMG and Startle Eyeblink

In addition to measures of autonomic peripheral activity, the social or behavioral scientist can also measure somatic nervous system activity by recording the activity of striated muscles. Electromyography (EMG) measures muscle activity with sensors sensitive to the electrical activity that precedes and accompanies muscle contractions. As motor neurons stimulate the muscle, muscle fibers contract and the electrical activity associated with this stimulation propagates across the muscle. The electrical activity, called the muscle action potential, permeates the surrounding tissue including the skin surface. EMG sensors detect this electrical activity at the skin surface. From the sensors, the electrical activity is amplified and recorded for later analysis (for details about muscle physiology and the processing of EMG, see Tassinary, Cacioppo, & Vanman, 2016; Van Boxtel, 2010). The voltages of a muscle's action potential are commensurate with the magnitude of a muscle's contraction: Larger contractions produce greater voltages. EMG is particularly useful because it provides a precise index of muscle activity that is quantifiable in real time, and sensitive enough to detect even the smallest amounts of muscle activity.

EMG has been used in a wide range of applications, including to quantify muscle tension in ergonomics (e.g., Delisle, Lariviere, Plamondon, & Salazar, 2009), to study mimicry (e.g., Korb, Malsert, Strathearn, Vuilleumier, & Nideenthal, 2016), to study the muscles that maintain posture and coordinate gait in highly skilled acts (e.g., Trepman, Gellman, Solomon, Murthy, Micheli, and de Luca, 1994), and to measure sociality effects of avatars in a virtual environment (e.g., Philipp, Storrs, & Vanman, 2012). In behavioral research, EMG has been largely used to measure facial activity, as illustrated in Figure 10.2. It can be used to measure incipient emotional reactions (e.g., in response to

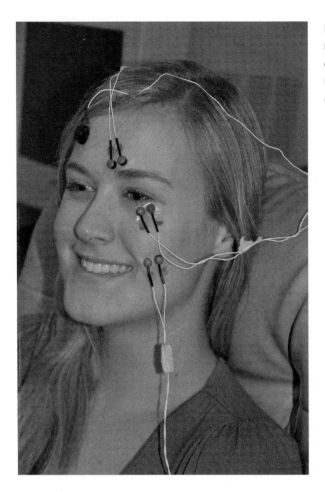

FIGURE 10.2 An example of a facial EMG recording setup. In the picture, pairs of electrodes are placed over the zygomaticus major (cheek), corrugator supercilii (brow), and orbicularis oculi (eyeblink) regions

pictures of a disliked person or a feared stimulus such as a snake). EMG can also be used to record activity that is associated with the cognitive processing of stimuli, such as subvocal reading (i.e., muscle activity from the lip region) or the time course of early attention (i.e., the use of probes during a startle response). Below, we further differentiate between facial EMG to measure muscles of expression and that used to measure the startle eyeblink reflex.

10.3.1 Facial Expressions and EMG

Well before the modern era of social and behavioral science research, scientists were drawn to the importance of facial expressions as markers of aesthetic enjoyment (e.g., Bell 1877), taxonomies of mental states (e.g., Duchenne de Boulogne (1990/1862), and functional displays of internal motivations (e.g., Darwin, 1965/1872). Since the 1970s, the majority of research on facial expressions has examined the correspondence of expressions (e.g., smile) to internal states (e.g., happiness) and the subtle morphological differences that differentiate expressions. This pursuit led to a number of methods for coding facial movements like the Facial Action Coding System (FACS; Ekman & Friesen, 1978), in which recorded expressions are meticulously coded using a sophisticated scoring system. Although FACS is still used widely, EMG has become another popular method for understanding how facial expressions map to internal states.

Facial EMG can measure a variety of expression muscles from the levator labii superior (crinkling of the nose accompanied by feelings of disgust) to the lateral frontalis (pulling up of eyebrows when surprised). However, most EMG research has focused on two muscles in particular: the zygomaticus major (i.e., the muscle that pulls the corners of the mouth into a smile) and the corrugator supercilii (i.e., the muscle that furrows the brow and pulls the eyebrows together into a frown). Activity of these muscles is typically indicative of real-time changes in emotional feelings. For example, zygomaticus major activity increases and corrugator supercilii activity reduces when participants report more positive affect; the inverse holds for experiences of negative affect (e.g., Cacioppo, Petty, Losch, & Kim, 1986).

When using facial EMG to measure affective states, participants quickly become accustomed to the sensors during an experimental session. As long as the researcher is careful not to draw attention to the precise purpose of the sensors, participants rarely suspect that EMG sensors are measuring their facial muscles. This means that facial EMG is far less susceptible to social desirability biases compared to other methods. This can be especially useful in assessing intergroup prejudice (e.g., Vanman, Paul, Ito, & Miller, 1997), where self-reports of attitudes toward outgroups may be more indicative of social norms than personal attitudes.

10.3.2 Startle Eyeblink

Another common use of facial EMG to index emotional and attentional states is by startling a person and measuring the muscle activity associated with a startle-evoked eyeblink initiated by the sudden contraction of the muscles around the eyes that cause the eyelids to close. Researchers measuring a startle eyeblink use a loud (i.e., 105 decibels) noise or puff of air to startle participants and measure the eyeblink startle using EMG sensors over the orbicularis oculi muscle – just below the lower eyelid (see Figure 10.2). The greater a startle reflex triggered in the participant, the larger the EMG activity. The amplitude of the startle reflex can be modified by whatever the participant was attending to just before the startle noise. For example, if a person views a picture of a cockroach, the presentation of a startle noise will elicit a larger blink reflex than if that person is viewing a picture of a flower (Bradley, Cuthbert, & Lang, 1993; Mallan & Lipp, 2007). More generally, startle eyeblinks are smaller for stimuli that the participant finds especially salient or interesting but are greater for stimuli toward which the participant is deliberately directing his or her attention (Filion, Dawson, & Schell, 1998). The exact time courses of affective as well as attentional modulation of the eyeblink reflex are sensory modality and task dependent (Neumann, Lipp, & Pretorius, 2004; Vanman, Boehmelt, Dawson, & Schell, 1996). Startle eyeblink modification has yielded a long line of fruitful research in psychophysiology, but it is important to keep in mind that, compared to other facial EMG methods, it is more intrusive (i.e., participants are frequently interrupted by a loud, aversive noise), less efficient (i.e., participants start to habituate to the startle noise), and possibly less sensitive, given that it can be simultaneously influenced by affect, attention, and sensory modality processes (Tassinary et al., 2016).

10.4 Hormones

Hormones are molecules released by glands or the brain which serve as messengers that travel at the speed of blood throughout the body's bloodstream. Specialized receptors at the organs receive these messages, which can trigger different responses depending on the organ's function. A single type of hormone (e.g., vasopressin) can, therefore, have multiple effects on the body. One major effect of hormones is they can be organizing – that is, they can cause permanent structural differences to develop gradually in the body (e.g., the changes that occur during puberty). Another major effect is that they can be activating – they can cause temporary changes in behavior that occur over seconds, minutes, hours, or days following the release of the hormone.

As our knowledge of endocrinology has grown, social and behavioral scientists have been increasingly interested in the activating effects of hormones. Some of this research has looked at correlations between naturally occurring levels of hormones and different psychological variables. For example, higher levels of endogenous testosterone have been found to be associated with lower levels of accuracy when people try to infer other's thoughts and feelings (Ronay & Carney, 2013). Another way to study the role of hormones in behavior is to measure them in response to some manipulation or otherwise known stimulus. For example, cortisol was found to increase when people participated in an extreme dance ritual associated with bondage/discipline, dominance/submission, and sadism/masochism (BDSM, Klement et al., 2016). In another study, people who gave up Facebook for five days showed a drop in their cortisol levels (Vanman, Baker, & Tobin, 2018). Finally, some hormones can be directly manipulated through injections or nasal sprays. For example, some limited evidence suggests that oxytocin sprays might reduce some of the social deficits associated with autism (Guastella & Hickie, 2016). Other research has shown that a single dose of testosterone administered under the tongue enhances responsiveness to socially threatening stimuli that are presented to women (Hermans, Ramsey, & van Honk, 2008).

The two most common ways social and behavioral scientists measure hormones are via blood samples or saliva samples. The advantages of drawing blood samples are that they can be collected throughout the day, and they can show immediate changes because the bloodstream where the hormones travel is being directly sampled. The primary disadvantage to this method is that it requires venipuncture. Needles have to be injected into veins, which can cause discomfort to the participant and may require a venipuncture-qualified research assistant to collect the samples. Saliva samples, therefore, can be an attractive alternative, as they are noninvasive and can also be taken frequently (see Schultheiss & Stanton, 2009). Saliva is usually collected through the use of a small plastic tube that contains a cotton roll. The participants place the cotton in their mouth for a couple of minutes and then put the roll back into the tube. Alternatively, they might try to deposit some saliva directly into the tube. This means that saliva sampling can be self-administered and, therefore, collected in any location, in or out of the laboratory. The primary disadvantage to saliva sampling, however, is that protein hormones (e.g., oxytocin, prolactin) are not found in saliva. In addition, enzymes and bacteria can change the structure of steroids that can be measured (e.g., testosterone, cortisol). Finally, if a little blood is mixed in the sample (perhaps due to bleeding gums) one can get false high values. Two other ways to measure hormones, which are not often used in human behavioral research, are via the collection of urine or fecal samples. Each of these methods also has advantages and disadvantages (see Schultheiss & Stanton, 2009).

Once the sample containing the hormone is collected, it is then typically stored in a freezer at −20°C or lower until it can be sent to a laboratory to be assayed. Some local universities or clinics may provide plasma or salivary assay services or the researcher may ship the samples off to other laboratories elsewhere in the world. Such assays can be expensive, costing between US $4.00 and $7.00 per sample, depending on the hormone that is to be assayed. The sample itself is not typically returned, but the laboratory will provide a file with the values measured for each sample.

Other important considerations regarding hormone sampling have to do with the timing of when measurements are made. For example, cortisol, a hormone associated with the stress response, has a diurnal rhythm. Levels vary widely during the day and, thus, time of day becomes critical in any study that might involve cortisol reactivity (Dickerson & Kemeny, 2004). For that reason, researchers tend to conduct their research with cortisol in the afternoon, when cortisol levels are relatively stable. Testing participants early in the morning can also be affected by the cortisol awakening response, which involves rapid increases in cortisol within the first hour of wakening. If the researcher is interested in hormone responses to a stimulus, such as a stressor, it can take several minutes for the hormone to show its peak reactivity, particularly when using saliva samples. For example, Dickerson and Kemeny (2004) found in their review of 208 stressor studies that it took twenty to forty minutes after the stressor for cortisol to reach peak levels.

10.5 EEG and ERPs

10.5.1 EEG

Hans Berger published the first paper on the human electroencephalogram (EEG) in 1929. What was most prominent in his early recordings of voltage changes from the surface of the head was a steady rhythm or wave that occurred around 8–12 Hz (cycles per second), which Berger designated "alpha" activity. Even today, it is estimated that about 70% of all EEG activity in adult humans is alpha activity. Perhaps surprisingly, the origins of this EEG activity are not yet completely understood. It is a small, oscillating voltage that reflects a large number of neurons somewhere beneath an electrode on the scalp. This electrical activity is *not* due to action potentials firing below in the cortex. Instead, some of the signal is partly due to glial cells and the changing potentials of inhibitory and excitatory postsynaptic potentials. Alpha activity appears to be the result of many of these potentials in synchronization with each other. They become desynchronized when the brain has new inputs that require further cortical processing.

EEG recording today is far more sophisticated than during Berger's time, of course, but the basic logic of the recording is still the same. To record EEG, researchers attach dozens of electrodes all over the scalp at precise locations, which allows one to make topographic maps of the distributions of the waveforms. The participant usually wears an EEG cap to standardize the placements, and the skin is typically lightly abraded before gel is added to conduct the signal from the scalp to the electrode. Recording of the very weak EEG signal is susceptible to many artifacts, such as the electrical noise in the room, muscle movement (e.g., the participant chews gum), and eye movements. The latter artifact occurs because the human eyeball is an electrical dipole that can be picked up by the EEG electrodes near the front of the head. For this reason, eye movements are simultaneously recorded during EEG using the electrooculogram. Any ocular deflections, including eyeblinks, can then be corrected in the relevant EEG channels.

EEG can be recorded with eyes open or closed. While the participant performs a task (e.g., imagining a scene, reading some text), EEG is continuously recorded and bands of activity can be detected in the ongoing signal, which are categorized by their amplitudes and frequencies. For example, compared to alpha, beta waveforms occur between 13 and 30 Hz and have a smaller amplitude. The relationship between alpha and beta is largely reciprocal, so some EEG research measures alpha activity suppression as an index of increased activation in a brain area. For example, Harmon-Jones and Sigelman (2001) recorded EEG activity over the right and left prefrontal cortex while participants were experimentally induced to feel angry while performing a task. By measuring alpha suppression, the researchers found that anger was associated with increased left-prefrontal and decreased right-prefrontal EEG activity, which suggested that prefrontal asymmetry activity reflects motivational direction and not differences in emotional valence.

10.5.2 Event-Related Potentials (ERPs)

As the technology behind EEG recording improved, researchers discovered that one could also measure small changes in the EEG signal in response to the presentation of discrete stimuli. Event-related potentials (ERPs) are patterned voltage changes that are embedded in the ongoing EEG but which reflect a process in response to a particular event, like the presentation of a visual or an auditory stimulus. ERPs are recorded using the EEG, but they are usually detected by presenting the same stimulus repeatedly over dozens, if not hundreds, of trials. The ongoing activity of the EEG signal is considered the background "noise," whereas the activity that reflects processing of the stimulus becomes the "signal." By averaging across many trials, the signal-related activity can be extracted because it is time-locked to the presentation of the stimulus and the random noise averages to zero. Typically, researchers focus on the 1-second period after stimulus presentation and then average together the responses across all like stimuli. A somewhat idealized waveform emerges that can reflect sensory, motor, and/or cognitive events in the brain. This waveform is an aggregate of numerous ERP components, as depicted in Figure 10.3.

ERP components are defined in one of three ways: 1) by the positivity/negativity of the peaks, 2) by aspects of the ERP that covary across participants, manipulations, and locations, and 3) neural structures that are believed to generate the ERP. The first method depends on identifying peaks, which are theorized to represent the sum of several functional or structure sources. The same group of neurons may contribute to more than one of the peaks of the ERP. The earliest components occur in less than 100 ms following the stimulus onset and occur in an obligatory fashion that reflects how a stimulus is processed automatically (e.g., edge detection, light). The later components (100–1000 ms) are sensitive to changes in the state of the organism, the meaning of the stimulus, or the information processing demands of the task. It is these components that ERP researchers focus on the most. The amplitudes and the latency of the components will presumably vary as a function of the independent variables of the experiment. For example, Ito and Urland (2005) measured the amplitude of these components as they varied as a function of the gender and race of target faces that participants viewed. The location of the peaks on the scalp is also important. ERP researchers often use topographical maps of the components as they change over the time and location, giving the appearance of a brain map that one might obtain in neuroimaging. One should be cautious in interpreting such maps, however. The scalp distribution does not necessarily reflect cortical specialization nor does that area where the components are at

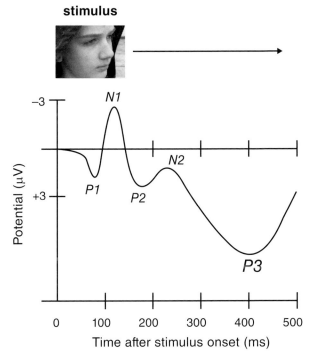

FIGURE 10.3 A typical EEG waveform showing several ERP components. By convention, positive values are plotted on the bottom. Common ERP components are labeled P1, N1, P2, N2, and P3. The letter of each component refers to whether the amplitude is positive (P) or negative (N) in microvolts. The number roughly corresponds to timing of the component in hundreds of milliseconds. So a P2 component is a positive deflecting wave that occurs around 200 ms after stimulus exposure. The earliest components (e.g., P1) can occur in less than 100 ms following the stimulus onset and reflect automatic processes (e.g., edge detection, light). Later components (e.g., N2) are sensitive to changes in the state of the organism, the meaning of the stimulus, or the information processing demands of the task.

the maximum amplitude suggest that the neural generator lies underneath.

10.6 Functional MRI and PET

When neuroimaging became a readily available tool for social and behavioral science research, it seemed that scientists would finally have a direct line into the workings of the human brain without the need of surgery or electrodes inserted into the cortex. Positron emission tomography (PET) was the first of these techniques to examine functional activity in the living brain, beginning in the 1970s. It was the dominant method for imaging brain function until the early 1990s, when functional magnetic resonance imaging (fMRI), which was an extension of structural MRI, became widely available. PET research requires the production of a radiotracer, which contains a positron emitting radioisotope. The radiotracers have short half-lives, which means they have to be produced relatively close to the PET imaging system. A local cyclotron is, thereby, required to synthesize radiotracers, which themselves are large, highly specialized pieces of equipment requiring dedicated staff trained in radiochemistry. The radioisotope is then injected or inhaled by the participant, who then performs the task. The PET imaging system measures the distribution of the radioisotope in the blood during brain activation, giving a representation of which areas of the brain are most active during the task.

Functional MRI, because it was cheaper and safer, became the dominant way to image brain function in the 1990s. Not only did it not require an off-site cyclotron, fMRI had no limit to the number of times someone could be studied safely. By contrast, PET involves a significant amount of radiation, so people can participate in only one PET study per year. Perhaps more important, the spatial resolution is superior in fMRI. Higher-strength magnets can increase the spatial resolution down to less than 1 mm, whereas PET's resolution is much worse than that (i.e., 8–15 mm). In the MRI scanner, a large magnetic field is used to align the protons (hydrogen molecules) in the body to the field. A radio frequency (RF) coil emits a pulse that knocks the protons over. When they realign, they emit energy that gradient coils built into the scanner receive, which are then used to encode 3-D spatial information. The MR signal changes depending on the amount of oxygenated blood in a specific area. Thus, one is not directly measuring the neurons firing in the brain (or elsewhere), but instead fMRI measures BOLD (blood oxygenation level-dependent) signals that result from a combination of blood flow changes, blood volume changes, and oxygen consumption (see Figure 10.4). Much like EEG, the precise physiological meaning of what fMRI is measuring is not fully understood, but it is assumed to reflect changes in how much a particular brain area is activated (or deactivated) during a task.

It is common that, in both PET and fMRI studies, stimuli are presented in blocks, which are then compared to a baseline condition. Although a great deal of behavioral phenomena can be studied in this manner, it does impose limits on the realism of the phenomena being studied. For example, engaging in an oral conversation with a friend is practically impossible to study in a scanner. The precision and timing of the scanner necessitate that participants remain perfectly still.

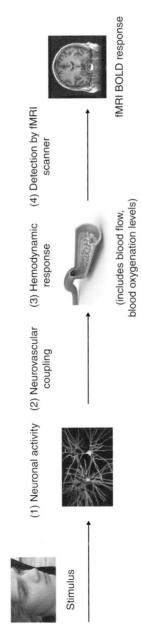

Stimulus   (1) Neuronal activity   (2) Neurovascular coupling   (3) Hemodynamic response   (4) Detection by fMRI scanner

(includes blood flow, blood oxygenation levels)

fMRI BOLD response

FIGURE 10.4 From stimulus to the BOLD response, as recorded by fMRI (Arthurs & Boniface, 2002)

The slightest movements can create significant artifacts that might mean the whole session has to be thrown out. Further, the powerful magnet used in an fMRI scanner precludes any metal to be in the same room – whether implanted in the participant (e.g., a pacemaker) or part of some other recording equipment. Functional MRI scanners are also expensive to use – with many universities charging their researchers US $400–600 per hour, including the time of the expert technologist running the system. As a result, fMRI research is often statistically underpowered (Button et al., 2013) because the sample sizes are relatively small.

Perhaps more than other methods reviewed in this chapter, the use of a neuroimaging technique like fMRI requires advanced knowledge of both its anatomical basis and the physics behind the MR signal itself. When designing the experiment, just like when using other physiological measures, the researcher should carefully consider the temporal dynamics of the signal among other factors, as the BOLD response can take 1–2 seconds to be produced and detected by the MRI scanner. Excellent references to consult while developing fMRI research are Buxton (2009) and Huettel, Song, and McCarthy (2014), which provide comprehensive coverage of the physics of MRI, the neurochemistry of the BOLD response, experimental design, and analysis of fMRI data.

10.7 Conclusion

Participating in any experiment, but especially one with electrical recording devices attached, is often a novel experience for research participants. Moreover, the close proximity of the experimenter, who is cleaning the skin and attaching electrodes, can heighten anxiety. In the fMRI scanner, lying perfectly still in a hospital gown in a very large machine can be similarly anxiety-provoking. Thus, the testing session becomes a social situation itself, which can affect the participants' physiological responses in unintended ways (see Strohmetz, Chapter 8 for more on this). Tassinary et al. (2016) and Blascovich et al. (2011) have reviewed some of the following procedures that can minimize participant stress and anxiety:

1. *Before the experiment begins, provide an introductory session and/or a tour of the laboratory facilities.* This could be done virtually or on a website. In the presentation, show participants what will happen when they arrive at the experiment. Show them exactly what the electrodes will look like, what the experimenter will do to prepare the skin, etc.
2. *Use a less emotional word instead of "electrode" when speaking to the participants.* "Electrode" easily connotes electricity and can remind participants of other more infamous studies involving electric shock. In our laboratories, we try to use "biosensors" or "sensors" in our written documentation, instructions, and consent forms.
3. *Reduce the feeling of being watched.* If possible, have the experimenter sit in a separate room and monitor the participant using a concealed camera. In addition, avoid making participants aware that they are being closely monitored by refraining from asking them about their movements (e.g., "Did you just yawn?").

4. *Use a cover story and dummy electrodes, when needed.* As mentioned, facial EMG researchers rarely explicitly tell their participants that their facial muscles are being measured. Instead, one might say "we are recording activity that comes from the head, sort of like an EEG." If participants know their facial muscles are being recorded, it might invoke social desirability problems (e.g., "I should not be smiling at that, should I?"). The attachment of dummy electrodes (i.e., ones that don't record any physiological activity) to the back of the neck can help diminish the sense that facial muscles are of interest. See Nichols and Edlund, Chapter 3 for more on cover stories.

Finally, we cannot stress enough that the researcher who is about to embark on the recording of physiological activity should refer to more comprehensive guides and reviews about their measure of interest. Although Erasistratus in 294 BC was able to draw a strong inference about love sickness based on a fairly rudimentary understanding of physiological changes, the modern scientist has many subtler, yet more powerful and accurate, tools at hand. More interesting research using these physiological measures is forthcoming, and it will continue to yield deeper knowledge in social and behavioral science.

KEY TAKEAWAYS

- The use of physiological measures requires that the scientist considers a multiple-levels-of-analysis approach to understanding behavior, where the physiological processes are just one level of analysis that should be considered in conjunction with others, such as the social situation or the structure of society.
- Physiological measures can be used to investigate cognitive, emotional, and/or social phenomena unobtrusively and continuously during the research study.
- Choosing which physiological measure to use depends the phenomenon of interest, but also the expertise of the investigator and the costs of using the measure.
- Advances in technology should enable researchers to use more wearable, lightweight physiological equipment in the coming decade. In addition, the use of virtual and/or augmented reality will enable physiological research to have greater ecological validity and experimental realism.

IDEAS FOR FUTURE RESEARCH

- How reliable and valid are new, wearable technologies to measure physiological changes during the normal course of a person's day?
- Does the use of wearable physiological measures change the normal behavior and physiological response of participants?
- Will less intrusive measures of a person's physiology improve external validity or will the lack of control hinder their usefulness?

SUGGESTED READINGS

Blascovich, J., Vanman, E., Mendes, W. B., & Dickerson, S. (2011). *Social Psychophysiology for Social and Personality Psychology.* London: Sage.

Cacioppo, J. T., Berntson, G. G., Sheridan, J. F., & McClintock, M. K. (2000). Multilevel integrative analyses of human behavior: Social neuroscience and the complementing nature of social and biological approaches. *Psychological Bulletin, 126,* 829–843.

Cacioppo, J. T., Tassinary, L. G., & Berntson, G. G. (2016). *Handbook of Psychophysiology* (4th ed.). Cambridge: Cambridge University Press.

Huettel, S. A., Song, A. W., & McCarthy, G. (2014). *Functional Magnetic Resonance Imaging* (3rd ed.). Sunderland, MA: Sinauer.

Rohrbaugh, J. W. (2016). Ambulatory and non-contact recording methods. In J. T. Cacioppo, L. G. Tassinary, & G. G. Berntson (Eds.), *Handbook of Psychophysiology* (4th. ed., pp. 300–338). Cambridge: Cambridge University Press.

REFERENCES

Arthurs, O. J., & Boniface, S. (2002). How well do we understand the neural origins of the fMRI BOLD signal? *Trends in Neurosciences, 25,* 27–31. doi:10.1016/S0166-2236(00)01995-0

Bell, C. (1877). *The Anatomy and Philosophy of Expression: As Connected with the Fine Arts* (7th ed.). London: Williams Clowes and Sons.

Berger, H. (1929). Über das Elektrenkephalogramm des Menschen. *Archiv für Psychiatrie und Nervenkrankheiten, 87*(1), 527–570.

Blascovich, J., Vanman, E., Mendes, W. B., & Dickerson, S. (2011). *Social Psychophysiology for Social and Personality Psychology.* London: Sage.

Bradley, M. M., Cuthbert, B. N., & Lang, P. J. (1993). Pictures as prepulses: Attention and emotion in startle modification. *Psychophysiology, 30,* 541–545.

Button, K. S., Ioannidis, J. P. A., Mokrysz, C., Nosek, B. A., Flint, J., Robinson, E. S. J., & Munafò, M. R. (2013). Power failure: Why small sample size undermines the reliability of neuroscience. *Nature Reviews Neuroscience, 14*(5), 365–376. http://doi.org/10.1038/nrn3475

Buxton, R. B. (2009). *Introduction to Functional Magnetic Resonance Imaging: Principles and Techniques.* Cambridge: Cambridge University Press.

Cacioppo, J. T., & Berntson, G. G. (1992). Social psychological contributions to the decade of the brain: The doctrine of multilevel analysis. *American Psychologist, 47,* 1019–1028.

Cacioppo, J. T., Berntson, G. G., Sheridan, J. F., & McClintock, M. K. (2000). Multilevel integrative analyses of human behavior: Social neuroscience and the complementing nature of social and biological approaches. *Psychological Bulletin, 126,* 829–843.

Cacioppo, J. T., Petty, R., Losch, M., & Kim, H. (1986). Electromyographic activity over facial muscle regions can differentiate the valence and intensity of affective reactions. *Journal of Personality and Social Psychology, 50*(2), 260–268.

Cacioppo, J. T., Tassinary, L. G., & Berntson, G.G. (2016). *Handbook of Psychophysiology* (4th ed.). Cambridge: Cambridge University Press.

Cella, M., Okruszek, L., Lawrence, M., Zarlenga, V., He, Z., & Wykes, T. (2018). Using wearable technology to detect the autonomic signature of illness severity in schizophrenia. *Schizophrenia Research, 195,* 537–542.

Darwin, C. (1965). *The Expression of the Emotions in Man and Animals.* Chicago, IL: University of Chicago Press. [Originally published 1872.]

Dawson, M., Schell, A., & Filion, D. (2016). The electrodermal system. In J. T. Cacioppo,

L. G. Tassinary, & G. G. Berntson (Eds.), *Handbook of Psychophysiology* (4th ed., pp. 217–243). Cambridge: Cambridge University Press. doi:10.1017/9781107415782.010

Delisle, A., Lariviere, C., Plamondon, A., & Salazar, E. (2009). Reliability of different thresholds for defining muscular rest of the trapezius muscles in computer office workers. *Ergonomics, 52,* 860–871.

Demaree, H. A., & Everhart, D. E. (2004). Healthy high-hostiles: Reduced parasympathetic activity and decreased sympathovagal flexibility during negative emotional processing. *Personality and Individual Differences, 36,* 457–469.

Dickerson, S. S., & Kemeny, M. E. (2004). Acute stressors and cortical responses: A theoretical integration and synthesis of laboratory research. *Psychological Bulletin, 130*(3), 355–391.

Duchenne, G. B. A. (1990). *The Mechanism of Human Facial Expression.* R. A. Cuthbertson (Ed. and Trans.). Cambridge: Cambridge University Press. [Originally published 1862.]

Ekman, P., & Friesen, W. V. (1978). *The Facial Action Coding System.* Palo Alto, CA: Consulting Psychologists Press.

Feldman Barrett, L., Quigley, K. S., Bliss-Moreau, E., & Aronson, K. R. (2004). Interoceptive sensitivity and self-reports of emotional experience. *Journal of Personality and Social Psychology, 87,* 684–697.

Filion, D. L., Dawson, M. E., & Schell, A. M. (1998). The psychological significance of human startle eyeblink modification: A review. *Biological Psychology, 47,* 1–43.

Fowles, D., Christie, M. J., Edelberg, R., Grings, W. W., Lykken, D. T., & Venables, P. H. (1981). Publication recommendations for electrodermal measurements. *Psychophysiology, 18,* 232–239.

Fraser, P. M. (1969). The career of Erasistratus of Ceos. *Istituto Lombardo, Rendiconti, 103,* 518–537.

Guastella, A. J., & Hickie, I. B. (2016). Oxytocin treatment, circuitry, and autism: A critical review of the literature placing oxytocin into the autism context. *Biological Psychiatry, 79,* 234–242. doi:10.1016/j.biopsych.2015.06.028

Harmon-Jones, E., & Sigelman, J. (2001). State anger and prefrontal brain activity: Evidence that insult-related relative left-prefrontal activation is associated with experienced anger and aggression. *Journal of Personality and Social Psychology, 80,* 797–803.

Hermans, E. J., Ramsey, N. F., & van Honk, J. (2008). Exogenous testosterone enhances responsiveness to social threat in the neural circuitry of social aggression in humans. *Biological Psychiatry, 63,* 263–270. doi:10.1016/j.biopsych.2007.05.013

Huettel, S. A., Song, A. W., & McCarthy, G. (2014). *Functional Magnetic Resonance Imaging* (3rd ed.). Sunderland, MA: Sinauer.

Ito, T. A., & Urland, G. R. (2005). The influence of processing objectives on the perception of faces: An ERP study of race and gender perception. *Cognitive, Affective, and Behavioral Neuroscience, 5,* 21–36.

Johnson, R. R., Prentice, T. W., Bridegam, P., Young, C. R., Steelman, A. J., Welsh, T. H., ... & Meagher, M. W. (2006). Social stress alters the severity and onset of the chronic phase of Theiler's virus infection. *Journal of Neuroimmunology, 175,* 39–51.

Klement, K. R., Lee, E. M., Ambler, J. K., Hanson, S. A., Comber, E., Wietting, D., ... Sagarin, B. J. (2016). Extreme rituals in a BDSM context: The physiological and psychological effects of the "Dance of Souls." *Culture, Health & Sexuality, 19,* 453–469. doi:10.1080/13691058.2016.1234648

Korb, S., Malsert, J., Strathearn, L., Vuilleumier, P., & Niedenthal, P. (2016). Sniff and mimic – intranasal oxytocin increases facial mimicry in a sample of men. *Hormones and Behavior, 84,* 64–74. doi:10.1016/j.yhbeh.2016.06.003

Kuan, G., Morris, T., Kueh, Y. C., & Terry, P. C. (2018). Effects of relaxing and arousing music during imagery training on dart-throwing performance, physiological arousal indices, and competitive state anxiety. *Frontiers in Psychology*, *9*. doi:10.3389/fpsyg.2018 .00014

Lacey, J. I. (1967). Somatic response patterning and stress: Some revisions of activation theory. In M. H. Appley, & R. Trumbull (Eds.), *Psychological Stress: Issues in Research* (pp. 14–37). New York: Appleton-Century-Crofts.

Mallan, K. M., & Lipp, O. V. (2007). Does emotion modulate the blink reflex in human conditioning? Startle potentiation during pleasant and unpleasant cues in the picture-picture paradigm. *Psychophysiology*, *44*, 737–748. doi:10.1111/j.1469-8986.2007 .00541.x

Mendes, W. B., Blascovich, J., Lickel, B., & Hunter, S. (2002). Challenge and threat during interactions with White and Black men. *Personality and Social Psychology Bulletin*, *28*, 939–952.

Millasseau, S., & Agnoletti, D. (2015). Non-invasive estimation of aortic blood pressures: A close look at current devices and methods. *Current Pharmaceutical Design*, *21*, 709–718.

Neumann, D. L., Lipp, O. V., & Pretorius, N. R. (2004). The effects of lead stimulus and reflex stimulus modality on modulation of the blink reflex at very short, short, and long lead intervals. *Perception and Psychophysics*, *66*, 141–151. doi:10.3758/BF03194868

Philipp, M. C., Storrs, K. R., & Vanman, E. J. (2012). Sociality of facial expressions in immersive virtual environments: A facial EMG study. *Biological Psychology*, *91*, 17–21.

Picard, R. W., Fedor, S., & Ayzenberg, Y. (2015). Multiple arousal theory and daily-life activity asymmetry. *Emotion Review*, *8*, 62–75.

Rodriguez-Paras, C., Susindar, S., Lee, S., & Ferris, T. K. (2017). Age effects on drivers' physiological response to workload. *Proceedings of the Human Factors and Ergonomics Society Annual Meeting*, *61*, 1886–1886.

Ronay, R., & Carney, D. R. (2013). Testosterone's negative relationship with empathic accuracy and perceived leadership ability. *Social Psychological and Personality Science*, *4*, 92–99. doi:10.1177/1948550612442395

Sarchiapone, M., Gramaglia, C., Iosue, M., Carli, V., Mandelli, L., Serretti, A., ... & Zeppegno, P. (2018). The association between electrodermal activity (EDA), depression and suicidal behaviour: A systematic review and narrative synthesis. *BMC Psychiatry*, *18*, 22. doi:10.1186/s12888-017-1551-4

Schultheiss, O. C., & Stanton, S. J. (2009). Assessment of salivary hormones. In E. Harmon-Jones & J. S. Beer (Eds.), *Methods in Social Neuroscience* (pp. 17– 44). New York: Guilford.

Tassinary, L. G., Cacioppo, J. T., & Vanman, E. J. (2016). The somatic system. In J. T. Cacioppo, L. G. Tassinary, & G. G. Berntson (Eds.), *Handbook of Psychophysiology* (4th ed., pp. 151–182). Cambridge: Cambridge University Press.

Trepman, E., Gellman, R. E., Solomon, R., Murthy, K. R., Micheli, L. J., & de Luca, C. (1994). Electromyographic analysis of standing posture and demi-plié in ballet and modern dancers. *Medicine and Science in Sports and Exercise*, *26*, 771–782.

van Boxtel, A. (2010). Facial EMG as a tool for inferring affective states. In A. J. Spink, F. Grieco, O. Krips, L. Loijens, L. Noldus, & P. Zimmerman (Eds.), *Proceedings of Measuring Behavior* (pp. 104–108). Wageningen: Noldus Information Technology.

Vanman, E. J., Baker, R., & Tobin, S. J. (2018). The burden of online friends: The effects of giving up Facebook on stress and well-being. *The Journal of Social Psychology*. doi:10.1080/00224545.2018.1453467

Vanman, E. J., Boehmelt, A. H., Dawson, M. E., & Schell, A. M. (1996). The varying time courses of attentional and affective modulation of the startle eyeblink reflex. *Psychophysiology*, *33*, 691–697.

Vanman, E., Paul, B., Ito, T. A., & Miller, N. (1997). The modern face of prejudice and structural features that moderate the effect of cooperation on affect. *Journal of Personality and Social Psychology, 73*(5), 941–959. http://doi.org/10.1037/0022-3514.73.5.941

Wilke, K., Martin, A., Terstegen, L., & Biel, S. S. (2007). A short history of sweat gland biology. *International Journal of Cosmetic Science, 29*(3), 169–179.

Zillman, D. (1984). *Connections Between Sex and Aggression*. Hillsdale, NJ: Lawrence Erlbaum Associates.

11 Eyetracking Research

Jeff B. Pelz

Erasmus Darwin dedicated his book, *Zoonomia; or, The Laws of Organic Life* (Darwin, 1794) to "all those who study the Operations of the Mind as a Science." Darwin was fascinated by the movements of the eyes, and devoted a section of his book to "The Motions of the Retina Demonstrated by Experiments." His descriptions of eye movements were, of course, not the first. Darwin referenced work done by William Porterfield, a physician in Edinburgh, who published "An essay concerning the motions of our Eyes" (Porterfield, 1735). In that essay, Porterfield described the dramatic anisotropy across the visual field and the need to make eye movements to inspect objects in detail (Porterfield, 1735, pp. 184–185). He also pointed out the fact that we are typically not aware of the lack of detailed vision away from central vision, nor of the eye movements we make to compensate for that loss, stating that,

> in viewing any large Body, we are ready to imagine that we see at the same Time all its Parts equally distinct and clear: But this is a vulgar Error, and we are led into it from the quick and almost continual Motion of the Eye, whereby it is successively directed toward all the Parts of the Object in an Instant of Time. (p. 186)

The "vulgar error" that Porterfield described, the commonly held belief (or "illusion") that our visual system supports high acuity in parallel across the visual field, is what makes eyetracking such a powerful tool. This apparent **parallel process** is, in fact, a rapid **serial process** – humans make over 100,000 rapid eye movements per day to successively bring objects to the high-acuity central region of the retina. The ability to track an observer's direction of gaze offers the researcher a unique window into behavior, attention, and performance. Because the acuity of the human visual system varies so dramatically across the visual field, with high acuity only available in the fovea (a small region in the center of the retina with the highest density of cones), humans have evolved a rich suite of oculomotor behaviors to move the eyes and reorient the fovea to regions requiring high acuity and to stabilize the retinal image on the fovea during those tasks. Perhaps more important for eyetracking research, humans also tend to move their eyes to **foveate** regions of interest, even when high acuity is not required for a given task. As a result, tracking the eyes offers a method of tracking visual attention because, while visual attention and gaze can be disassociated, they rarely are (Spering & Carrasco, 2015). As a result, eyetracking provides a powerful tool for monitoring performance, exploring task strategies, and understanding human behavior.

Most people are completely unaware of the hundreds of rapid eye movements they make each minute to bring objects of interest to the high-acuity fovea (which covers less than 1/100 of 1% of the visual field) (Clarke, Mahon, Irvine, & Hunt, 2017; Foulsham & Kingstone, 2013). Despite the fact that it is impossible not to attend to the target of an eye movement (Shepherd, Findlay, & Hockey, 1986), observers perform poorly when asked to report on their own eye movements. Marti, Bayet, and Dehaene (2015) found that, in addition to being unaware of many of their fixations, observers also reported many fixations in locations that were not actually fixated. The value of eyetracking is, therefore, evident; a record of observers' gaze patterns provides a record of objects and regions attended that is even more accurate than known to the observers. Monitoring these eye movements also provides an external marker of visual attention that is more accurate than alternate methods, such as verbal report and post-tests intended to infer attention, which are often

unreliable and can change the behavior one is trying to understand. Their use across the social and behavioral sciences continues to grow, with applications spanning the fields of anthropology, criminal justice, economics, human factors, linguistics, and sociology (see "Sample Applications," below).

11.1 Background

The oculomotor system of humans (and non-human primates) consists of control architecture and six extraocular muscles attached to each eye. The muscles are arranged in three agonist-antagonist pairs, as seen in Figure 11.1. The first pair, attached on the left and right of each eye, are the **lateral rectus** (on the temporal, or outer side of the eye) and the **medial rectus** (on the nasal, or inner side of the eye). Together, the lateral and medial recti rotate the eyes outward and inward. The second pair, attached on the top and bottom of each eye, are the **superior rectus** (on the top of the eye) and the **inferior rectus** (on the bottom). Together, the superior and inferior recti rotate the eyes up and down. The final pair, also attached on

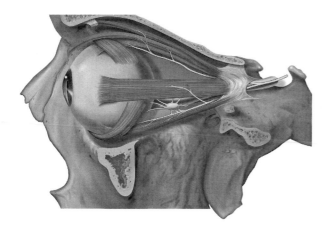

FIGURE 11.1 Extraocular muscles.
Adapted from Patrick Lynch, Yale University

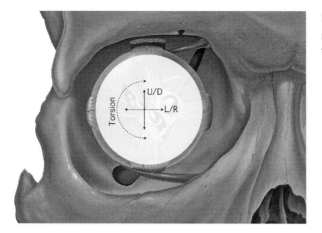

FIGURE 11.2 Directions of eye movements. Adapted from Patrick Lynch, Yale University

the top and bottom of each eye, are the **superior oblique** (on the top) and the **inferior oblique** (on the bottom). The oblique muscles are so named because they are attached in such a manner that they rotate the eye primarily about the optical axis rather than redirect that axis. Together, the superior and inferior obliques rotate the eyes about the optical axis ("torsional" eye movements), as illustrated in Figure 11.2.

The extraocular muscles can be seen as serving two primary purposes: 1) stabilizing the image on the retina to optimize the information available to the visual system (especially in central vision), and 2) moving the retina to a new location to gather different information.

It is useful to parse oculomotor events into categories based on motion of the observer and of the target being viewed. In the simplest case, both observer and target are stationary. When the eyes are stable in the head, the static target (or scene) is stable on the retina, and the eyes are said to be in **fixation**. When an observer shifts the orientation of the eyes to bring a new location to the retina to gather different information, optimal performance

requires that the shift take place as quickly as possible to minimize the time during which the retinal image is unstable. The rapid movement of the eyes from one place to another is known as a **saccade**, a French term introduced by Javal in 1878 to describe the intermittent "jerks" or "twitches" he observed in movements of the eyes (Tatler & Wade, 2003), but the term was adopted widely in the literature decades later when Raymond Dodge reintroduced it in a review of international literature and suggested that it was preferable to the then-current term "Type I" eye movement (Dodge, 1916, pp. 422–423).

Figure 11.3 shows a record of the horizontal (solid line) and vertical (dashed line) orientation of gaze over a 1-second interval. The vertical axis is scaled in degrees of visual angle. Fixations are evident as periods where the orientation is constant (e.g., the first 0.17 seconds). Saccades, in contrast, are periods where the orientation changes at a high rate (e.g., between 0.18 and 0.23 seconds). Saccadic eye movements are distinguished from other movements by their velocity, reaching rotational velocities of over 500°/sec (Becker, 1989). The relationship between saccadic

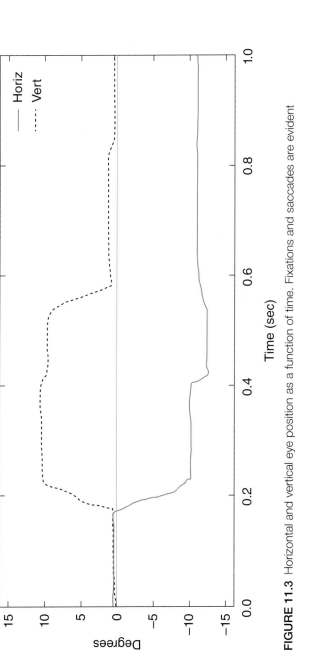

FIGURE 11.3 Horizontal and vertical eye position as a function of time. Fixations and saccades are evident

amplitude and peak velocity has been characterized over a wide range, from a small fraction of a degree to over 45 degrees. Bahill, Brockenbrough, and Troost (1981) showed that there is a smooth relationship between amplitude and peak velocity and termed the relationship the **main sequence**. Over a relatively large range of saccadic amplitudes (e.g., 5–45°), one can estimate the duration of a saccade with the simple linear relationship $D = D_0 + dA$ – where D is the duration of the saccade, D_0 is the minimum duration for small saccades, d is the rate of increase in duration per degree of saccadic amplitude, and A is saccadic amplitude (Becker, 1989). Values for D_0 vary from 20 to 30 milliseconds, and for d from 1.5 to 3.0 milliseconds/degree, but a simple approximation can be obtained as $D = 20 + 2A$. The first saccade in Figure 11.3 is ~ 10 degree in amplitude, so its estimated duration would be $20 + 2(15) = 50$ milliseconds. Inspection of the plot verifies the estimate.

In the simplest case, when both observer and target are stationary, the two oculomotor events are *fixations* and *saccades*. The next case we consider is when the observer is stationary, but the target is moving within the observer's field of view. If gaze did not follow the target, vision would suffer both because the target of interest would drift away from the high-acuity fovea and because the relative motion of the image on the retinal surface would result in blur. Both potential faults are compensated by matching the eyes' rotation to the angular position of the target in the visual field with **smooth pursuit** eye movements. The result is a stable (or nearly stable) **retinal image**. In the case of fixations, it is static because both the object and eye are stationary. Note that smooth pursuit

movements are rarely perfectly matched to object motion, resulting in some **retinal slip** (Grossberg, Srihasam, & Bullock, 2012). Figure 11.4 shows a record of the horizontal (solid line) and vertical (dashed line) orientation of gaze over a 1-second interval. The vertical axis is scaled in degrees of visual angle. Smooth pursuit is evident throughout the 1-second period, with angular velocity ranging over ~ 15–$20°$/sec. Because smooth pursuit eye movements are often unable to exactly compensate for target motion, retinal errors can accumulate, moving the target away from the center of the retina. Small **catch-up** saccades are frequently executed to re-center the target, as evident between 0.5 and 0.75 seconds in Figure 11.4.

Smooth pursuit eye movements compensate for motion of the target when the observer is static. The opposite case would also degrade visual acuity. If the target is static, but the head rotates, vision would again suffer because the retinal image would be blurred and it would drift away from the center of the retina. **Vestibular-ocular response** (VOR) eye movements compensate for head rotations by counterrotating the eyes to stabilize the retinal image. Figure 11.5 shows a series of three saccadic eye movements surrounded by VOR. This eye-movement record is the result of an observer inspecting a printed object, making three small, predominantly leftward saccadic eye movements, while simultaneously rotating the head to the left (and slightly upward). While gaze was fixed on a target location (between the saccades), the eyes rotated slowly to the right (to increasingly positive horizontal values). The saccadic eye movements rapidly brought gaze to the left (more negative horizontal values).

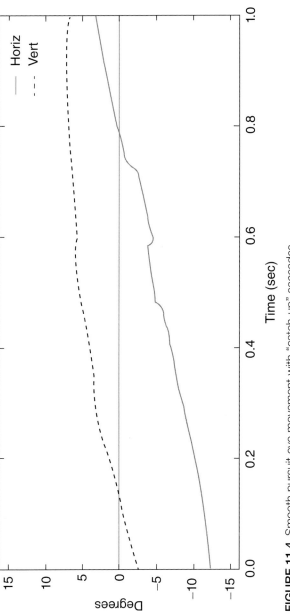

FIGURE 11.4 Smooth pursuit eye movement with "catch-up" saccades

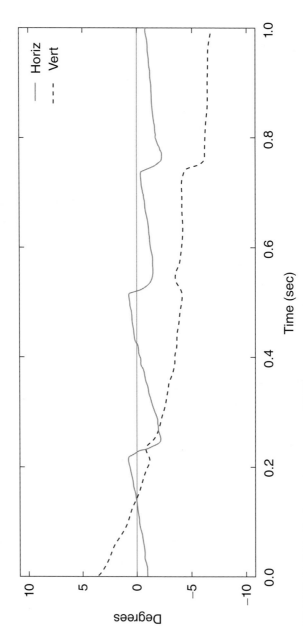

FIGURE 11.5 Vestibular-ocular response (VOR) eye movements

These are the most important classifications of **conjugate** eye movements, movements that affect both eyes equally. Other classifications include **vergence** eye movements (Kowler, 1995), which affect the two eyes differently to vary the angle between the eyes, **miniature eye movements** (or **movements of fixation**), made up of tremor, drift, and microsaccades (Rucci & Victor, 2015), and the **optokinetic reflex** (OKR), which is related to smooth pursuit but driven by large-field motion instead of a small target (Leigh & Zee, 2015). It is important to note that, while all of these classifications are convenient and valuable for discussion, they are often seen together in complex combination during a real task, as seen in Figure 11.6, which shows a 5-second segment from a task in which an observer was inspecting a printed document. All of the oculomotor events discussed so far – fixations, saccades, smooth pursuit, and VOR eye movements – are evident, and they are not easily separated because the classifications are blurred in complex tasks.

Many methods have been used to monitor eye movements over the years, from relatively crude mechanical linkages, such as the "… mirror, attached to the cocainized cornea by an aluminum capsule" described by Dodge (1911, p 383), to modern high-speed video devices that use computer vision algorithms to track features of the eye. The goal has always been the same: To obtain high-quality, low-noise measures of eye position (or velocity) over time without unduly affecting the underlying behavior. Even as we move past systems that require mechanical linkages to the eyes, one must consider the influence that knowledge that one's gaze is being monitored may have on visual performance.

Modern, video-based eyetrackers can be classified into five broad categories: 1) **remote eyetrackers** that can monitor an observer's gaze position from a distance without requiring any contact with the observer; 2) **tower eyetrackers** – a subclass of remote eyetrackers that place the camera(s) near the observer and limit head movements; 3) **headband-mounted eyetrackers** in which cameras are worn by the observer and are tethered to a laboratory computer; 4) **wearable eyetrackers** that are self-contained and can be used outside of the laboratory; and 5) **head-mounted display eyetrackers** that are integrated into head-mounted displays. The category boundaries can be blurred, but the categories are useful to introduce the hardware and systems in common use today. All of these systems rely on one or more video cameras to capture images of the observer's eye(s) and some method of representing the scene that the observer is viewing.

11.2 Remote Eyetrackers

Eyetrackers of this class are typically built into (or mounted below) computer displays (Figure 11.7), though they can be used without displays to monitor the gaze position of observers viewing other scenes. The primary advantages of remote eyetrackers are that nothing needs to be attached to the observer, and output gaze position can be given directly in the fixed reference frame of display coordinates. The disadvantage is that the observer must typically remain within a relatively constrained position (the "headbox") throughout the data collection if the camera(s) capturing the observer's eye(s) are static. Some systems address this issue by using dynamic eye cameras that follow the observer's

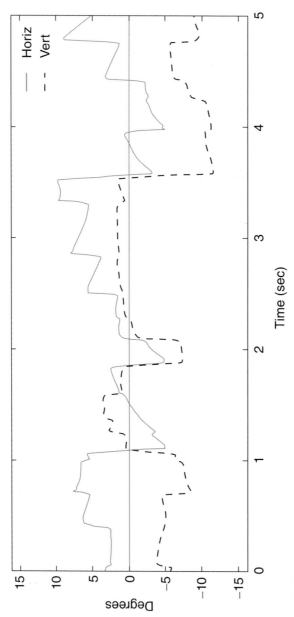

FIGURE 11.6 Five-second gaze orientation record

FIGURE 11.7 Remote eyetracker mounted below computer monitor (Tobii Pro Spectrum)

FIGURE 11.8 Remote eyetracker mounted below computer display (LC Technologies EyeGaze Edge EYEFOLLOWER)

movements, extending the effective headbox (Figure 11.8).

In addition to in research, common uses for remote eyetrackers are market research and assistive technologies for individuals with mobility or communication challenges.

11.3 Tower Eyetrackers

Tower eyetrackers are designed to optimize performance at the expense of observer movement and flexibility. Figure 11.9 shows an EyeLink 1000 Tower eyetracker. The illuminator and eye camera are mounted on the "tower" above the observer, and a "hot mirror" is placed between the observer and the stimulus monitor. The hot mirror serves as a beam splitter, allowing most of the visible light from the computer monitor to pass to the observer while reflecting most of the near infrared power to be reflected to illuminate and image the observer's eye. Because the participant's position is more strictly limited than in remote eyetrackers, the system design

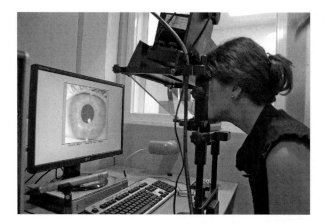

FIGURE 11.9 Tower eyetracker (EyeLink 1000 Tower)

allows careful illumination and image capture of the observer's eye at high magnification and good focus without concern that the image will drift. As a result, tower eyetrackers can provide some of the highest-quality gaze data among video-based eyetrackers. The tradeoff, of course, is that tasks are limited to those that can be performed under these restricted conditions, and motion of the observer can limit data quality. Like remote eyetrackers, the tower eyetracker has the advantage that gaze position output is directly in the fixed reference frame of computer display coordinates because the tower eyetracker, display, and observer are stationary in that space.

11.4 Headband-Mounted Eyetrackers

Headband-mounted eyetrackers build the illumination and eye-capture hardware into a device that is worn by the observer and tethered to a laboratory computer. They offer some of the advantages of a tower eyetracker (high-speed and high-magnification eye images) without placing severe limitations on head movements (and useful field of view). Because the illuminators and eye cameras are attached to the headband, they move with the participant's head, extending the useful field of view to 360°. Unlike remote and tower eyetrackers, however, there is no fixed reference frame for the gaze output, so gaze output from headband-mounted eyetrackers is limited to the relative angle of the headband and observer's head (assuming no slippage). Headband-mounted eyetrackers typically include a **scene** (or **world**) **camera** mounted to the headband that captures a video of the scene from the observer's (dynamic) viewpoint. Because the scene camera moves with the observer (and eye camera), gaze can be output in the reference frame of the scene camera. Note that motion of the scene camera represents head (and full-body) motion of the observer, so mapping from scene-camera coordinates to world coordinates is a non-trivial problem.

While scene and eye cameras are mounted to the headband, processing and/or recording of the video streams is typically done on a tethered laboratory computer, limiting the range of motion, mobility, and tasks that can be performed with headband-mounted eyetrackers.

11.5 Wearable Eyetrackers

Wearable eyetrackers are designed to allow the entire system required for recording eye and scene videos to be worn by the participant. The systems are lighter and less obtrusive than typical headband-mounted eyetrackers and are typically built into eyeglass-style frames, making it possible to take tasks outside of the laboratory. Figures 11.10–11.12 show wearable systems from Positive Science, Tobii, and Pupil Labs. Like headband-mounted eyetrackers, eye- and scene-cameras move with the participant's head, so all eye-movement records are in an "eye-in-head" reference frame, not in a world reference frame. The systems typically output a video record captured from the scene camera with a cursor overlaid that indicates the observer's point of gaze on each frame, as shown in Figure 11.13 for a system from Positive Science. Gaze data from wearable (and headband-mounted) eyetrackers can be analyzed by examining the output video stream (Vansteenkiste, Cardon, Philippaerts, & Lenoir, 2015), mapping the eye-in-head gaze to gaze-in-world coordinates by incorporating head position measurements (Evans, Jacobs, Tarduno, & Pelz, 2012) or by

FIGURE 11.10 Wearable eyetracker (Positive Science)

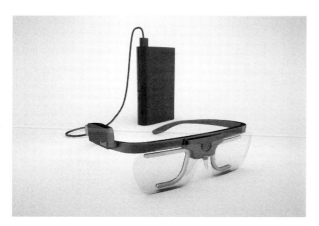

FIGURE 11.11 Wearable eyetracker (Tobii)

FIGURE 11.12 Wearable eyetracker (Pupil Labs)

FIGURE 11.13 Frame of output video from wearable eyetracker (Positive Science)

incorporating markers in the environment to automatically map the scene by calculating the transformation necessary to map gaze position in the scene-camera reference frame to the reference frame defined by the markers (Pfeiffer & Renner, 2014).

11.6 Head-Mounted Display (HMD) Eyetrackers

Using HMDs to display virtual-reality (VR) and augmented-reality (AR) is becoming more widespread. While eyetracking in HMDs has been done since the 1990s (Duchowski, Shiva-shankaraiah, Rawls, Gramopadhye, Melloy, & Kanki, 2000; Geisler & Perry, 1998; Pelz, Hayhoe, Ballard, Shrivastava, Bayliss, & Von der Heyde, 1999), the cost of HMDs has plummeted in the last five years with the advent of commodity HMDs such as the HTC Vive and the Oculus Rift (see Figure 11.14). Their arrival brought the cost of high-quality, binocular HMDs down by an order of magnitude and dramatically broadened the field of eyetracking in VR/AR. In its simplest form, gaze "inside" a virtual environment can be tracked in eye-in-head coordinates and analyzed manually by viewing the video record showing gaze position overlaid on the virtual scene. However, VR systems require head tracking to update the display based on observer movements, so it is possible to map the eye-in-head signal onto the virtual environment, yielding a "gaze-in-(virtual-)world" signal (Diaz, Cooper, Kit, & Hayhoe, 2013). Figure 11.15 shows the gaze output overlaid on the virtual environment being viewed by a participant in a virtual environment (Diaz et al., 2013, Figure 2). The angular distance between the observer's gaze and the bullseye target can be calculated because gaze is known in the reference frame of the head, head position is measured by the motion-capture system, and the bullseye target is displayed in the HMD at a known position with respect to the head.

11.7 Selecting an Eyetracker

Selecting the type of eyetracker you will use is typically based on the accuracy and precision required, the type of task and stimulus to be presented to the observer, the degree to which

FIGURE 11.14 Oculus Rift with integrated eyetracker (Pupil Labs)

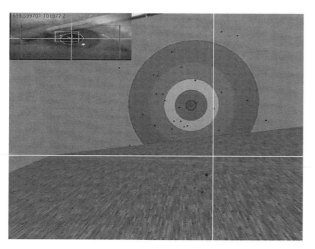

FIGURE 11.15 Frame from virtual environment (Diaz et al., 2013, Figure 2)

the experimenter is willing to limit the movements of the observer, and the display to be used. In general, the more restrictions that are placed on the observer (and task), the greater is the likelihood of capturing accurate and precise data. For example, tower eyetrackers are capable of providing very high-quality data at data rates exceeding 1000 samples per second (1 kHz), with the tradeoff that observers typically use a chinrest that limits movement, and the field of view is limited by the apparatus. At the other extreme, wearable eyetrackers can be worn by observers as they perform natural tasks outside of the laboratory with an unlimited field of view, but data rates are limited to ~250 Hz, data quality does not match that of a tower tracker, and the field of view over which data can be collected reliably is smaller than the total field of view. In some cases, the task requirements demand the flexibility of a wearable eyetracker, and the experimental paradigm must be adapted to the available data quality. In other cases, the required accuracy and precision may dictate a simpler task (and a more restrictive environment for the observers). It is ultimately up to the researcher to decide what is the best for his/her research question.

11.8 Data Quality

Whether data is collected in a laboratory (see Nichols & Edlund, Chapter 3), in "the wild" (see Elliott, Chapter 4), or in a virtual environment (see Guadagno, Chapter 5), it is critical to understand the effects of eyetracking data quality. The primary metrics of data quality are **accuracy** and **precision** (see Wagner & Skowronski, Chapter 2). A useful definition of accuracy in the context of eyetracking data quality is the agreement between an averaged gaze position and the actual gaze position during that period. A useful definition of precision is the dispersion of reported gaze position over time.

If an observer is asked to fixate a known target for 0.5 seconds to validate a calibration, and the average of the reported gaze position differs from the known target position by 1° of visual angle, the system can be said to have an accuracy (or more formally an *inaccuracy*) of 1°. Figure 11.16 shows the scene from an observer's perspective during a calibration routine. The left panel of Figure 11.17 illustrates the average gaze position over a 500-millisecond period while the observer was looking at the central target. This record is said to have "high accuracy" (or low

FIGURE 11.16 View from scene camera

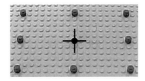

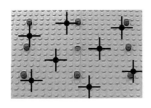

FIGURE 11.17 Average gaze location with 0° (left) and 2° (right) inaccuracy

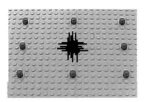

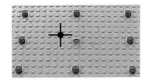

FIGURE 11.18 Gaze locations with 0° average inaccuracy and high (left) and low (right) precision

inaccuracy) – approximately 0° of visual angle in this illustration. The right panel of Figure 11.17 illustrates a case where the reported average position over the same duration deviates from the actual target position by approximately 2° of visual angle. Accuracy is typically reported in degrees of visual angle and is usually specified as a single value, presumed to be the maximum deviation of the reported average value from the actual value. Unfortunately, there is not wide agreement in the field about how to define, measure, or report accuracy in eyetracking research, and it is not unusual for publications to appear that simply report the manufacturers' claims of accuracy that are often best-case values measured under ideal conditions (Reingold, 2014). This would be like reporting the reliability of a scale from its testing phase and not providing any information regarding its reliability in your sample. Ideally, researchers would report accuracy ranges based on position in the visual field and over trial duration.

Measures of accuracy made during an experiment end up measuring the accuracy of the entire system, including the eyetracker and the observer. To measure the eyetracker without including the influence of an observer's ability to hold fixation during calibration and validation, an artificial eye can be used. Wang, Mulvey, Pelz, and Holmqvist (2017) evaluated several eyetrackers using a number of artificial eyes provided by eyetracker manufacturers and adapted a commercially available model eye designed for ophthalmic image training.

Individual gaze records of a system with "high accuracy" (or low inaccuracy) may still vary widely on a short timescale; accuracy only refers to the time-averaged position, not the instantaneous position or the dispersion of reported values. Figure 11.18 shows two cases that both have high accuracy. The left panel has 0° average inaccuracy and low dispersion, while the right panel has the same 0° average inaccuracy but a much larger dispersion.

Accuracy and precision are independent; it is possible for an eyetracker to be accurate and imprecise, inaccurate and precise, etc. as illustrated in Figure 11.19.

Reports of precision in eyetracking research are complicated by the fact that there are several different definitions of precision in use in the community. Blignaut and Beelders (2012) reviewed many of the precision measures in use and proposed a new measure. Two

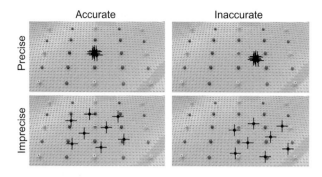

FIGURE 11.19 Accuracy and precision in eyetracker data

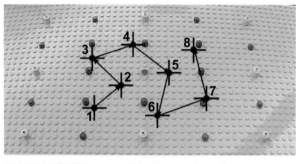

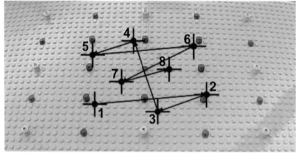

FIGURE 11.20 Temporal order effects on precision measures

commonly reported measures of precision are the **standard deviation** of position over time (σ) and the **sample-to-sample root-mean-square** distance (RMS-S2S). The two measures share a computational form (both are root-mean-square measures) but measure different characteristics of an eyetracker signal. Most important, σ ignores the temporal sequence (and data rate) of the gaze data (with the resultant value being a function only of the spatial layout of the samples) while RMS-S2S is very sensitive to temporal sequences and data rate. The two panels in Figure 11.20 show the same spatial distribution of eight gaze positions reported over a short period of time. The top panel shows the reported gaze "drifting" from left to right. The bottom panel shows a more random pattern between the same points. The precision of the two cases is identical when reported in terms of standard deviation (5.5°), but the RMS-S2S is dramatically different – 4.8° for the "drifting" in the top panel of Figure 11.20 and 8.4° for the more random pattern in the bottom panel.

Spatial accuracy and precision are the most commonly reported measures of eyetracker data quality, but other measures are important. Holmqvist, Nyström, and Mulvey (2012) reviewed other measures including **temporal precision** (the eyetracking system's ability to accurately report the timing and duration of events), **data loss** (number of observers excluded and the fraction of data loss in included participants), and **robustness** to a number of factors that can affect data quality.

Holmqvist et al. (2012) list factors that influence data quality including the **physical characteristics of the participants** (e.g., spectacle correction, complexion, makeup), **operator** (e.g., experience), **task** (e.g., difficulty, stimulus), **environment** (e.g., lighting, vibration), and **eyetracker design** (e.g., camera temporal and spatial resolution, algorithm design). Researchers should also consider the influence of gaze position within the visual field. Feit et al. (2017) reported the effect of position on accuracy and precision in the context of user-interface design, but the results are at least as important to researchers.

11.9 Sample Applications

Once a specialized tool available only to a small number of experts, eyetracking is now being applied across many disciplines, and its reach is increasing as advances in technology make the hardware and software tools more accessible in terms of cost, ease of use, and data quality. A few sample applications from the social and behavioral sciences are highlighted below.

In recent work by Mühlenbeck, Jacobsen, Pritsch, and Liebal (2017), a remote eyetracker was used to explore the origins of man-made markings on Paleolithic era artifacts. The researchers recorded the eye-movement patterns of human observers as they viewed images of archeological artifacts (hand axes and sticks) with and without markings and incisions. The human observers were from two groups – adolescent "city dwellers" (from Hamburg and Berlin) and an age-matched group from a tribe of hunter-gatherers in Namibia. The groups differed in their cultural backgrounds and in the visual surroundings in which they spend most of their time, as the indigenous hunter-gatherers lived on the South African savannah and the German adolescents in industrialized cities. The results of the eyetracking study showed that the marked objects were fixated longer by both groups, though neither group judged the marked objects as more attractive than the unmarked objects.

In addition to the human subjects, the researchers also measured the eye movements of a group of orangutans who were trained to perform a similar task while their eye movements were tracked. Unlike the two human groups, the orangutans showed no increased fixation on the marked objects, leading the researchers to conclude that the function of early markings was to direct attention.

While Mühlenbeck and colleagues used eyetracking to study the intersection of anthropology, archeology, psychology, and art, others have applied the tools of eyetracking to the study of business practices and economic theory. Criminal justice and crime-scene analysis are other areas where eyetracking has recently offered insights into long-standing questions. Lasky, Fisher, and Jacques (2017) recruited two "active shoplifters" and used a wearable eyetracker to monitor their eye movements while they simulated shoplifting in retail stores. The results provide a unique opportunity to examine the efficacy of security measures. Watalingam, Richetelli, Pelz, and Speir (2017) explored criminal behavior from the perspective of the crime-scene analyst. They used a wearable eye-tracker to monitor the eye-movement patterns of thirty-two crime-scene analysts (categorized as "experts" or "trained novices") as they examined a mock crime scene. The experts were more efficient and effective (achieving higher scores in the same time). Comparison of the eye movements revealed some reasons for the difference. Comparisons of total contiguous fixation time in a region ("dwell time") and time in areas of interest (AOIs) revealed significant differences based on expertise. Notably, experts had more variation in dwell time on critical evidence (e.g., blood spatter and cartridge case) but had less variation in their search patterns.

Linguistics and language acquisition are two other application areas where eyetracking has proved to be a valuable tool. Eberhard, Spivey-Knowlton, Sedivy, and Tanenhaus (1995) started a rich line of research in this area with their paper titled, "Eye movements as a window into real-time spoken language comprehension in natural contexts." In their paradigm, now known as the "Visual World Paradigm" (Salverda & Tanenhaus, 2017), observers listen to (or produce) spoken language while their gaze is tracked as they view objects related to the audio stream. The fixations reveal a detailed record of the temporal process of spoken language comprehension (or production). This record can be at the level of a phrase ("the brown dog brush") or a single word ("carpet"). In the first case, if a scene contains a brown box, a brown dog, and a brown dog brush, it is instructive to watch the pattern of eye movements of an observer viewing the scene as the phrase is spoken. Theories about when the meaning of the phrase can be understood can be tested by examining when gaze moves reliably to the brush. In the second case, if a scene contains a camera, a car, a card, and a carpet, the time course of word disambiguation can be understood by monitoring gaze over the short time period of a single word's utterance. In addition to the established literature in linguistics, researchers in language acquisition and second-language research are also making use of eyetracking (Conklin & Pellicer-Sánchez, 2016).

Eyetracking has also been applied to the study of survey methodology in sociological research. Surveys continue to play an important role in such research, despite our awareness that survey results can be affected by subjects' concern over how their answers will appear to others – the "social desirability bias" (Belli, Traugott, & Beckmann, 2001). Kaminska and Foulsham (2016) used a wearable eye-tracker to monitor the gaze of observers as they filled out surveys in one of three modes: **web**, **self-administered questionnaire**, and **face-to-face**. By measuring the relative time observers spent looking at the questions, the response options, and the interviewer (in the face-to-face mode), the researchers were able to examine the influence of mode, response option position, and question time.

11.10 Conclusion

Eyetracking is a powerful tool that provides quantitative measures of behavior, performance, stress, and attention that are otherwise difficult to obtain (Kovesdi, Spielman, LeBlanc, & Rice, 2018; Macatee, Albanese, Schmidt, & Cougle, 2017). Gaze offers an external marker of attention, and by guiding gaze it may be possible to guide cognition (Grant & Spivey, 2003). Recent advances have led to miniature, robust systems that can be used in real and virtual environments, extending the reach of eyetracking into new domains. The price of entry into eyetracking has fallen significantly in the last decade. While it used to cost tens of thousands of dollars for an entry-level system, it is now possible to equip a laboratory with a basic system for less than $5000. The use of eyetracking across a broad range of application areas demonstrates the value of the methodology. Predicting future research directions is always dangerous, but it is evident that three areas are ripe for methodological development that will allow new insights in the social and behavioral sciences. Further miniaturization of the instrumentation to capture high spatial and temporal resolution gaze data will allow more natural behaviors to be studied. Wireless and/or high-density portable storage will support long-term monitoring, allowing "experiments" extending hours or days instead of seconds. Finally, development of sophisticated data-analysis tools must be prioritized to support the resultant complex datasets.

KEY TAKEAWAYS

- Eyetracking technology allows understanding human cognition in ways that are not accessible in other manners. You can learn about conscious, semiconscious, and unconscious cognitive processes through the careful use of this technology.
- There are several classes of eyetrackers, including remote, tower, headband-mounted, and wearable. The classifications are based largely on whether and how the eyetracker is attached to the observer (or vice versa). In general, the larger, more restrictive systems provide faster temporal rates and superior spatial accuracy and/or resolution but limit task complexity and reality by inhibiting natural behaviors.
- Accuracy and precision are much more than manufacturer specifications. Everything from eyetracker design to eye makeup can influence data quality. Calibration followed by validation is an important part of any eyetracking study and allows researchers to include meaningful accuracy and precision metrics in publications.

IDEAS FOR FUTURE RESEARCH

- What similarities and differences occur between people from different cultures in terms of what we process and how we process visual events?
- How can eyetracking technology evolve to allow for high accuracy and precision measurements of eye movements while still allowing for natural behavior?
- How can research fields work together using eyetracking technology to create impactful interdisciplinary research?

SUGGESTED READINGS

Becker, W. (1989). The neurobiology of saccadic eye movements. Metrics. *Reviews of Oculomotor Research*, *3*, 13.

Kowler, E. (1995). Cogito ergo moveo: Cognitive control of eye movement. In M. Landy, L. Maloney, & M. Pavel (Eds.), *Exploratory Vision: The Active Eye* (pp. 51–77). New York: Springer-Verlag.

Leigh, R. J., & Zee, D. S. (2015). *The Neurology of Eye Movements* (Vol. 90). New York: Oxford University Press.

Wade, N., & Tatler, B. W. (2005). *The Moving Tablet of the Eye: The Origins of Modern Eye Movement Research*. New York: Oxford University Press.

Watalingam, R. D., Richetelli, N., Pelz, J. B., & Speir, J. A. (2017). Eye tracking to evaluate evidence recognition in crime scene investigations. *Forensic Science International*, *280*, 64–80.

REFERENCES

Bahill, A. T., Brockenbrough, A., & Troost, B. T. (1981). Variability and development of a normative data base for saccadic eye movements. *Investigative Ophthalmology & Visual Science*, *21*(1), 116–125.

Becker, W. (1989). The neurobiology of saccadic eye movements. Metrics. *Reviews of Oculomotor Research*, *3*, 13.

Belli, R. F., Traugott, M. W., & Beckmann, M. N. (2001). What leads to voting overreports? Contrasts of overreporters to validated voters and admitted nonvoters in the American National Election Studies. *Journal of Official Statistics*, *17*(4), 479.

Blignaut, P., & Beelders, T. (2012). The precision of eye-trackers: A case for a new measure. In *Proceedings of the Symposium on Eye Tracking Research and Applications* (pp. 289–292). New York: ACM.

Clarke, A. D., Mahon, A., Irvine, A., & Hunt, A. R. (2017). People are unable to recognize or report on their own eye movements. *The Quarterly Journal of Experimental Psychology*, *70*(11), 2251–2270.

Conklin, K., & Pellicer-Sánchez, A. (2016). Using eye-tracking in applied linguistics and second language research. *Second Language Research*, *32*(3), 453–467.

Darwin, E. (1794). *Zoonomia; or, The Laws of Organic Life* (Vol. 1). London: J. Johnson.

Diaz, G., Cooper, J., Kit, D., & Hayhoe, M. (2013). Real-time recording and classification of eye movements in an immersive virtual environment. *Journal of Vision*, *13*(12), 1–14.

Dodge, R. (1911). Visual motor functions. *Psychological Bulletin*, *8*(11), 382–385. (1916). Visual motor functions. *Psychological Bulletin*, *13*(11), 421–427.

Duchowski, A. T., Shivashankaraiah, V., Rawls, T., Gramopadhye, A. K., Melloy, B. J., & Kanki, B. (2000, November). Binocular eye tracking in virtual reality for inspection training. In *Proceedings of the 2000 Symposium on Eye Tracking Research & Applications* (pp. 89–96). New York: ACM.

Eberhard, K. M., Spivey-Knowlton, M. J., Sedivy, J. C., & Tanenhaus, M. K. (1995). Eye movements as a window into real-time spoken language comprehension in natural contexts. *Journal of Psycholinguistic Research*, *24*(6), 409–436.

Evans, K. M., Jacobs, R. A., Tarduno, J. A., & Pelz, J. B. (2012). Collecting and analyzing eye tracking data in outdoor environments. *Journal of Eye Movement Research*, *5*(2), 6.

Feit, A. M., Williams, S., Toledo, A., Paradiso, A., Kulkarni, H., Kane, S., & Morris, M. R. (2017). Toward everyday gaze input: Accuracy and precision of eye tracking and implications for design. In *Proceedings of the 2017 CHI Conference on Human Factors in Computing Systems* (pp. 1118–1130). New York: ACM.

Foulsham, T., & Kingstone, A. (2013). Fixation-dependent memory for natural scenes: An experimental test of scanpath theory. *Journal of Experimental Psychology: General, 142*(1), 41.

Geisler, W. S., & Perry, J. S. (1998). Real-time foveated multiresolution system for low-bandwidth video communication. In *Photonics West '98 Electronic Imaging* (pp. 294–305). Bellingham, WA: International Society for Optics and Photonics.

Grant, E. R., & Spivey, M. J. (2003). Eye movements and problem solving: Guiding attention guides thought. *Psychological Science, 14*(5), 462–466.

Grossberg, S., Srihasam, K., & Bullock, D. (2012). Neural dynamics of saccadic and smooth pursuit eye movement coordination during visual tracking of unpredictably moving targets. *Neural Networks, 27*, 1–20.

Holmqvist, K., Nyström, M., & Mulvey, F. (2012). Eye tracker data quality: What it is and how to measure it. In *Proceedings of the Symposium on Eye Tracking Research and Applications* (pp. 45–52). New York: ACM.

Kaminska, O., & Foulsham, T. (2016). Eye-tracking social desirability bias. *Bulletin of Sociological Methodology/Bulletin de Méthodologie Sociologique, 130*(1), 73–89.

Kovesdi, C., Spielman, Z., LeBlanc, K., & Rice, B. (2018). Application of eye tracking for measurement and evaluation in human factors studies in control room modernization. *Nuclear Technology, 202*(2–3), 1–10.

Kowler, E. (1995). Cogito ergo moveo: Cognitive control of eye movement. In M. Landy, L. Maloney, & M. Pavel (Eds.), *Exploratory Vision: The Active Eye* (pp. 51–77). New York: Springer-Verlag.

Lasky, N. V., Fisher, B. S., & Jacques, S. (2017). "Thinking thief" in the crime prevention arms race: Lessons learned from shoplifters. *Security Journal, 30*(3), 772–792.

Leigh, R. J., & Zee, D. S. (2015). *The Neurology of Eye Movements* (Vol. 90). New York: Oxford University Press.

Macatee, R. J., Albanese, B. J., Schmidt, N. B., & Cougle, J. R. (2017). Attention bias towards negative emotional information and its relationship with daily worry in the context of acute stress: An eye-tracking study. *Behaviour Research and Therapy, 90*, 96–110.

Marti, S., Bayet, L., & Dehaene, S. (2015). Subjective report of eye fixations during serial search. *Consciousness and Cognition, 33*, 1–15.

Mühlenbeck, C., Jacobsen, T., Pritsch, C., & Liebal, K. (2017). Cultural and species differences in gazing patterns for marked and decorated objects: A comparative eye-tracking study. *Frontiers in Psychology, 8*, 6.

Pelz, J. B., Hayhoe, M. M., Ballard, D. H., Shrivastava, A., Bayliss, J. D., & von der Heyde, M. (1999). Development of a virtual laboratory for the study of complex human behavior. In *Electronic Imaging '99* (pp. 416–426). Bellingham, WA: International Society for Optics and Photonics.

Pfeiffer, T., & Renner, P. (2014, March). EyeSee3D: A low-cost approach for analyzing mobile 3D eye tracking data using computer vision and augmented reality technology. In *Proceedings of the Symposium on Eye Tracking Research and Applications* (pp. 369–376). New York: ACM.

Porterfield, W. (1735). An essay concerning the motions of our eyes. Part I. Of their external motions. *Edinburgh Medical Essays and Observations, 3*, 160–261.

Reingold, E. M. (2014). Eye tracking research and technology: Towards objective measurement of data quality. *Visual Cognition, 22*(3–4), 635–652.

Rucci, M., & Victor, J. D. (2015). The unsteady eye: An information-processing stage, not a bug. *Trends in Neurosciences, 38*(4), 195–206.

Salverda, A. P., & Tanenhaus, M. K. (2017). 5 The Visual World Paradigm. *Research Methods in Psycholinguistics and the Neurobiology of Language: A Practical Guide, 9*, 89.

Shepherd, M., Findlay, J. M., & Hockey, R. J. (1986). The relationship between eye movements and spatial attention. *The Quarterly Journal of Experimental Psychology Section A, 38*(3), 475–491.

Spering, M., & Carrasco, M. (2015). Acting without seeing: Eye movements reveal visual processing without awareness. *Trends in Neurosciences, 38*(4), 247–258.

Tatler, B. W., & Wade, N. J. (2003). On nystagmus, saccades, and fixations. *Perception, 32*(2), 167–184.

Vansteenkiste, P., Cardon, G., Philippaerts, R., & Lenoir, M. (2015). Measuring dwell time percentage from head-mounted eye-tracking data – comparison of a frame-by-frame and a fixation-by-fixation analysis. *Ergonomics, 58*(5), 712–721.

Wade, N., & Tatler, B. W. (2005). *The Moving Tablet of the Eye: The Origins of Modern Eye Movement Research.* New York: Oxford University Press.

Wang, D., Mulvey, F. B., Pelz, J. B., & Holmqvist, K. (2017). A study of artificial eyes for the measurement of precision in eye-trackers. *Behavior Research Methods, 49*(3), 947–959.

Watalingam, R. D., Richetelli, N., Pelz, J. B., & Speir, J. A. (2017). Eye tracking to evaluate evidence recognition in crime scene investigations. *Forensic Science International, 280*, 64–80.

12 Questionnaire Design

Heather M. Stassen

Heather J. Carmack

Questionnaires are a common data collection method used by researchers, pollsters, news reporters, market researchers, and campaign workers to capture individuals' attitudes, retrospective behaviors, and knowledge (Brace, 2013). Morning news outlets report the findings from the most recent Gallup poll, students are stopped on the sidewalk to answer a few questions about the new student union food options, consumers rate the customer service of their cable provider – questionnaires are one of the most popular and most familiar research methods, in part because we encounter them on a daily basis. This chapter discusses issues to consider when developing questionnaires and writing questions and response options as well as limitations and concerns related to the use of questionnaires.

12.1 Considerations Prior to Questionnaire Development

Crafting an effective questionnaire requires careful thought. Although developing a series of questions may seem rather simple, there are numerous considerations prior to beginning questionnaire development. As Hague, Hague, and Morgan (2013) noted, "It is true that anyone can string some questions together, and normally you will get an answer in some shape or form, but it may not be the answer you are looking for" (p. 106). While the construction of a survey or questionnaire "allows license to the creator," initial key considerations help ensure that investigators are able to garner relevant information. These forethoughts include the distribution method, identifying extant measures, developing a plan for a pilot test, and thinking about the limitations of questionnaires (Hague et al., 2013, p. 106).

12.1.1 Distribution Method

Survey and questionnaire distribution and completion are either self-administered (e.g., field, mail, and online surveys) or investigator-administered (face-to-face and telephone surveys). Each mechanism has advantages and disadvantages. For example, while online surveys are often preferred due to ease of distribution, online distribution may not always be feasible or advantageous for reaching a specific participant pool (e.g., senior citizens). Similarly, mail surveys traditionally have low response rates but allow participants time to complete questions, which is particularly useful when looking for detailed description in open-ended questions. For example, in response to a question asking a participant to describe their most recent workplace conflict, the lack of perceived time constraints and the absence of an investigator could enable participants to provide more detail and context to their answer.

Self-Administered Questionnaires. Broadly, self-administration occurs when an individual is asked to complete a questionnaire without an investigator reading the questions to the participant. This requires that participants be able to read well and, if open-ended questions are present, have adequate writing skills. Although mail surveys are self-administered, not all self-administered questionnaires are done through mail. For example, many academic researchers have access to a sample population through university research pools where students receive credit or other incentives for participating in research studies. These questionnaires are often administered in classroom or lecture hall settings. In other cases, a researcher may be able to identify a physical location of a sample group and would bring questionnaire copies to a location for participants to complete and return in that setting. Often known as field self-administered surveys, this particular mode is known to have a higher return rate than mail self-administered questionnaires (Dunning & Calahan, 1974). Although a high return rate is common for field self-administered questionnaires, at times, participants rush to complete them, often skipping questions or providing incomplete responses to open-ended questions.

Mail surveys often suffer from poor response rates and an overall delay in responses, particularly if there is not an incentive (Dillman, 2000). Mail surveys with a prepaid incentive of as little as $1 will garner a response rate 6% higher than a mail survey without an incentive. A $10 incentive will yield a response rate 22% higher (Mercer, Caporaso, Cantor, & Townsend, 2015). However, because mail surveys are self-administered, they "can provide a sense of privacy for participants," and, without the influence of an investigator asking the questions, may be optimal for questionnaires that request private or sensitive information from participants (Salant & Dillman, 1994, p. 36). Mail surveys were once considered cost effective given that materials and postage were the primary cost. However, the advent of online questionnaire programs has, in contrast, made mail questionnaires seemingly expensive.

Online questionnaires completed through a survey software, such as SurveyMonkey or Qualtrics, also fit the criteria of self-administered questionnaires. Online administration, while incredibly time and cost effective, can be difficult in that it is easy for participants to overlook requests distributed via email or on social media. Moreover, online surveys suffer from non-response error. As Manfreda, Batagelj, and Vehovar (2002) noted, "non-response is measured with the percentage of respondents who prematurely abandon the Web questionnaire" and also the "proportion of unanswered items." In other words, mail surveys that are returned tend to be more complete than online surveys in which participants often skip questions or abandon the questionnaire, leaving researchers with partial data.

Investigator-Administered Questionnaires. Unlike self-administered questionnaires, investigator-administered questionnaires entail significant additional costs, as a trained investigator is needed to read the questions to the participants and record responses. Telephone, once a primary mechanism for data collection, has become much more complex with the increase in cellphones and the reduction in landlines. Although random-digit dialing can be a way to reach participants, refusal rates are often high (Brick, Collins, & Chandler, 1997). Marketing research often utilizes automated calling for consumer feedback. In contrast, social and behavioral science research has continued to move away from this channel, given the lack of availability of phone numbers and low response rates yielded from random-digit dialing (Brace, 2013).

Similarly, a face-to-face survey, in which an investigator asks a participant questions, may lead to answers that will reflect favorably on the participant. As Pedhazur and Schmelkin (1991) indicated, "Of the various aspects of respondent role-enactment, probably the most important and pervasive is that of self-preservation, which refers to the desire on the part of the respondent to present him or herself to the research in a particular light" (p. 140). For instance, a participant may be embarrassed to admit behaviors or attitudes that might be judged negatively (e.g., not exercising regularly, consuming fatty foods, or recreational drug use), and subsequently not provide truthful responses to the researcher.

Design Considerations. The way in which a researcher intends to distribute the questionnaire significantly impacts the design of the survey itself. In other words, the intended

distribution method must be taken into consideration when designing the questionnaire. In his discussion of Tailored Design, Dillman (2000) noted that commonalities to good questionnaires include "the development of survey procedures that create respondent trust and perception of increased rewards and reduced costs for being a respondent, that take into account features of the survey situation, and that have their goal the overall reduction of survey error" (p. 4). The survey situation undoubtedly includes the mode of distribution. All modes have some limitations, but these can be mitigated by considering the mode of distribution when designing the survey.

The general consensus is that the length of a survey has a negative correlation with response rate. In other words, the longer the survey, the lower the response rate (Dillman, 2000). Dillman (2000) also noted that "there is more to length than a simple count of pages, and that the frequent tendency of questionnaire designers to cram more questions into fewer pages to make a questionnaire shorter is *not* likely to accomplish its intended purpose" (p. 306). Some of these factors mitigating response burden, other than length, include the number of prompts participants are asked to consider, organization of questions, and the level of thought required to answer the questions. Asking a participant over the phone to participate in a survey taking thirty to forty minutes would not yield excellent response rates – though the same questionnaire administered by phone, with advanced notice through mail, will undoubtedly yield a higher rate. Similarly, online surveys suffer from high breakoff rates, where participants stop engaging, when the survey starts fast and then becomes slow (e.g., more difficult or open-ended questions coming later) or when they become discouraged by a progress bar indicator (Tourangeau, Conrad, & Couper, 2013). Including interactive grids may make the questions easier to answer or make the questionnaire appear shorter in online formats, but "it may promote non-differentiation or straightlining, in which respondents select the same response for every (or nearly every) item" (Tourangeau et al., 2013, p. 113). The length of time needed for completion, often used to identify response burden rather than page length in the online era, should be a reflection of the mode of distribution. Modes in which breakoff is more socially acceptable because of anonymity, including telephone, should be transparent about the investment of time and forewarn participants when considerable time is needed for completion. This will increase response rates, reduce non-response error, and maintain a positive relationship with the respondent.

In addition to the length of the questionnaire, the mode of distribution ought to significantly influence the wording of questions. A question read by a participant should be phrased differently than a question heard by a respondent. Garnering the participant's attention is also crucial in the first few questions of a self-administered survey. Salant and Dillman (1994) argued that the first question should be of interest to the participant and, while easy to answer, be at the intersection of the participant's information and the topic. Dillman (2000) and others discouraged inclusion of demographic questions at the start of self-administered questionnaires unless there is a high chance of breakoff or non-response before the completion of the survey due to a high response burden from length of the instrument or intensity of questions. The importance of the wording of the first question may not ultimately be as critical when an investigator is able to communicate the importance of the questionnaire with verbal and nonverbal cues. The reliance on text in self-administered surveys necessitates both clarity and interest to prevent confusion and non-response error. The mode of questionnaire distribution and collection can and will impact response rate, threshold for perceived response burden, desire to answer in a socially acceptable way, potential breakoff, and tolerance for level of personal disclosure in a question.

12.1.2 Exploring Extant Measures

Questionnaires are ubiquitous in social and behavioral science research. Although many are proprietary and cannot be used without consent or payment to the initial instrument designers, many others are available for free use. It is appropriate to attempt to identify existing measures, when available, to reduce time and costs, while also increasing validity by using a standard measure that has undergone significant testing. In other words, do not reinvent the wheel if items or an entire instrument have been crafted and tested examining the variables in your study. For example, in the field of communication studies, measures created by James C. McCroskey, a notable and prolific scholar, are available free for use on his personal website and each contains a history of the scale development (and each scale has been used in at least one academic journal article, which helps investigators see previous uses of each scale). Likewise, numerous survey items are available to use freely through the ESCR Question Bank, now hosted by UK Data Service. When an initial search does not find extant measures, investigators may wish to seek out information through professional listservs. The key is to evaluate existing measures prior to use, as Hyman, Lamb, and Bulmer (2006) noted, "the use of pre-existing survey questions is unfortunately not free of drawbacks; researchers ought to be aware of these when undertaking any question 'recycling'" (p. 5).

12.1.3 Testing the Instrument

The terms "pre-test" and "pilot study" are, at times, used interchangeably in the literature and, at other times, differentiated. For example, the word "pre-test" could be used to identify a questionnaire being used to garner a baseline measure of participants' behavior and attitudes with a subsequent "post-test" measure to be administered after some type of intervention. In this case, a full sample would be used. Similarly, the term "pilot study" is, at times, used to identify a complete study, generally exploratory in nature, which would come before a secondary study. Our use of the terms pre-test and pilot is meant to reference a trial of the questionnaire to a select group of individuals – not with the goal of data collection or identifying a baseline measure, but rather for understanding the clarity and reliability of the instrument.

Pre-testing of the instrument should emulate the conditions in which the questionnaire will be administered, which makes evaluation of the measures in interview and focus group formats less preferable to field testing (Fowler, 1995). For self-administered questionnaires, the instrument should be tested by participants similar in key characteristics to the sample population. Moreover, the instrument should be tested in the appropriate mode – mail, online, or in a physical setting. Regarding telephone interviews, Salant and Dillman (1994) noted, "Inevitably, problems arise as soon as questions are read to someone who is not within sight" (p. 133), which makes piloting an instrument in the channel it will be ultimately administered an imperative. Those who engage in the pre-test should be allowed not just to answer the questions, but be provided space to comment on question clarity as they complete the instrument. Moreover, it may be appropriate to ask participants – either in writing or verbally – about clarity of specific questions, the definition of variables they were asked about, and how they reached certain answers. For interviewer-administered questionnaires, it is important to garner feedback from the interviewers as well. The purpose of pre-testing is ultimately to "make sure that the questionnaire measures what it is supposed to, that people understand and can easily answer the questions, and that interviewers can work with it" (Salant & Dillman, 1994, p. 133). Fowler (1995) noted that, since a small sample is used for pre-testing, researchers ought to be cautious about making significant changes to an instrument based on the feedback from a few individuals, but that "if a question is problematic for 20% or 40% of the sample, that constitutes a serious problem" (p. 115).

Pre-testing should have two primary goals: Fixing errors – both for participants and investigators – and examining the quality of the data that is being collected (Presser et al., 2004). Investigators should plan resources and time for pre-testing. Not planning accordingly may make it tempting to skip over this important step and move directly into data collection. Pre-testing is imperative and essential (Presser et al., 2004). Bad survey items create bad questionnaires, which leads to bad data and unusable results. Although it may save a few days and dollars to not pre-test the instrument, it would take much longer and cost much more to start over after running the full study.

12.1.4 Self-Report Limitations

A questionnaire may ask about an individual, a close other (e.g., family member or romantic other), or a general situation (e.g., asking for an appropriate response to a situation). Many questionnaires engage in self-report, in that they are asking participants to report on their own behaviors, attitudes, and beliefs. In these types of questionnaires, the responses generated are significantly limited by the nature of self-reporting. Participants may be unable or unwilling to provide accurate information regarding their own behaviors for fear of a negative impression to the investigator or creating cognitive dissonance for themselves regarding contradictions in their own attitudes and behaviors. For example, in their study regarding media usage, Vraga, Bode, and Troller-Renfree (2016) noted, "Relying on self-reports of content exposure may lead to unreliable estimates, as people define content differently or inaccurately recall the frequency with which they view content topics and styles on Facebook" (p. 152). Similarly, in their study examining workplace absenteeism and healthcare utilization, Short et al. (2010) found that participants were able to accurately recall doctor appointments and missed work in the last month but were not as accurate in recalling these actions over the past year. Moreover, healthier individuals were able/willing to recall information regarding work absences more accurately than their unhealthy counterparts. Research illustrates that moving to a non-self-report measure, particularly in instances where actions or beliefs may not be deemed socially acceptable, may be particularly useful for garnering more truthful information and avoiding response distortion (Stewart, Bing, Davison, Woehr, & McIntyre, 2009).

12.2 Questionnaire Design

12.2.1 Instructions

Instructions for a questionnaire may ask for participants to provide a direct response, reflect on a previous situation, or provide their feedback to a vignette. For example, instructions seeking a direct response such as, "Please rate your agreement to the following set of statements" is an example of an instruction seeking a direct response. In contrast, "When answering the following questions, please keep in mind your most recent romantic relationship" is asking participants to reflect on a previous experience and to answer accordingly. Vignettes are "stories about individuals in situations which make reference to important points in the study of perceptions, beliefs, and attitudes ... Participants are typically asked to respond to these stories with what they would do in a particular situation" (Hughes, 1998, p. 381). Alexander and Becker (1978) stated that "it is frequently argued that questionnaires and interviews are not well suited for the study of human attitudes and behavior because they elicit unreliable and biased self-reports" and they subsequently noted that vignettes present a stimulus for response that "more closely approximate[s] a real-life decision-making or judgement-making situation" (p. 93). As vignettes serve as an anchor, "systematic variation across individuals in the rating of the vignettes can be attributed to differences in reporting behavior" (Rice, Robone, & Smith, 2011, p. 142). For example, a vignette may contain a description of a conflict scenario in the workplace followed by several closed-ended questions asking participants to identify the severity of the conflict and appropriateness of the ways in which the individuals engaged in the conflict.

Unless there is some level of deception warranted in the study (see also Nichols & Edlund, Chapters 3 and Blackhart & Clark, Chapter 9), questionnaire instructions should be clear and concise. Moreover, if the question type changes – moving from a Likert-scale matrix to a Guttman-style question, for example – a new set of directives for the participants must also be included. In the case of investigator-administered questionnaires, researchers ought to provide directions for both the individual administering the questionnaire as well as the participant answering the questionnaire items. In all cases, instructions and directives should be included during the pre-testing of the instrument.

12.2.2 Inclusion of Demographics

Demographic questions, such as age, level of education, race, ethnicity, sex, gender identification, and income, are of varying importance depending on the hypotheses and research questions driving the study. As noted previously, best practices indicate that demographic questions come at the end of the questionnaire unless specific questions are imperative to the study and there is a high likelihood of breakoff based on the length or level of disclosure in the questionnaire. Dillman (2000) also noted that it may be important, particularly for web-based surveys, where a convenience sample is used, to include demographic questions at the start of the instrument. For example, if a study is examining friendship in older age, a demographic question pertaining to age would help to exclude individuals who are not age-appropriate participants for the study.

Demographic questions are often recycled from pre-existing measures, and some survey software offers demographic question templates. However, these may be limited and not appropriate to a particular study – proceed with caution. For example, a prototype question on SurveyMonkey asks participants to identify their gender, but only provides "male" and "female" as options – this is limited and could lead to incomplete, or even unusable, data if gender is a key variable. Government resources, such as the United States Census Bureau's American Community Survey, may help to identify consistent demographic items with current language for race and ethnicity identification.

12.3 Question Considerations

Writing clear questions requires a concern for presenting concepts, questions, and responses in a way that is easy for participants to understand and answer. Researchers must consider the questions *and* answer options they include in a questionnaire. Whether researchers are using extant measures or drafting original questions, questionnaire items must be reliable and valid (Converse & Presser, 1986; see also Wagner & Skowronski, Chapter 2 for information regarding reliability and validity).

Questions must be consistent, meaning that they should be read and understood by participants in a similar way (Brace, 2013). For example, the question "Where are you from?" can be interpreted in several different ways – country, region, state, or city? This makes the question unreliable because participants may write in different levels of living in a location (some participants may write city, whereas some may write country, and everywhere in between). The question, "In what US state do you currently live?" is specific and can be consistently understood by participants. Questionnaire items must also be valid and reflect the concept or topic on which they focus (Converse & Presser, 1986). If researchers want to ask about participants' intention to leave their current jobs, they need to consider if they are interested in if participants are planning on leaving their current jobs or thinking about leaving their current jobs. Planning on leaving and thinking about leaving are different questions, so the researchers would have to be clear about which concept (planning or considering) they want to study.

Researchers writing their own questions must also consider the answer options they provide participants as well as the types of questionnaire scales they want to use to measure participants' answers. Part of these decisions will be guided by whether the researchers are studying nominal, ordinal, ratio, or interval data (see also Edlund & Nichols, Chapter 1 for information about data types) because certain variables are associated with certain questions and scales. Researchers who are not clear about how they write their questions, do not identify what types of variables they are studying, and do not select the best scales to measure participants' responses

will struggle with the reliability and validity of their questionnaires. There are four major considerations for researchers writing questionnaire items: mutually exclusive answer options, exhaustive options, reverse coded items, and scale options.

12.3.1 Mutually Exclusive Options

One of the challenges of writing questionnaire items is ensuring the answer options are mutually exclusive. Mutually exclusive options are response options which are distinct and can be clearly differentiated from each other (Bradburn, Sudman, & Wansink, 2004). Participants should be able to read the response options and see how they are different from each other. Take, for example, a questionnaire item about sleep deprivation in college. The researcher is interested in how many hours, on average, participants sleep at night.

Q: On average, how many hours do you sleep at night?
   A. 3 hours or less
   B. 3–4 hours
   C. 4–6 hours
   D. 6–8 hours
   E. 8 or more hours

The problem with this item and response options is that they are not mutually exclusive. If a participant sleeps, on average, 6 hours a night, which option should they select: C or D? One participant may select C while another may select D. This makes it difficult for the researcher to report how much sleep college students get. Additionally, the question and response options would not meet the reliability requirement. This kind of question is easy to correct (and identify before sending participants the questionnaire) if researchers are vigilant. The researcher in this scenario can easily correct the response options:

Q: On average, how many hours do you sleep at night?
  A. Less than 3 hours
  B. 3–4 hours
  C. 5–6 hours
  D. 7–8 hours
  E. 9 or more hours

There are times, however, when participants may want the opportunity to select multiple options or not be sure about which response option they want to select. Researchers can give participants multiple options and still ensure response options are mutually exclusive. Additionally, researchers can add an "other" or write-in option for participants. The "other" option addresses a concern of questionnaire items not being exhaustive. Using a write-in option, such as "On average, how many hours do you sleep a night? ___hours," also provides researchers flexibility to capture mutually exclusive answers while giving participants the opportunity to drive their answer. Researchers may direct participants to choose the option that best represents their opinion, attitude, or feeling about a topic or offer them the option of selecting multiple options by telling participants to "check all that apply" (Bradburn et al., 2004). For example, if the researcher studying college students and sleep wanted to know what keeps participants from getting the recommended 8 hours of sleep, she could ask:

Q: What are the major reasons you do not get 8 hours of sleep a night? Please select all the options that apply to you.

It is important, when providing participants multiple options, to list an "other" option in case the list provided does not include response options for participants. If researchers want more detail, they can also include a space for participants to list other response options.

12.3.2 Exhaustive Options

Researchers also need to ensure their questionnaire response options are exhaustive. When researchers provide lists of options for participants to choose, it is important those lists capture all the possible ways a participant could answer a question. These lists need to be both exhaustive and equivalent or equal to each other (Keyton, 2011). As discrete data, no response option should have more meaning than another. The order of responses does not communicate ranking or hierarchy, just categories. Demographic questions are prime examples where exhaustive response options are important. Owens (2011) used marital status to illustrate the importance of exhaustive options, listing "married," "divorced," "widowed," "separated," and "never married." However, this list is not exhaustive: What about participants who are in committed relationships but are not married (p. 249)? Participants would not have a response option that represents them and would be forced to either leave the question blank, select an untruthful answer, or choose the "other" option, if provided.

As discussed earlier, it is not uncommon for researchers to use extant measures which have been validated. However, an important factor to consider with exhaustive responses is that they can change. The marital status demographic question is a good example of this. Until the 2015 US Supreme Court decision that same-sex marriage is legal, an exhaustive marital status response list would have to include "civil unions" for members of the LGBTQ+ community who were not able to marry. The exhaustive response options for sex and gender have also changed. "Transgender" is expected to appear alongside "male" and "female" when asking about participants' sex; gender options have also recently expanded to include other options, such as "cisgender" and "genderfluid" (Bauer, Braimoh, Scheim, & Dharma, 2017).

12.3.3 Reverse Coded Questions

Researchers will often write questionnaire items as reverse coded questions, where the wording of questions varies in the use of positive and negative wording or by changing the direction of a scale's anchors (Weems & Onwuegbuzie, 2001). If a questionnaire includes positively and negatively worded questions, the negatively worded questions would be reverse coded. For example, in a questionnaire about the impact of speech and debate competition on graduate school success, Holm and Carmack (2012) used reverse coded questions. The questionnaire focused on successes and failures in graduate school and the failure questions, which were negatively worded questions, were reverse coded. Reverse coding is a popular, albeit not heavily used, questionnaire design practice since it can be difficult to write all questionnaire questions in positively keyed language. A good example is the Willingness to Communicate about Health Scale (Wright, Frey, & Sopory, 2007), a ten-item measure with questions such as "I am comfortable talking about health with health providers." Two questions from the scale are reverse coded questions: "When

I don't feel well, I don't want to talk" and "I experience difficulties communicating successfully with healthcare providers." In this case, the researchers had to write negatively worded questions because writing these questions in positively worded language would make the questions difficult to read.

Reverse coded questions are also a way to avoid response bias, which occurs when response patterns appear when there is not a pattern (van Sonderen, Sanderman, & Coyne, 2013). This can happen for a number of reasons, including not understanding the question, not reading questions closely enough to realize the question wording changed, not having a strong feeling about a response, or tending to respond at the extreme ends of a measure (Swain, Weathers, & Niedrich, 2008). However, changing the valence orientation of certain questions in questionnaires that require reverse coding can create problems for researchers. Alternating between positively and negatively worded questions can create confusion in participants and may negatively impact their responses (Magazine, Williams, & Williams, 1996). In a study of children in grades 4 through 6, Benson and Hocevar (1985) found the students were "less likely to indicate agreement by disagreeing with a negatively phrased item than to indicate agreement by agreeing with a positively phrased item" (p. 236).

12.3.4 Types of Questionnaire Scales

An important part of questionnaire design is selecting the correct measurement platforms for questions. Researchers have a variety of scale options from which to choose but must start with whether they want to ask **closed-ended** or **open-ended** questions. A closed-ended question is designed to

provide participants with previously designated response options. Open-ended questions ask participants to write in their own
answers to questions, with no predetermined
options for participants to choose. There are a
number of different types of questionnaire
scale options for researchers interested in
closed-ended questions, including Likert-type,
Guttman, and Osgood's semantic differential.

Likert-Type. The Likert-type scale (pronounced Lick-ert), named after Rensis Likert,
is perhaps the most commonly used measurement design to collect interval data (Keyton,
2011). Likert-type scales ask participants to
respond to questions or statements using
balanced, bipolar opposite anchors, the most
popular being "strongly agree" to "strongly
disagree," with a neutral option as the midpoint. For example, researchers studying the
use of social media platforms to search for
internship opportunities could ask participants about their attitudes about use:

Using social media platforms is useful for finding internships				
1	2	3	4	5
Strongly Disagree	Disagree	Neither Agree Nor Disagree	Agree	Strongly Agree

Likert-type measures typically use a five-
point scale of response choices, but can be
expanded to a seven-point scale. There is
assumed equal meaning between each
response choice. Likert-type designs can be
used with single-question items or with multiple items that can be summed and averaged
into one variable score. Researchers can also

choose to eliminate the numbers altogether
and only use the word anchors. Although there
is assumed equal meaning between response
choices, it may be difficult for participants to
distinguish between responses; for example,
what really is the difference between "Agree"
and "Strongly Agree" when it comes to emotional infidelity in relationships? How can and
do participants tell the difference in their feelings on the topic?

Guttman. There are times when researchers
may want to ask more complex questions
beyond the traditional binary (Yes/No) question. A Guttman question format, named after
Louis Guttman, expands binary questions by
allowing participants to identify variance in
their binary response (Guttman, 1944).
A popular example of a Guttman scale is the
Bogardus Social Distance Scale; participants
respond to questions on a five-point scale
ranging from "least extreme" to "most
extreme" (Parrillo & Donoghue, 2005). One
of the items asks, "Are you willing to permit
immigrants to live in your neighborhod?"
Guttman's major critique of Likert-type items
and scale is that it is problematic to assume
there is or should be equal meaning between
each response choice (Massof, 2004). In
a Guttman scale, each response is assigned a
different score and those scores are used to
compare participants.

Semantic Differential. Osgood's semantic
differential scale is another common interval
data measure. Named after Charles Osgood,
semantic differential scales present a topic or
phenomenon anchored by pairs of polar
opposite adjectives (Osgood, Suci, & Tannenbaum, 1957). Participants use a five-item or

seven-item scale to express their feeling about the topic. For example, Ohanian (1990) used semantic differential scales to ask participants about source credibility. One of the three dimensions of source credibility, attractiveness, gives participants five different adjective pairs to rate a speaker's attractiveness:

Attractive–Unattractive
Classy–Not classy
Beautiful–Ugly
Elegant–Plain
Sexy–Not sexy

Semantic differential questions are different from Likert and Guttman measures because they allow researchers to get at the evaluation, potency, and activity feelings related to a topic because of the use of adjectives (Osgood et al., 1957).

Open-Ended Questions. The biggest critique of closed-ended questions is that researchers select the answers for participants. There are, however, many times when researchers would want participants to report their own answers to questions. When this happens, researchers turn to open-ended questions (Fink, 1995). Open-ended questions are especially helpful when researchers are studying topics which are new or when views on a topic may have changed. Unlike closed-ended questions and measures which identify how to score answers, open-ended questions do not have a built-in scoring system; instead, researchers must identify ways, such as content analysis or thematic analysis, to identify meaning (Hetsroni, 2007). Of important concern for researchers using open-ended questions is question wording. Keyton (2011) showed how the question wording is important

through a simple example: "How would you describe North Carolina State University?" (p. 173). Participants could answer in different ways, none of which could actually get at the purpose of the question. When asking open-ended questions, researchers must be clear about what they mean and, if appropriate, need to make sure the question is directed. In the aforementioned example, a researcher could instead ask, "What positive words do you use to describe North Carolina State University? What negative words do you use to describe North Carolina State University?"

12.4 Limitations to Questionnaire Design

A number of issues can hamper the success of questionnaire design which researchers need to be aware of in order to prevent the questionnaire from being unreliable and invalid. These can be divided into two categories: Concerns related to self-reporting participants and concerns related to design. When researchers ask participants to self-report answers, they open the door for participants to control the questionnaire and their "true" responses. First, researchers need to worry about respondent refusal, attrition, and carelessness in answering questions. The length of a questionnaire can negatively impact individuals' participation. Refusal rate, the refusal to start a questionnaire, increases with the length of the questionnaire (Hoerger, 2010); participants are willing to complete a five-minute questionnaire, but may be less likely to complete a forty-five-minute questionnaire. Long questionnaires also have attrition problems: Participants are more likely to not complete a survey after starting if they feel it takes too long to complete the questionnaire (Brace, 2013). Finally, participants may be careless in their answering of questions. This is a common problem when individuals feel the questionnaire is too long. Rather than stop their participation, individuals may instead answer every Likert-type question as "neither agree nor disagree" or form a pattern or design with their answers (a common practice among students). Researchers need to be thoughtful about the length of their questionnaires and ask themselves if they have to ask all the questions listed or if there are better, more efficient ways to ask questions. An easy way to address this limitation is to pre-test questionnaires to determine time length and question clarity. A discussion with pre-test participants in a focus group about the questions' intended and perceived meaning, validity, and reliability can help to further refine a questionnaire (Schwartz, 1999).

Another concern for questionnaire design is social desirability response. Participants may feel pressure to answer questions in a way that is socially appropriate or favorable, even if they do not feel that way (De Maio, 1984). Self-reported data are fallible (Schwartz, 1999), and, coupled with perceived pressure to answer in a certain way, make the data more problematic. This is especially a concern when researchers ask about sensitive topics or ask for personal information. However, participants will even give socially desirable answers to questions about mundane activities, such as not paying parking tickets or returning an overdue library book, if they feel it might make them look less favorable (Oliphant, 2014). Social desirability is a problem whether respondents are completing questionnaires with researchers present or alone (or online).

Even when participants are anonymous and are alone when completing the questionnaire, some answer questions the way they feel makes them look better (Brace, 2013). Tied to social desirability, participants may feel a tendency to publicly agree with questions (called acquiescence) even when they do not privately agree with them (Krosnick, 1999). This may be related to social desirability, but it could also be because they want to get through the questionnaire or because the question is leading and there appears to be a "correct" answer and they want to show they know it. If a researcher suspects participants could respond to a question in a certain way because of social desirability, he or she could attempt to control for this by writing additional questions to counter this or use the question as a way to detect problematic participants.

The second major concern researchers need to work to address or avoid are limitations related to survey design. Some issues, such as exhaustive response options and valid questionnaire items, have already been discussed. There are several other design issues researchers need to consider. First, researchers need to be aware of question order effects, which happens when the ordering of questions can implicitly or explicitly influence how participants answer later questions (Schuman & Presser, 1981). Order effects assume that Question A may be answered differently if asked before Question B than if it were asked after Question B (Strack, 1992). A common order effects example is stereotype threat. A researcher interested in if gender identity influences participants' math competency would want to ask the math competency questions first because asking about gender identity may influence whether participants feel competent in their math abilities. If researchers can control the sequencing of questions, they need to consider the organization of questions to prevent this from happening.

Researchers also need to focus on writing questions that are neutral, focused, and clear. Unless researchers are reading questions to participants, participants will not have a chance to ask for clarification. When writing questions, one of the biggest problems researchers encounter is writing neutral questions. If researchers want to capture participants' "true" feelings about a topic, they cannot ask leading questions. Leading questions can be questions that lead participants to a specific answer or that are emotionally charged (Brace, 2013). For example, when studying the abortion debate, language such as "pro-choice," "anti-choice," "anti-abortion," and "pro-abortion" are all emotionally charged, and questions using that language could bias participants' answers (Davis & Lachlan, 2017). Researchers must also avoid assumptions in their wording, especially with questions which assume knowledge about a topic. The question, "How likely are you to vote for the Democratic nominee for the 5th Congressional district" can only be clearly answered if participants 1) know who the Democratic nominee is, 2) know where the 5th Congressional district is, and 3) know if they are in the 5th Congressional district. Finally, researchers should avoid double-barreled questions. A questionnaire question becomes double-barreled when researchers try to ask about multiple topics in one question (Bassili & Scott, 1996). For example, when presented with the question "Did you enjoy NBC's coverage of the Winter and Summer Olympics?," is the participant answering about the Winter Olympics coverage or the Summer Olympics coverage? If the participant says "Yes," researchers would not be able to determine which they enjoyed and the results would not be valid. Guttman scales can be used if researchers need to ask a double-barreled question; however, double-barreled questions should mostly be avoided.

Finally, researchers need to consider the different ways to determine if their survey is valid and reliable (see Wagner & Skowronski, Chapter 2). Two of the most common ways to do this is to compute the reliability estimates of summated variables (e.g., variables created from summing multiple items to create an aggregate variable measure) and to run a factorial analysis. Reliability estimates (also known as Cronbach's alpha scores) tell researchers how reliable their variables are. It is important to note, however, that reliability cannot be computed for individual items, so this is not always the most appropriate method. By comparison, a confirmatory factor analysis helps researchers to identify if their measures are consistent with the researchers' understanding of the construct. This is typically done by creating a hypothesis model (Campbell & Fiske, 1959; Thompson, 2004). For example, McEwan, Fletcher, Eden, and Sumner (2014) used confirmatory factor analysis to test whether their Facebook relational maintenance scale was valid. Lastly, researchers should consult a psychometrics framework, such as item response theory (IRT – see Hauenstein & Embretson, Chapter 19), to determine if participants' performance on the overall measure is consistent with their responses to individual survey items (Embretson & Hershberger, 1999).

12.5 Conclusion

While questionnaires are ubiquitous, often offer an expedient way to collect data, and provide myriad options for question types and administration, there are many considerations for administration and construction of survey instruments. The questions researchers seek to ask ought to be reflected in the survey items the instrument comprises. The pitfalls surrounding questionnaire construction are plentiful, but careful thought and planning can help researchers successfully navigate the terrain. Moving toward the future, there are a number of theoretical frameworks, such as Theory of Planned Behavior, that are in need of a valid and/or reliable measure for researchers to use. Scholars have recently critiqued the lack of reliable measures directly related to theory (e.g., Carmack & Heiss, 2018), but more work is needed to create instruments that are theoretically grounded.

KEY TAKEAWAYS

- Always think about the logistics of your survey. Length, time to completion, and how the survey will be administered all need to be planned as you write the survey items.
- Take the stance of the participants to consider the different ways people could interpret questions. This will ensure that you write questions to actually measure what you want to measure.
- Be thoughtful about the type of questionnaire scales used in your survey and pick the appropriate scale types that help you accurately measure your topic and run the appropriate statistical tests.
- Don't neglect your demographic questions. Be inclusive and exhaustive in order to get a complete picture of participants and be able to use demographics as variables in your study.
- Pilot test your survey. It will help you identify problems with the survey and make sure you are measuring what you want to measure.

IDEAS FOR FUTURE RESEARCH

- Research should continue to build and test scales that reliably measure commonly used theories and concepts in the social sciences (as was recently done with the Theory of Planned Behavior).
- What is the best way to distribute surveys in an increasingly digital era?
- How does the changing social landscape influence how we assess demographic questions in order to provide truly inclusive options?

SUGGESTED READINGS

Brace, I. (2013). *Questionnaire Design* (3rd ed.). London: Kogan Page.

Bradburn, N. M., Sudman, S., & Wansink, B. (2004). *Asking Questions: The Definitive Guide to Questionnaire Design*. San Francisco, CA: Wiley.

Converse, J. M., & Presser, S. (1986). *Survey Questions: Handcrafting the Standardized Questionnaire* (Sage University Paper 63: Series: Quantitative Application in the Social Sciences). Thousand Oaks, CA: Sage.

Lavrakas, P. J. (Ed.) (2011) *Encyclopedia of Survey Research Methods*. Thousand Oaks, CA: Sage.

Schwartz, N. (1999). Self-reports: How the questions shape the answers. *American Psychologist*, *54*, 93–105. Retrieved from www.apa.org/pubs/journals/amp/

REFERENCES

Alexander, C. S., & Becker, H. J. (1978). The use of vignettes in survey research. *Public Opinion Quarterly*, *42*(1), 93–104.

Bassili, J. N., & Scott, B. S. (1996). Response latency as a signal to question problems in survey research. *Public Opinion Quarterly*, *60*, 390–399.

Bauer, G. R., Braimoh, J., Scheim, A. I., & Dhrama, C. (2017). Transgender-inclusive measures of sex/gender for population surveys: Mixed-methods evaluation and recommendations. *PLoS One*, 1–28. [Epub ahead of print]

Benson, J., & Hocevar, D. (1985). The impact of item phrasing on the validity of attitude scales for elementary school children. *Journal of Educational Management*, *22*, 231–240.

Brace, I. (2013). *Questionnaire Design* (3rd ed.). London: Kogan Page.

Bradburn, N. M., Sudman, S., & Wansink, B. (2004). *Asking Questions: The Definitive Guide to Questionnaire Design*. San Francisco, CA: Wiley.

Brick, J. M., Collins, K., & Chandler, K. (1997). *An Experiment in Random-Digit Dial Screening*. Washington, DC: US Department of Education/National Center for Educational Statistics.

Campbell, D. T., & Fisk, D. W. (1959). Convergent and discriminant validation by the multitrait-multimethod matrix. *Psychological Bulletin*, *56*, 81–105. http://dx.doi.org/10.1037/h0046016

Carmack, H. J., & Heiss, S. N. (2018). Using Theory of Planned Behavior to predict college students' intent to use LinkedIn for job searches and professional networking. *Communication Studies*, *69*(2), 145–160. https://doi.org/10.1080/10510974.2018.1424003

Converse, J. M., & Presser, S. (1986). *Survey Questions: Handcrafting the Standardized Questionnaire* (Sage University Paper 63: Series: Quantitative Application in the Social Sciences). Thousand Oaks, CA: Sage.

Davis, C. S., & Lachlan, K. A. (2017). *Straight Talk about Communication Research Methods* (3rd ed.). Dubuque, IA: Kendall Hunt.

De Maio, T. J. (1984). Social desirability in survey measurement: A review. In C. F. Turner & E. Martin (Eds.), *Surveying Subjective Phenomena* (Vol. 2, pp. 257–282). New York: Russell Sage.

Dillman, D. A. (2000). *Mail and Internet Surveys: The Tailored Design Method* (2nd ed.). New York: John Wiley & Sons.

Dunning, B., & Cahalan, D. (1974). By-mail vs. field self-administered questionnaires: An armed forces survey. *Public Opinion Quarterly*, *37*(4), 618–624.

Embretson, S. E., & Hershberger, S. L. (Eds.) (1999). *The New Rules of Measurement*. Mahwah, NJ: Lawrence Erlbaum Associates.

Fink, A. (1995). *How to Ask Survey Questions* (Vol. 2). Thousand Oaks, CA: Sage.

Fowler, F. J., Jr. (1995). *Improving Survey Questions: Design and Evaluation*. Thousand Oaks, CA: Sage.

Guttman, L. (1944). A basis for scaling qualitative data. *American Sociological Review*, *9*, 139–150.

Hague, P., Hague, N., & Morgan, C. (2013). *Market Research in Practice* (2nd ed.). London: Kogan Page.

Hetsroni, A. (2007). Open or closed – this is the question: The influence of question format on the cultivation effect. *Communication Methods and Measures*, *1*, 215–226.

Hoerger, M. (2010). Participant dropout as a function of survey length in Internet-mediated university studies: Implications for survey design and voluntary participation in psychological research. *Cyberpsychology, Behavior, and Social Networking*, *13*, 697–700.

Holm, T. T., & Carmack, H. J. (2012). Forensics as a correlate of graduate school success. *Journal of the Association for Communication Administration, 31*(1), 29–45.

Hughes, R. (1998). Considering the vignette technique and its application to a study of drug injecting and HIV risk and safer behavior. *Sociology of Health & Illness, 20*(3), 381–400.

Hyman, L., Lamb, J., & Bulmer, M. (2006). The use of pre-existing survey questions: Implications for data quality. Paper presented at the European Conference on Quality in Survey Statistics, Cardiff, UK.

Keyton, J. (2011). *Communication Research: Asking Questions, Finding Answers* (3rd ed.). New York: McGraw-Hill.

Krosnick, J. A. (1999). Maximizing questionnaire quality. In J. P. Robinson, P. R. Shaver, & L. S. Wrightsman (Eds.), *Measures of Political Attitudes* (pp. 37–57). San Diego, CA: Academic Press.

Magazine, S. L., Williams, L. J., & Williams, M. L. (1996). A confirmatory factor analysis examination of reverse coding effects in Meyer and Allen's affective and continuance commitment scales. *Educational and Psychological Measurement, 56*, 241–250.

Manfreda, K. L., Batagelj, Z., & Vehovar, V. (2002). Design of web survey questionnaires: Three basic experiments. *Journal of Computer-Mediated Communication, 7*(3), np.

Massof, R. W. (2004). Likert and Guttman scaling of visual function rating scale questionnaires. *Ophthalmic Epidemiology, 11*, 381–399.

McEwan, B., Fletcher, J., Eden, J., & Sumner, E. (2014). Development and validation of a Facebook relational maintenance measure. *Communication Methods and Measures, 8*, 244–263. doi:10.1080/19312458.2014.967844

Mercer, A., Caporaso, A., Cantor, D., & Townsend, R. (2015). How much gets you how much? Monetary incentives and response rates in household surveys. *Public Opinion Quarterly, 79*(1), 105–129.

Ohanian, R. (1990). Construction and validation of a scale to measure celebrity endorser's perceived expertise, trustworthiness and attractiveness. *Journal of Advertising, 19*, 39–52.

Oliphant, T. (2014). "I'm a library hugger!": Public libraries as valued community assets. *Public Library Quarterly, 33*, 348–361.

Osgood, C. E., Suci, G., & Tannenbaum, P. (1957). *The Measurement of Meaning*. Urbana, IL: University of Illinois Press.

Owens, L. (2011). Exhaustive. In P. J. Lavrakas (Ed.), *Encyclopedia of Survey Research Methods* (p. 249). Thousand Oaks, CA: Sage.

Parrillo, V. N., & Donoghue, C. (2005). Updating the Bogardus social distance studies: A new national survey. *The Social Science Journal, 42*, 257–271.

Pedhazur, E. J., & Schmelkin, L. P. (1991). *Measurement, Design, and Analysis: An Integrated Approach*. Hillsdale, NJ: Lawrence Erlbaum Associates.

Presser, S., Rothgeb, J. M., Couper, M. P., Lessler, J. T., Martin, E., Martin, J., & Singer, E. (2004). *Methods for Testing and Evaluating Survey Questionnaires*. New York: John Wiley & Sons.

Rice, N., Robone, S., & Smith, P. (2011). Analysis of the validity of the vignette approach to correct for the heterogeneity in reporting health system responsiveness. *European Journal of Health Economics, 12*, 141–162.

Salant, P., & Dillman, D. A. (1994). *How to Conduct Your Own Survey*. New York: John Wiley & Sons.

Schuman, H., & Presser, S. (1981). *Questions and Answers in Attitude Surveys*. San Diego, CA: Academic Press.

Schwartz, N. (1999). Self-reports: How the questions shape the answers. *American Psychologist, 54*, 93–105. Retrieved from www.apa.org/pubs/journals/amp/

Short, M. E., Goetzel, R. Z., Pei, X., Tabrizi, M. J., Ozminkowski, R. J., Gibson, T. B., … Wilson, M. G. (2010). How accurate are self-reports? An analysis of self-reported healthcare utilization and absence when compared to administrative data. *Journal of Occupational & Environmental Medicine, 51*(7), 786–796.

Stewart, S. M., Bing, M. N., Davison, H. K., Woehr, D. J., & McIntyre, M. D. (2009). In the eyes of the beholder: A non-self-report measure of workplace deviance. *Applied Journal of Psychology, 94*(1), 207–215.

Strack, F. (1992). "Order effects" in survey research: Activation and information functions of proceeding questions. In N. Schwartz & S. Sudman (Eds.), *Context Effects in Social and Psychological Research* (pp. 23–34). New York: Springer-Verlag.

Swain, S. D., Weathers, D., & Niedrich, R. W. (2008). Assessing three sources of misresponse to reversed Likert items. *Journal of Marketing Research, 45*, 116–131.

Thompson, B. (2004). *Exploratory and Confirmatory Factor Analysis: Understanding Concepts and Applications*. Washington, DC: American Psychological Association.

Tourangeau, R., Conrad, F. G., & Couper, M. P. (2013). *The Science of Web Surveys*. Oxford: Oxford University Press.

van Sonderen, E., Sanderman, R., & Coyne, J. C. (2013). Ineffectiveness of reverse wording in questionnaire items: Let's learn from cows in the rain. *PLoS One, 8*, e68967.

Vraga, E., Bode, L., & Troller-Renfree, S. (2016). Beyond self-reports: Using eye tracking to measure topic and style difference in attention to social media content. *Communication Methods and Measures, 10*(2), 149–164.

Weems, G. H., & Onwuegbuzie, A. J. (2001). The impact of midpoint responses and reverse coding of survey data. *Measurement and Evaluation in Counseling and Development, 34*, 166–176.

Wright, K. B., Frey, L., & Sopory, P. (2007). Willingness to communicate about health as an underlying trait of patient self-advocacy: The development of the willingness to communicate about health (WTCH) measure. *Communication Studies, 58*(1), 35–51.

13 Reaction Time Measures in the Social and Behavioral Sciences: Speed Matters

Jeremy D. Heider

Jason T. Reed

I think it necessary to mention that my assistant, Mr. David Kinnebrook, who observed the transits of the stars and planets very well ... began, from the beginning of August last, to set them down half a second of time later than he should do, according to my observations; and in January of the succeeding year, 1796, he increased his error to eight-tenths of a second.

(Maskelyne, as quoted in Howse, 1989, p. 169)

Eight-tenths of a second. That is all it took to set in motion a series of events that would eventually lead to the widespread use of **reaction time** measures among social and behavioral scientists. What is reaction time? How is it measured? And why should such measures be included in the social and behavioral scientist's toolbox? To address these key questions, this chapter will unfold in four major sections: 1) an overview of the evolution of reaction time research; 2) a discussion of the state of reaction time knowledge today; 3) a review of some of the most common reaction time measures employed by contemporary social and behavioral scientists; and 4) a description of specific technological tools that can be used to administer those measures.

13.1 A Brief History of Reaktionszeit

The simple, yet elegant, term **reaction time** (RT) describes the interval between the initial appearance of a stimulus and an organism's response to that stimulus (Luce, 1986). Though they share certain basic characteristics, RTs are distinct from **reflexes**, which are involuntary reactions that protect organisms from harm. RT measures have a long history in the social and behavioral sciences, but their prominence began in astronomy and physiology. Decades before Austrian physiologist Sigmund Exner (1873) coined the term *reaktionszeit* (reaction time), the aforementioned Kinnebrook made a series of "errors" that cost him his job – and also inspired the study of RT as we know it (Schultz & Schultz, 2016).

13.1.1 On the "Incompetence" of Astronomers' Assistants

Kinnebrook was the assistant to Nevil Maskelyne, Royal Astronomer of England. When studying stellar transit, Kinnebrook's observations became consistently slower than Maskelyne's, and the assistant was fired (Boring, 1950). Approximately twenty years later, German astronomer Friedrich Bessel determined such "errors" were not the result of incompetence but of normal variation among human observers. He developed the personal equation, in which systematic differences between observers are statistically adjusted (Boring, 1950). For example, presuming Kinnebrook's observations were consistently 0.8 seconds slower than Maskelyne's, their figures could be equated by subtracting 0.8 seconds from each observation made by Kinnebrook (Hergenhahn, 2013). The personal equation was important for two reasons: 1) it showed subjective psychological phenomena could be quantified, nullifying the philosophical argument that mental processes could not be studied experimentally; and 2) it provided evidence from a "real" (i.e., respected as objective) science that subjective perception is critical to understand, meaning the role of the human observer must be appreciated in all sciences (Schultz & Schultz, 2016).

13.1.2 On the "Nerve" of Nineteenth-Century Physiologists

Around the time astronomers were stumbling upon RT, physiologists were studying the rate of nerve conduction. Johannes Müller of Germany believed nervous transmission was virtually instantaneous – perhaps on the order of the speed of light – and, therefore, impossible to measure (Müller, 1833–1840). However, Müller's pupil, Hermann von Helmholtz, disagreed. Helmholtz used his newly invented **myograph** to determine that the rate of human nerve conduction was

approximately 50–100 meters per second (slower than sound, let alone light; Boring, 1950). Unfortunately, substantial variability in conduction rates both between and within persons led Helmholtz to abandon the study of RT. Nonetheless, his work was refined by other physiologists, including Franciscus Donders of the Netherlands. In line with earlier work, Donders first studied **simple reaction time**, gauging the basic lag between stimulus and response. Later, he conducted innovative investigations of **recognition reaction time**, presenting numerous stimuli to participants at once but with instructions to only respond to one of them, and **choice reaction time**, presenting several different stimuli but with instructions to respond to each one differently. He established that simple RTs are the shortest followed by recognition and choice RTs, respectively (Hergenhahn, 2013).

13.1.3 Wundt Brings RT to Social and Behavioral Science

Germany's Wilhelm Wundt, often considered the founding father of psychology, was himself a trained physiologist. Thus, being influenced by Müller, Helmholtz, and Donders, it is no surprise that Wundt incorporated RT into many of his experiments (Hergenhahn, 2013). Overall, Wundt and his students detailed seven types of reactions ranging from the fastest (**reflexes**) to the slowest (**judgement reactions**). By subtracting one type of reaction from another, they could establish the time it took for a person to engage in various cognitive processes (Boring, 1950). Wundt so enjoyed this line of work that he used the term **mental chronometry** to designate it as an entire subdiscipline of psychology. Of course, much like Helmholtz before him, Wundt encountered considerable variability in RTs across participants, time, sensory modality, stimulus intensity, and other variables, and, therefore, eventually stopped studying RT altogether. Still, by this point, it was clear that RT was fully infused into the new psychology and was there to stay (Hergenhahn, 2013).

13.2 The State of Reaction Time Research Today

Why is the study of reaction time important to social and behavioral scientists? There are many answers to such a question. From an information processing perspective, RT studies allow us to examine the human cognitive system during optimal performance. Other cognitive processes, such as memory, are often only understood when they fail (e.g., Atkinson & Shiffrin, 1968). Also, as Wundt first demonstrated, RT studies are valuable in determining the relative length of various cognitive processes and the manner in which they are temporally organized (i.e., whether they occur serially or in parallel). Of course, there are also many real-world implications of RT, many of which have potentially serious consequences. RT impacts a driver's ability to appropriately recognize stimuli on the road, the speed with which a law enforcement officer determines the level of threat posed by a suspect, and the particular associations that are activated when a teacher encounters a student of another race. Each of these processes could, in turn, have behavioral implications. For example, the driver might turn the steering wheel to avoid an obstruction, the officer might draw his or her firearm to deter the suspect from further threatening

behaviors, or the teacher might be motivated to suppress negative stereotypes associated with the race of the student. Conversely, each of these individuals' cognitive systems might fail to initiate such behaviors.

Contemporary studies of RT can be classified in much the same way Donders classified them in the nineteenth century: Simple reaction time, recognition reaction time, and choice reaction time (Luce, 1986). Simple RT measures involve a single stimulus and a single response. For example, if a participant pressed the spacebar on a keyboard as quickly as possible each time a red "X" appeared on the computer screen – and the red "X" was the only type of stimulus presented – then he or she would be completing a simple RT task. Recognition RT measures incorporate some stimuli that require a response and other stimuli that require non-response. An example of this type of measure is the Go/No-Go Association Task (GNAT; Nosek & Banaji, 2001), in which participants respond with a key press (the "go" response) to stimuli representing a category (e.g., science) or an attribute (e.g., positivity) but do *not* press a key (the "no-go" response) for any stimuli not associated with the category or attribute. Finally, choice RT measures present participants with multiple potential active responses. An example is the Implicit Association Test (IAT; Greenwald, McGhee, & Schwartz, 1998), in which participants categorize stimuli by pressing one key in response to stimuli from social and/or evaluative categories on the left side of a screen and a different key in response to stimuli from categories on the right side of the screen.

As was the case in Donders' laboratory, RT studies in the twentieth century and beyond have shown RTs vary reliably according to the nature of the task, such that Simple RTs < Recognition RTs < Choice RTs. Importantly, these differences are the result of differential cognitive processing time, not other potential mechanisms such as motor reactions (Miller & Low, 2001). There are also well-established differences in RT across sensory modalities, such that Auditory RTs < Touch RTs < Visual RTs (taste and smell RTs are rarely studied, as they involve inherently less precision in both stimulus onset and detection; see Kling & Riggs, 1971; Robinson, 1934). Importantly, these differences in RT across the three senses emerge regardless of whether one employs simple, recognition, or choice tasks (Sanders, 1998).

Of course, many other variables besides task type moderate the length of RTs. Although an exhaustive review of all such variables is beyond the scope of this chapter, we will nonetheless review several key moderators. Our primary goal in presenting this information is to equip researchers with sufficient knowledge to design RT studies that enhance experimental control, thereby reducing unwanted error variability. RT data are notorious for revealing substantial individual differences and also for exhibiting a degree of positive skew. Awareness of variables that moderate RT should allow researchers to minimize the impact of such factors.

13.2.1 Stimulus Intensity and Complexity

Aside from task type and sensory modality, perhaps the best understood moderators of reaction time are **stimulus intensity** and **stimulus complexity**. Numerous lines of evidence converge on two complementary

conclusions: 1) increased stimulus intensity decreases RTs, and 2) increased stimulus complexity increases RTs (Luce, 1986). In other words, people respond more quickly to intense stimuli (e.g., a bright light rather than a dim one) and more slowly to more complex stimuli (e.g., five words on a screen rather than a single word). However, the intensity–RT relationship is curvilinear, such that the tendency for intensity to decrease RT levels off after the stimulus reaches a sufficient degree of intensity. The complexity–RT relationship, on the other hand, seems to vary according to the type of task. In recognition RT tasks, the complexity–RT relationship is often linear (e.g., Sternberg, 1969). Conversely, it is often curvilinear in choice RT tasks. Specifically, RTs tend to increase along with the number of stimuli, up to a point, after which the amount of increase in RT with each additional stimulus gets smaller and smaller (e.g., Hick, 1952). The important lesson here is that researchers need to control for intensity and complexity by either holding such variables constant or by carefully manipulating them.

13.2.2 Arousal

Physiological arousal has also received considerable attention in the RT literature, with many findings paralleling the **Yerkes–Dodson Law** (Yerkes & Dodson, 1908) regarding the relationship between arousal and performance. According to this law, the arousal–performance relationship follows an inverted-U pattern, whereby performance is best at moderate levels of arousal and worst at very low or very high levels of arousal. With occasional exceptions in which some participants exhibit a pattern opposite that of the traditional inverted-U (e.g., VaezMousavi, Barry, & Clarke, 2009), the vast majority of RT studies show participants respond with optimum speed to a wide range of stimuli when arousal is moderate (e.g., Arent & Landers, 2003; see Levitt & Gutin, 1971 for a demonstration of optimal heart rate for RT tasks). Durlach, Edmunds, Howard, and Tipper (2002) found moderately increased arousal via doses of caffeine decreased RTs and increased attentional focus. In the opposite direction, it is well known that sleep deprivation worsens RT on both simple and choice RT tasks (Jaśkowski & Włodarczyk, 1997). Because of these patterns, we recommend creating conditions of moderate arousal to optimize participant performance. In other words, unless one is purposely studying varying levels of arousal, it is important to avoid conditions in which arousal might be too low (e.g., due to sleep deprivation or other forms of fatigue) or too high (e.g., a highly stressful situation).

13.2.3 Focus of Attention Effects

Numerous studies have documented substantially faster visual reaction times to stimuli presented foveally (i.e., focused centrally on the retina) compared to those presented parafoveally (i.e., in one's peripheral vision; Babkoff, Genser, & Hegge, 1985). This effect can be attenuated through practice, whereby practice with parafoveal stimuli decreases RT to foveal stimuli and vice versa (Ando, Kida, & Oda, 2002; 2004). In addition, Johns, Crowley, Chapman, Tucker, and Hocking (2009) presented evidence that visual RTs increase significantly – sometimes by as much as two-tenths of a second – when blinks and/or saccadic eye movements occur either immediately before or immediately after the presentation of a stimulus. However, other studies clearly show that distractor stimuli increase response latency (e.g., Elliott, Morey, Morey, Eaves, Shelton, & Lufti-Proctor, 2014). For these reasons, studies of visual RT (and other forms of RT, for that matter) typically utilize a "warning" system to orient the participant's attention prior to stimulus presentation. A common feature of such systems is the use of a visual fixation point. Participants are instructed to look directly at that point, as it is the exact location where the target stimulus will appear. Though nothing is guaranteed with this approach, it does increase the likelihood that participants will orient their attention to the right place and time compared to approaches that do not use such a warning system (see Pelz, Chapter 11 for a detailed discussion of eyetracking).

13.2.4 Practice Effects

As noted previously, Ando et al. (2002; 2004) found practice with foveal and parafoveal stimuli improved reaction time. Ando et al. (2004) further documented that the practice gains were still present after a three-week interval. However, these studies hardly represent the only evidence of practice effects (see Dutilh, Vandekerckhove, Tuerlinckx, & Wagenmakers, 2009 for a model of the processes underlying practice effects). Dutilh et al. (2009) used a lexical decision task (i.e., distinguishing words from nonwords) to study practice effects, but others have used experience with a specific activity as a proxy for practice and found similar patterns of results. For example, Fontani, Lodi, Felici, Migliorini, and Corradeschi (2006) found more experienced practitioners of karate were faster on a simple RT task than those with less experience. However, this pattern did not replicate among volleyball players. Less experienced volleyball players actually had faster RTs, although it is important to note that they also committed more errors. Still, this lends support to the notion that practice is a crucial component of overall performance.

Practice effects are most noticeable when the RT task is complex and the practice is extended over a period of days or weeks, but such effects can also be accomplished with brief "warmup" periods within a single session. Such "warmups" are common to the specific RT measures we will review later in this chapter. For example, the IAT (Greenwald et al., 1998), which typically takes participants only ten to fifteen minutes to complete, incorporates blocks of both practice and critical trials. The practice blocks are intended to orient the participant to the nature of the categorization task, thus reducing error variability during the critical blocks. Of course, there is also a downside to practice effects. For example, if a participant becomes "too good" at a particular RT task, it may cease to be a valid indicator of how members of the population at large process information.

Before we move on to our discussion of widely used RT measures in the social and behavioral sciences, we should reiterate that our treatment of potential moderators in this section was not meant to be exhaustive. Thus, we encourage the reader to examine other possible moderators, such as age (e.g., slower and more variable RTs among very young and very old participants; Welford, 1981), sex (e.g., slightly slower and more variable RTs among female participants; Der & Deary, 2006), learning disorders (e.g., slower RTs among dyslexic children; Vakil, Lowe, & Goldfus, 2015), brain injury (e.g., delayed RTs

among concussed athletes; Lavoie, Dupuis, Johnston, Leclerc, & Lassonde, 2004), personality traits (e.g., neuroticism positively correlating with RT variability; Robinson & Tamir, 2005), and hemispheric specialization (e.g., minor differences if visual stimuli are only processed in the left or right visual field; Hernandez, Nieto, & Barroso, 1992). Again, the primary goal of familiarizing oneself with such patterns is to aid in designing RT studies that reduce unwanted noise.

13.3 Commonly Employed Reaction Time Measures in the Social and Behavioral Sciences

The ubiquity of reaction time measures in social and behavioral science research makes it difficult to describe all possible measures. In fact, many modern software applications, including those described in the next section of this chapter, make it possible to record RTs for literally every response a participant provides during a computer-based procedure. One possibility, then, is to simply use RT as a dependent variable to gauge cognitive processing time for any kind of response. As noted above, the practical applications of such information are numerous (e.g., speed of response when law enforcement officers encounter suspects). However, many specific RT tasks have gained immense popularity among social and behavioral scientists in their examinations of cognitive processing. In this section, we examine five such tasks: The Stroop test, the Eriksen flanker task, evaluative priming, the Implicit Association Test, and the Go/No-Go Association Task. Also, although alternative methods of administration are possible, we will focus on computer-based administration as it is by far the most frequent manner in which all five tasks are executed.

13.3.1 The Stroop Color and Word Test

The **Stroop Color and Word Test** (SCWT), more commonly known as the **Stroop test** (Stroop, 1935), is a choice reaction time task that examines participants' ability to process two separate stimulus features simultaneously – typically the semantic meaning of a color word (e.g., "blue") and the color in which the word is presented (which may or may not match the semantic meaning). However, unlike many other choice RT tasks, the speed of participants' cognitive responding is typically not measured on a per-stimulus basis. Rather, the total amount of time participants take to process an entire set of stimuli is the focus because participants typically state their responses verbally rather than responding with a mechanical key press. Such verbal responses are inherently more difficult to measure with millisecond-level precision.

Though there are many variations, Stroop's (1935) original test involves the participant processing three stimulus sets as quickly as possible: 1) color words presented in black lettering; 2) basic shapes (e.g., squares) presented in different colors; and 3) color words presented in lettering of an incongruent color (e.g., the word "blue" presented in red lettering). The first two sets are considered congruent, as there is no interference between stimulus features. However, the third stimulus set requires participants to inhibit a relatively automatic process (reading the word) in order to engage a more effortful process

(naming the color of the lettering; see MacLeod & Dunbar, 1988). Stroop (1935) found only a 5.6% increase in processing time of the third stimulus set compared to the first set when participants read the color words themselves, but a 74.3% increase when participants had to name the color of the lettering (thus having to inhibit the automatic process in favor of the effortful one). This increase in processing time has been termed the **Stroop effect**. An interesting application of the Stroop test, typically known as the **emotional Stroop test**, involves naming the colors of words that are emotionally negative or neutral. Gotlib and McCann (1984) found that depressed participants reacted more slowly when naming colors of the negative emotional words. Thus, although the Stroop test was originally designed to assess cognitive processing, it can be used to gauge emotional processing as well.

13.3.2 The Eriksen Flanker Task

The **Eriksen flanker task**, alternatively called the **Eriksen task** or **flanker task**, is a choice RT task that examines the effects of peripheral "flanker" stimuli that surround target stimuli (Eriksen & Eriksen, 1974). Much like the Stroop test, the flanker task involves interference between competing cognitive processes. However, two key differences are that the flanker task records RTs on a per-stimulus basis, and it avoids the potentially confounding influence of other variables like verbal ability. In a typical procedure, participants are presented with a group of stimuli and instructed to press either a left key or right key depending on the nature of the central or target stimulus. For example, imagine the letter "R" is the target stimulus. If "R" appears in the center of the stimulus group, the participant is to press the left key. If any other letter appears centrally, he or she is to press the right key. In either case, the central stimulus will be "flanked" by irrelevant stimulus letters. All else being equal, the task is easier in terms of both RT and accuracy when flanker stimuli are consistent with the target (e.g., "RRR") than when they are inconsistent (e.g., "TRT"). The difference in average per-trial RT between consistent and inconsistent trials is known as the **Eriksen effect** or **flanker effect**.

A large number of variations of the flanker task have been implemented in the literature. In most variations, the target and flanker stimuli are of the same type, such as letters. However, other types of stimuli, such as shapes or symbols, have been used to similar effect (e.g., Peschke, Hilgetag, & Olk, 2013). Some researchers have also experimented with placing flankers in varying patterns around the target stimulus and found the same basic flanker effect (e.g., Eriksen & St. James, 1986). Importantly, the flanker task has a number of potential applications, including as an indicator of the integrity of cognitive functioning in populations with neurodegenerative conditions (e.g., Parkinson's disease; Wylie et al., 2009).

13.3.3 Evaluative Priming

The remaining measures in this section have primarily been used by social and behavioral scientists to study implicit attitudes (i.e., relatively non-conscious evaluations) of various social groups in domains such as race/ethnicity, gender, sexual orientation, and religion, though they do have various other applications (e.g., the measurement of implicit self-esteem; see Greenwald & Farnham, 2000). The first of these techniques we will review is primarily known as **evaluative priming** (Fazio, Sanbonmatsu, Powell, & Kardes, 1986). Evaluative priming is a choice RT task in which participants make a series of decisions regarding the evaluative connotation of words (i.e., whether each word is positive/good or negative/bad in meaning). On each trial, the target word is preceded by a prime. Depending on the researcher's goals, the primes can be presented either supraliminally (e.g., Fazio, Jackson, Dunton, & Williams, 1995) or subliminally (e.g., Dovidio, Kawakami, & Gaertner, 2002; Dovidio, Kawakami, Johnson, Johnson, & Howard, 1997).

One common application of evaluative priming has been in the assessment of implicit racial attitudes. In this context, the primes are typically facial photographs of African American and Caucasian targets (though targets of any other social group can easily be substituted). Among samples that consist of entirely or at least primarily Caucasian participants, responses to negative adjectives are often facilitated (i.e., faster RTs) following an African American prime relative to a Caucasian prime. In addition, to some extent, responses to positive adjectives are inhibited following an African American prime. These differential RTs, depending on prime type, have, in turn, been related to discriminatory behavior in interracial interactions (e.g., Fazio et al., 1995).

13.3.4 The Implicit Association Test

In terms of sheer frequency of use, the undisputed king of modern-day reaction time measures is the **Implicit Association Test**, or IAT. As a testament to its widespread use, the original article reporting the development of the IAT (Greenwald et al., 1998) has been cited over 3500 times in the PsycINFO database, at the time of this writing, making it the most-cited paper in the history of the *Journal of Personality and Social Psychology*. The IAT is a choice RT task that, much like evaluative priming, has mostly been used to assess implicit attitudes (implicit racial attitudes, in particular). However, unlike evaluative priming, the IAT is a relative task that gauges differences in average RT between "compatible" trials (i.e., those involving judgements that are attitude-consistent for most respondents) and "incompatible" trials (i.e., those involving judgements that are attitude-inconsistent for most). The basic reasoning is that participants should be able to provide the same response to two concepts that are similar in evaluative connotation more easily than providing the same response to two concepts with contrasting evaluative connotations. In the context of race, an implicit preference for Caucasians over African Americans is indicated when the participant exhibits faster average RTs when responding with one key to Caucasian stimuli (typically facial photographs) and positive words while responding with another key to African American stimuli and negative words. Conversely, an implicit preference for African Americans over Caucasians is indicated by faster average RTs when the participant uses one key in response to African American names and positive words and a different key in response to Caucasian names and negative words.

Importantly, the enormous popularity of the IAT should not be taken to mean it is universally accepted as a measure of implicit attitudes. As is often the case when something becomes so widespread, criticisms of the IAT abound. One of the common criticisms is that differences in RTs between the compatible and incompatible portions of the test are relative, not absolute (Brendl, Markman, & Messner, 2001). For example, a result suggesting an implicit preference for African Americans could mean one of three things: 1) the participant has a positive implicit attitude toward African Americans and a negative implicit attitude toward Caucasians; 2) the participant has a positive implicit attitude toward both groups but said positive attitude toward African Americans is more pronounced; or 3) the participant has a negative implicit attitude toward both groups but said negative attitude toward African Americans is less pronounced. Whatever the case, the only meaningful conclusion one can draw is that the participant has an evaluative preference for one group over the other. Other criticisms of the IAT are more extensive, such as questioning whether the associations it measures are actually rooted within the person taking the test (Olson & Fazio, 2004). Nonetheless, the IAT has been shown to relate to a wide range of important outcomes, many of which have potential practical significance. For example, Senholzi, Depue, Correll, Banich, and Ito (2015) found IAT scores were positively related to amygdala activation (indicative of a fear reaction) in response to images of armed African Americans.

13.3.5 The Go/No-Go Association Task

Largely due to the aforementioned relative nature of the IAT, Nosek and Banaji (2001) developed the Go/No-Go Association Task, or GNAT, a similar computer-based task that assesses implicit attitudes without requiring a contrast category. Rather, the GNAT can gauge the positivity or negativity of the participant's implicit evaluation of any one social group, independent of his or her implicit evaluation of any other group. As its name implies, the GNAT is a recognition RT task in which participants actively respond to some stimuli (the "go" response) but do *not* respond to other stimuli (the "no-go" response). In a standard GNAT procedure, measuring implicit attitudes, participants are presented with stimuli from both the social category (e.g., African American) and positive or negative words. In some trial blocks, participants press the designated "go" key if the stimulus belongs to the social or positive word category; no response is given if any other stimulus appears. In other trial blocks, participants respond "go" to stimuli from the social and negative word categories but do not respond to any other stimuli.

Aside from the fact that a contrast category is unnecessary in the GNAT, its scoring also distinguishes it from the IAT. First, though the GNAT involves recording per-trial RTs, and such RTs can certainly be used (see Nosek & Banaji, 2001), its primary scoring method is grounded in signal detection theory (SDT; Green & Swets, 1966). To the extent that the social category and attribute dimension (positive or negative) are more strongly associated, sensitivity (i.e., discriminating signal from noise, indicated by d' – d-prime) should be enhanced. Therefore, differences in sensitivity between the two critical trial types indicate the nature of the participant's implicit evaluation of the social category (Nosek & Banaji, 2001). In our example, greater sensitivity

when African American and positive stimuli are paired together suggests a positive implicit attitude toward African Americans; greater sensitivity when African American and negative stimuli are paired suggests a negative implicit attitude toward African Americans.

13.4 Instruments for the Reaction Time Researcher's Toolbox

Reaction time tasks necessitate the use of equipment capable of presenting stimuli and recording responses with millisecond-level precision. In most social and behavioral scientific research, this equipment takes the form of computer hardware and software. Because hardware is often specific to certain tasks that involve RT and the continued development and advancement of the available software, this chapter focuses on software for the presentation and recording of RTs commonly available for use on most personal computers.

The rapid development of RT software is a double-edged sword. Advances in the capabilities of such software provide researchers with a myriad of options for how to best design and implement their studies. This amount of choice also highlights the importance of making informed decisions about which software is most appropriate for one's particular research. Ideally, the researcher wields the tool instead of the tool driving the researcher. Thankfully, discussion of RT software can focus on some important commonalities among the available options. Most of the well-known and widely used examples of the software are capable of use on most personal computers and with multiple and varied

RT tasks mentioned earlier in this chapter. The developers of the software also give explicit consideration to the critical issues of the reliability and validity of RT presentation and response. Use of the software also does not require specialized knowledge of programming languages and, instead, offers more user-friendly interfaces.

Software programs used in RT research with these key features are often commercially available products. Companies, as compared to individuals, are likely to have the available time and resources to focus on the timely development, distribution, and maintenance of such software. Although a variety of such software exists, it is most illustrative to focus on some specific examples that have been extensively used in published research and professional presentations: DirectRT, E-Prime, and SuperLab. We will also review some online and open-source options for collecting RT data.

13.4.1 DirectRT

DirectRT (Empirisoft Corp, New York, 2016a) is software that allows researchers to quickly and easily create tasks that require precision timing. The primary use of DirectRT is for tasks like cognitive/perception "blocks of trials" tasks that require measures of stimulus presentation and response times reliable and accurate to the millisecond (e.g., priming/lexical decision style). With basic understanding of spreadsheets like those used in Microsoft Excel, DirectRT allows users to create, copy, paste, and modify single or multiple trials in research designs. DirectRT uses DirectX to present sound, video, images, and text with exacting precision. Using

either standard hardware, such as a USB keyboard, serial mouse, joystick, or sound card microphone, or specialized millisecond-accurate keyboards and button boxes from the Empirisoft corporation, DirectRT obtains accurate, millisecond-level responses from participants. DirectRT also can be used in conjunction with other software, such as MediaLab (Empirisoft Corp, New York, 2016b), to add its RT-related capabilities to larger research designs (e.g., surveys and computer-based experiments). Output files are saved in a .txt format compatible with most spreadsheet and data-analysis software; data files can also be manipulated and merged using the Data Restructuring utility of DirectRT.

13.4.2 E-Prime

E-Prime (Psychology Software Tools, Pittsburgh, PA) uses a collection of applications to create and present RT tasks. Researchers can create different tasks involving reaction time using a graphical drop-and-drag interface. Specifically, E-Prime uses "objects" to control different functions such as the presentation of stimuli, creation of blocks of trials, presentation of feedback, and the order of presentation of these events. E-Prime collects input using both standard (i.e., keyboard and mouse) and specialized (i.e., button boxes) peripheral devices. Among its extensive documentation, E-Prime provides users with detailed information about the testing procedure to ensure the reliability and validity of its RT-based functions (https://support.pstnet.com/hc/en-us/articles/360008833253-INFO-E-Prime-3–0-Timing-Data-24332-). Like DirectRT, E-Prime provides users with tools for debugging problems that might arise while testing program files. Output files are saved in a .txt format compatible with most spreadsheet and data-analysis software; data files can also be viewed and cleaned in the E-DataAid application of E-Prime.

13.4.3 SuperLab

SuperLab (Cedrus Corp, San Pedro, CA) uses a point-and-click interface that allows researchers to control the presentation of stimuli such as images, sound, and video files. Software users can create a variety of tasks in blocks of trials that feature millisecond-level presentation of stimuli and recording of response input. SuperLab collects input using both standard (i.e., keyboard and mouse) and specialized (i.e., button boxes) peripheral devices. Timing of stimulus presentation is coordinated with the refresh rate of the device display, while timing of inputs is recorded to the nearest millisecond from various user-defined events like trial onset or stimulus presentation. Output files are saved in a .txt format compatible with most spreadsheet and data-analysis software; data files can also be viewed and cleaned in the Data Viewer function of SuperLab.

13.4.4 Online and Open-Source Tools

One feature common to the three examples of software used in reaction time research is that their use is primarily confined to the machine on which the software is installed. Understandably, adding anything between the actual hardware of the computer and the software used to run RT tasks introduces

possible variability in the execution of said tasks. Despite this valid concern, some software companies have taken up this challenge and successfully developed software to conduct RT research online. Projects like Project Implicit (www.projectimplicit .net/index.html) allow for online versions of specific tasks involving reaction time similar to the Implicit Association Test. Inquisit (Millisecond Software, Seattle, WA) is software with many of the same functions for RT research, such as described in our previous examples. However, Inquisit also has a version for use on the internet for different research tasks involving RT. A primary difference is that creation of these tasks in Inquisit "is based on an easy to use declarative scripting language similar to HTML" (www.millisecond.com/products/inquisit5/ features.aspx). Although the programming and creation of tasks in Inquisit may be more involved, the ability to run the tasks online is an attractive feature.

Use of the internet to implement RT tasks has also contributed to the creation and proliferation of open-source software available to researchers. Open-source software is attractive to researchers for its great potential to include numerous capabilities and features, often at little to no monetary cost to the researcher. Importantly, the open-source software has evidence of reliable and accurate timing functions (De Leeuw & Motz, 2016). However, these open-source programs are likely to use more complex programming languages to create reaction time tasks and may be troublesome for researchers without proficiency in computer programming. Open-source software is also heavily dependent on the development teams and users to continually support the software; the software is only as good as the effort its users put into its maintenance and maturation. This recent addition to the toolboxes of RT researchers and its potential contributions bears watching.

13.5 Conclusion

In many ways, the history of reaction time research is the history of social and behavioral science. Both borrowed and refined the study of physical sensations from traditional "hard sciences" for the study of human behavior. Both occupy a centrally situated role in contemporary empirical research of human behavior that is unlikely to go away any time soon. Both underwent a paradigm shift with the invention and proliferation of computer technology. Of course, RT research faces challenges in further defining its role. For example, how generalizable is existing RT research to cultures beyond the WEIRD (Western, Educated, Industrialized, Rich, and Democratic; Henrich, Heine, & Norenzayan, 2010) samples traditionally used in the social and behavioral sciences? Also, advances in technology, such as wearable devices (e.g., smartwatches and fitness trackers), that change the placement of displays from the traditional location of eye level, may affect RTs (Harrison, Lim, Shick, & Hudson, 2009). Researchers using such devices will need to take explicit care to compare and validate new technology with the existing body of RT research.

Future RT research should also consider the relationship between implicit and explicit thoughts and/or behaviors. The large body of empirical research on implicit and explicit thoughts and behaviors clearly demonstrates its variations for a number of topics of study such as attitudes (Nosek, 2007), emotion regulation (Gyurak, Gross, & Etkin, 2011), and self-esteem (Krizan & Suls, 2008). It also demonstrates the centrality of RT in such research using the internet instead of more traditional methods (Nosek, Banaji, & Greenwald, 2002). Researchers studying RT should be aware of possible moderators and mediators – including the RT measure itself – of the phenomena under investigation.

Finally, because the current volume focuses on methodology, we have not provided any discussion of approaches to statistical analysis when dealing with RT data. Thankfully, other authors have written extensive treatments of this subject (e.g., Whelan, 2008). We encourage researchers to examine such sources when dealing with their own RT datasets.

We hope this chapter illuminates the richness of the study of reaction time. The history of reaction time research demonstrates how a variable can progress from background noise to a signal worth study. The development and implementation of the variety of tasks used in reaction time research exemplify that the heart of the scientific process is the acquisition, investigation, correction, and integration of knowledge. The rapid creation and adoption of technologies used in reaction time research illustrate that the social and behavioral sciences are not sciences apart, but fields of study in step with other scientific disciplines concerned with human behavior. Reaction time research will undoubtedly continue to advance and develop at a rate we can only imagine – just don't blink or you might miss it.

- Reaction time is a measure of physical responses to stimuli. It can occur at conscious (measured in seconds) or unconscious (measured in milliseconds) thresholds.
- As used in social and behavioral science research, reaction time frequently represents the strength of cognitive associations.
- Key variables that moderate reaction times, such as the type of task, sensory modality, stimulus intensity and complexity, arousal, attentional focus, and practice effects, should be carefully controlled when designing RT studies.
- Social and behavioral science research should consider how its use of reaction time relates to the explicit or implicit nature of the thoughts and behavior under investigation. It should also consider the expected relationship between implicit and explicit measures as relevant to the thoughts and behaviors being studied.
- Measurement of reaction time in contemporary research primarily uses computer hardware and software. Researchers should give careful consideration to the particular features of the available software and hardware before deciding which to use for a particular research project.

IDEAS FOR FUTURE RESEARCH

- What factors mediate and moderate the relationship between implicit and explicit attitudes across different domains?

- What other applications of reaction time measures can be employed across the social and behavioral sciences?
- How can reaction time measures be more widely implemented in applied settings?
- How does culture impact the results obtained from reaction time measures?

SUGGESTED READINGS

Fazio, R. H., Sanbonmatsu, D. M., Powell, M. C., & Kardes, F. R. (1986). On the automatic activation of attitudes. *Journal of Personality and Social Psychology*, *50*, 229–238. http://dx.doi.org/10.1037/0022-3514.50.2.229

Greenwald, A. G., McGhee, D. E., & Schwartz, J. L. K. (1998). Measuring individual differences in implicit cognition: The Implicit Association Test. *Journal of Personality and Social Psychology*, *74*, 1464–1480. http://dx.doi.org/10.1037/0022-3514.74.6.1464

Luce, R. D. (1986). *Response Times: Their Role in Inferring Elementary Mental Organization*. New York: Oxford University Press.

Nosek, B. A., Banaji, M. R., & Greenwald, A. G. (2002). Harvesting implicit group attitudes and beliefs from a demonstration web site. *Group Dynamics: Theory, Research, and Practice, 6*, 101–115.

Whelan, R. (2008). Effective analysis of reaction time data. *Psychological Record, 58*, 475–482.

REFERENCES

Ando, S., Kida, N., & Oda, S. (2002). Practice effects on reaction time for peripheral and central visual fields. *Perceptual and Motor Skills*, *95*, 747–752.

Ando, S., Kida, N., & Oda, S. (2004). Retention of practice effects on simple reaction time for peripheral and central visual fields. *Perceptual and Motor Skills*, *98*, 897–900.

Arent, S. M., & Landers, D. M. (200). Arousal, anxiety, and performance: A reexamination of the inverted-U hypothesis. *Research Quarterly for Exercise and Sport, 74*, 436–444. doi:10.1080/02701367.2003.10609113

Atkinson, R. C., & Shiffrin, R. M. (1968). Human memory: A proposed system and its control processes. In K. W. Spence & J. T. Spence (Eds.), *The Psychology of Learning and Motivation* (Vol. 2, pp. 89–195). New York: Academic Press.

Babkoff, H., Genser, S., & Hegge, F. W. (1985). Lexical decision, parafoveal eccentricity and visual hemifield. *Cortex: A Journal Devoted to the Study of the Nervous System and Behavior, 21*, 581–593. http://dx.doi.org/10.1016/S0010-9452(58)80006-4

Boring, E. G. (1950). *A History of Experimental Psychology* (2nd ed.). New York: Appleton-Century-Crofts.

Brendl, C. M., Markman, A. B., & Messner, C. (2001). How do indirect measures of evaluation work? Evaluating the inference of prejudice in the Implicit Association Test. *Journal of Personality and Social Psychology, 81*, 760–773. doi:10.1037//0022-3514.81.5.760

Cedrus Corporation [SuperLab 5]. (2016a). Retrieved from https://cedrus.com/superlab/download.htm

de Leeuw, J. R., & Motz, B. A. (2016). Psychophysics in a Web browser? Comparing response times collected with JavaScript and Psychophysics Toolbox in a visual search task. *Behavior Research Methods, 48*, 1–12. http://dx.doi.org/10.3758/s13428-015-0567-2

Der, G., & Deary, I. J. (2006). Age and sex differences in reaction time in adulthood: Results from the United Kingdom health and lifestyle survey. *Psychology and Aging, 21*, 62–73. http://dx.doi.org/10.1037/0882-7974.21.1.62

Dovidio, J. F., Kawakami, K., & Gaertner, S. L. (2002). Implicit and explicit prejudice and interracial interaction. *Journal of Personality and Social Psychology, 82*, 62–68. http://dx.doi.org/10.1037/0022-3514.82.1.62

Dovidio, J. F., Kawakami, K., Johnson, C., Johnson, B., & Howard, A. (1997). On the nature of prejudice: Automatic and controlled processes. *Journal of Experimental Social Psychology, 33*, 510–540. http://dx.doi.org/10.1006/jesp.1997.1331

Durlach, P. J., Edmunds, R., Howard, L., & Tipper, S. P. (2002). A rapid effect of caffeinated beverages on two choice reaction time tasks. *Nutritional Neuroscience, 5*, 433–442.

Dutilh, G., Vandekerckhove, J., Tuerlinckx, F., & Wagenmakers, E. J. (2009). A diffusion model decomposition of the practice effect. *Psychonomic Bulletin and Review, 16*, 1026–1036. doi:10.3758/16.6.1026

Elliott, E. M., Morey, C. C., Morey, R. D., Eaves, S. D., Shelton, J. T., & Lutfi-Proctor, D. A. (2014). The role of modality: Auditory and visual distractors in Stroop interference. *Journal of Cognitive Psychology, 26*, 15–26. http://dx.doi.org/10.1080/20445911.2013.859133

Empirisoft Corporation [DirectRT 2016.1.104]. (2016a). Retrieved from www.empirisoft.com/Download.aspx?index = 4

Empirisoft Corporation [MediaLab 2016.1.104]. (2016b). Retrieved from www.empirisoft.com/Download.aspx?index = 4

Eriksen, B. A., & Eriksen, C. W. (1974). Effects of noise letters upon the identification of a target letter in a nonsearch task. *Perception and Psychophysics, 16*, 143–149. http://dx.doi.org/10.3758/BF03203267

Eriksen, C. W., & St. James, J. D. (1986). Visual attention within and around the field of focal attention: A zoom lens model. *Perception and Psychophysics, 40*, 225–240. http://dx.doi.org/10.3758/BF03211502

Exner, S. (1873). Experimentelle untersuchung der einfachsten psychischen processe: Erste abhandlung. *Archiv für die Gesammte Physiologie des Menschen und der Thiere, 7*, 601–660.

Fazio, R. H., Jackson, J. R., Dunton, B. C., & Williams, C. J. (1995). Variability in automatic activation as an unobtrusive measure of racial attitudes: A bona fide pipeline? *Journal of Personality and Social Psychology, 69,* 1013–1027. http://dx.doi.org/10.1037/0022-3514.69.6.1013

Fazio, R. H., Sanbonmatsu, D. M., Powell, M. C., & Kardes, F. R. (1986). On the automatic activation of attitudes. *Journal of Personality and Social Psychology, 50,* 229–238. http://dx.doi.org/10.1037/0022-3514.50.2.229

Fontani, G., Lodi, L., Felici, A., Migliorini, S., & Corradeschi, F. (2006). Attention in athletes of high and low experience engaged in different open skill sports. *Perceptual and Motor Skills, 102,* 791–805. http://dx.doi.org/10.2466/PMS.102.3.791-805

Gotlib, I. H., & McCann, C. D. (1984). Construct accessibility and depression: An examination of cognitive and affective factors. *Journal of Personality and Social Psychology, 47,* 427–439. http://dx.doi.org/10.1037/0022-3514.47.2.427

Green, D. M., & Swets, J. A. (1966). *Signal Detection Theory and Psychophysics.* Oxford: John Wiley.

Greenwald, A. G., & Farnham, S. D. (2000). Using the Implicit Association Test to measure self-esteem and self-concept. *Journal of Personality and Social Psychology, 79,* 1022–1038. http://dx.doi.org/10.1037/0022-3514.79.6.1022

Greenwald, A. G., McGhee, D. E., & Schwartz, J. L. K. (1998). Measuring individual differences in implicit cognition: The Implicit Association Test. *Journal of Personality and Social Psychology, 74,* 1464–1480. http://dx.doi.org/10.1037/0022-3514.74.6.1464

Gyurak, A., Gross, J. J., & Etkin, A. (2011). Explicit and implicit emotion regulation: A dual-process framework. *Cognition and Emotion, 25,* 400–412.

Harrison, C., Lim, B. Y., Shick, A., & Hudson, S. E. (2009, April). Where to locate wearable displays? Reaction time performance of visual alerts from tip to toe. In *Proceedings of the SIGCHI Conference on Human Factors in Computing Systems* (pp. 941–944). Maui, Hawaii: ACM.

Henrich, J., Heine, S. J., & Norenzayan, A. (2010). The weirdest people in the world? *Behavioral and Brain Sciences, 33,* 61–135. doi:10.1017/S0140525X0999152X

Hergenhahn, B. R. (2013). *An Introduction to the History of Psychology* (7th ed.). Belmont, CA: Wadsworth.

Hernandez, S., Nieto, A., & Barroso, J. (1992). Hemispheric specialization for word classes with visual presentations and lexical decision task. *Brain and Cognition, 20,* 399–408. http://dx.doi.org/10.1016/0278-2626(92)90029-L

Hick, W. E. (1952). On the rate of gain of information. *The Quarterly Journal of Experimental Psychology, 4,* 11–26. http://dx.doi.org/10.1080/17470215208416600

Howse, D. (1989). *Nevil Maskelyne: The Seaman's Astronomer.* Cambridge: Cambridge University Press.

Jaśkowski, P., & Włodarczyk, D. (1997). Effect of sleep deficit, knowledge of results, and stimulus quality on reaction time and response force. *Perceptual and Motor Skills, 84,* 563–572. http://dx.doi.org/10.2466/pms.1997.84.2.563

Johns, M., Crowley, K., Chapman, R., Tucker, A., & Hocking, C. (2009). The effect of blinks and saccadic eye movements on visual reaction times. *Attention, Perception, and Psychophysics, 71,* 783–788. http://dx.doi.org/10.3758/APP.71.4.783

Kling, J. W., & Riggs, L. A. (1971). *Woodworth & Schlosberg's Experimental Psychology* (3rd ed.). New York: Holt, Rinehart, & Winston.

Krizan, Z., & Suls, J. (2008). Are implicit and explicit measures of self-esteem related? A meta-analysis for the Name-Letter test. *Personality and Individual Differences, 44,* 521–531. http://dx.doi.org/10.1016/j.paid.2007.09.017

Lavoie, M. E., Dupuis, F., Johnston, K. M., Leclerc, S., & Lassonde, M. (2004). Visual P300 effects beyond symptoms in concussed college athletes. *Journal of Clinical and Experimental*

Neuropsychology, 26, 55–73. http://dx.doi.org/10.1076/jcen.26.1.55.23936

Levitt, S., & Gutin, B. (1971). Multiple choice reaction time and movement time during physical exertion. *Research Quarterly of the American Association for Health, Physical Education, and Recreation, 42,* 405–410.

Luce, R. D. (1986). *Response Times: Their Role in Inferring Elementary Mental Organization.* New York: Oxford University Press.

MacLeod, C. M., & Dunbar, K. (1988). Training and Stroop-like interference: Evidence for a continuum of automaticity. *Journal of Experimental Psychology: Learning, Memory, and Cognition, 14,* 126–135. http://dx.doi.org/10.1037/0278-7393.14.1.126

Miller, J. O., & Low, K. (2001). Motor processes in simple, go/no-go, and choice reaction time tasks: A psychophysiological analysis. *Journal of Experimental Psychology: Human Perception and Performance, 27,* 266–289. http://dx.doi.org/10.1037/0096-1523.27.2.266

Millisecond Software [Inquisit 5.0.9]. (2017). Retrieved from www.millisecond.com/download/

Müller, J. (1833–1840). *Handbuch der physiologie des menschen für vorlesungen.* Coblenz, Germany: Verlag von J. Hölscher.

Nosek, B. A. (2007). Implicit–explicit relations. *Current Directions in Psychological Science, 16,* 65–69.

Nosek, B. A., & Banaji, M. R. (2001). The Go/No-go Association Task. *Social Cognition, 19,* 625–666. http://dx.doi.org/10.1521/soco.19.6.625.20886

Nosek, B. A., Banaji, M. R., & Greenwald, A. G. (2002). Harvesting implicit group attitudes and beliefs from a demonstration web site. *Group Dynamics: Theory, Research, and Practice, 6,* 101–115.

Olson, M. A., & Fazio, R. H. (2004). Reducing the influence of extrapersonal associations on the Implicit Association Test: Personalizing the IAT. *Journal of Personality and Social Psychology, 86,* 653–667. http://dx.doi.org/10.1037/0022-3514.86.5.653

Peschke, C., Hilgetag, C. C., & Olk, B. (2013). Influence of stimulus type on effects of flanker, flanker position, and trial sequence in a saccadic eye movement task. *The Quarterly Journal of Experimental Psychology, 66,* 2253–2267. doi:10.1080/17470218.2013.777464

Psychology Software Tools, Inc. [E-Prime 3.0]. (2016). Retrieved from www.pstnet.com

Robinson, E. S. (1934). Work of the integrated organism. In C. Murchison (Ed.), *A Handbook of General Experimental Psychology* (pp. 571–650). Worcester, MA: Clark University Press.

Robinson, M. D., & Tamir, M. (2005). Neuroticism as mental noise: A relation between neuroticism and reaction time standard deviations. *Journal of Personality and Social Psychology, 89,* 107–114. http://dx.doi.org/10.1037/0022-3514.89.1.107

Sanders, A. F. (1998). *Elements of Human Performance: Reaction Processes and Attention in Human Skill.* Mahwah, NJ: Lawrence Erlbaum Associates.

Schultz, D. P., & Schultz, S. E. (2016). *A History of Modern Psychology* (11th ed.). Boston: Cengage Learning.

Senholzi, K. B., Depue, B. E., Correll, J., Banich, M. T., & Ito, T. A. (2015). Brain activation underlying threat detection to targets of different races. *Social Neuroscience, 10,* 651–662. http://dx.doi.org/10.1080/17470919.2015.1091380

Sternberg, S. (1969). Memory-scanning: Mental processes revealed by reaction-time experiments. *American Scientist, 57,* 421–457.

Stroop, J. R. (1935). Studies of interference in serial verbal reactions. *Journal of Experimental Psychology, 18,* 643–662. http://dx.doi.org/10.1037/h0054651

VaezMousavi, S. M., Barry, R. J., & Clarke, A. R. (2009). Individual differences in task-related activation and performance. *Physiology and Behavior, 98,* 326–330. http://dx.doi.org/10.1016/j.physbeh.2009.06.007

Vakil, E., Lowe, M., & Goldfus, C. (2015). Performance of children with developmental dyslexia on two skill learning tasks – Serial

Reaction Time and Tower of Hanoi Puzzle: A test of the specific procedural learning difficulties theory. *Journal of Learning Disabilities*, *48*, 471–481. http://dx.doi.org/10.1177/0022219413508981

Welford, A. T. (1981). Signal, noise, performance, and age. *Human Factors*, *23*, 97–109.

Whelan, R. (2008). Effective analysis of reaction time data. *Psychological Record*, *58*, 475–482.

Wylie, S. A., van den Wildenberg, W. P. M., Ridderinkhof, K. R., Bashore, T. R.,

Powell, V. D., Manning, C. A., & Wooten, G. F. (2009). The effect of speed-accuracy strategy on response interference control in Parkinson's disease. *Neuropsychologia*, *47*, 1844–1853. doi:10.1016/j.neuropsychologia.2009.02.025

Yerkes, R. M., & Dodson, J. D. (1908). The relation of strength of stimulus to rapidity of habit formation. *Journal of Comparative Neurology and Psychology*, *18*, 459–482. http://dx.doi.org/10.1002/cne.920180503

Part Four: Emerging Issues in Social and Behavioral Science Research

14 Replications and the Social and Behavioral Sciences

Courtney K. Soderberg

Timothy M. Errington

The scientific method provides a framework to formulate, test, and modify hypotheses. It is an iterative process that allows us to build knowledge in order to explain the world around us. Each finding is a piece of evidence in a continually growing body of knowledge that slowly reduces the uncertainty of a given phenomenon. A core feature of the scientific process is that each piece of evidence is reproducible – the findings and conditions necessary to produce it are accurately and transparently reported (Jasny, Chin, Chong, & Vignieri, 2011; Kuhn & Hacking, 2012; Merton, 1973; Popper, 1959/1934). Such reproducibility of methodology allows for a repetition of the test of an effect to determine if similar results can be obtained, also known as replication. Replications not only test if a piece of evidence is reproducible, but also provide a mechanism to better understand the conditions necessary to produce a given finding. Although replication is an essential aspect of the scientific process, replication studies do not appear as often as one would expect in the published literature (Makel & Plucker, 2014). Instead, replicable research is more of an assumption than a practice (Collins, 1992; Schmidt, 2009). However, in recent years, there

has been growing evidence from a variety of fields, including the social and behavioral sciences, medicine, and many others, that assumptions of the replicability of published research may not be well founded (Begley & Ellis, 2012; Camerer et al., 2016; Errington, Iorns, Gunn, Tan, Lomax, & Nosek, 2014; Mobley, Linder, Braeuer, Ellis, & Zwelling, 2013; Open Science Collaboration, 2015; Prinz, Schlange, & Asadullah, 2011). As a result, there is a growing body of replication studies appearing in the scientific literature that have brought replication to the forefront of scientists' minds, making the field much more open to replications than it used to be (Nelson, Simmons, & Simonsohn, 2018).

These studies provide examples of how replications can be conducted and published. In this chapter, we will briefly outline the different types of replications that researchers can conduct and the ways in which these different types of replications contribute to scientific knowledge. Building from the work of previous replications, we will outline the steps researchers should take when designing their own replications to make them as informative as possible, as well as discuss how these same steps can also improve the reproducibility of original research.

14.1 Defining Replication

What is a replication? At first, this may seem like an easy question to answer, but, over the years, replication has been defined very differently within and across different fields. Currently, there is no consensus definition of "replication" or the related terms of "reproduction" and "robustness" (National Academies of Sciences, Engineering, and Medicine, 2016). Clemens (2017), while attempting to better clarify and reclassify terms into four distinct categories, identified over fifty different terms that have been used across six disciplines to define various forms of replications – at times, the same term being used in conflicting fashions across disciplines. While some have tried to narrow the definition of replication, others take a broad view of the term, classifying a "replication" as any study that attempts to determine the validity of empirical results from a previous study (Berry, Coffman, Hanley, Gihleb, & Wilson, 2017; Duvendack, Palmer-Jones, & Reed, 2017). For the purposes of this chapter, we will adopt this broad view of the term, but will also distinguish between different categories or types of replications as needed.

14.1.1 Types of Replications

One category of replications reuses the original data, applying the same statistical protocol to the data to determine whether the original results and/or statistical conclusions can be reproduced. In many disciplines, this process of reproducing published results from original code and data is referred to as simply "reproducibility" or "computational reproducibility" (National Academies of Sciences, Engineering,

and Medicine, 2016; Peng, 2011). However, in economics, this type of reanalysis is commonly referred to as "replication" (Chang & Li, 2015), "pure replication" (Hamermesh, 2007), or "narrow replication" (Pesaran, 2003).

Another category of replications repeats the methodology of the original study to collect new data. Within this broad category, there is a continuum of similarity between the original and replication study methodologies. Traditionally, this continuum has been dichotomized into direct (or statistical) and conceptual (or scientific) replications. In direct replications, the replicators attempt to match the original protocol as closely as possible (Crandall & Sherman, 2016) or to at least employ a protocol that gives no a priori reason to expect a different outcome (Nosek & Errington, 2017; Schmidt, 2009). In a conceptual replication, some aspect of the protocol is purposefully changed (e.g., the operationalization of the dependent variable; Crandall & Sherman, 2016).

Though it is widely used, some have noted that the direct/conceptual dichotomy masks variability within each category. They have called for a more nuanced categorization with "exact," "very close," and "close" making up the continuum of direct replications, and "far" and "very far" making up the continuum of conceptual replications (LeBel, Berger, Campbell, & Loving, 2017). This continuum allows for increasing amounts of dissimilarity between the original and replication study. "Exact" to "close" replications increase in the level of theoretically irrelevant deviations (e.g., font of instructions, university where the study was conducted if no personality variables were a priori deemed as theoretically important), and "far" to "very far" replications increase in the level of theoretically relevant deviations (e.g., changing the operationalization of the independent or dependent variable).

14.1.2 Scales of Replications

Along with varying by type, replications can also vary in scale; a replication project may be done by one lab on one original study, or the project may involve multiple labs and/or multiple original studies. These small- and large-scale replication projects generally serve the same broad purpose – to build knowledge about phenomena – but differ in the amount of information they provide and the logistical complexity they entail. Large-scale projects also vary by breadth and depth, broadly falling into two categories: Wide but shallow, in which many studies from a particular field or set of journals are each replicated once, or narrow but deep, in which a smaller pool of studies is replicated many times across a large number of labs. The ways in which the scale and scope of replication projects affect the conclusions that can be drawn from them will be discussed further in later sections of this chapter.

14.2 How Replications Build Knowledge

Broadly, replications are concerned with verifying the empirical findings of earlier research. However, the category and scale of a replication has a large impact on the type of knowledge it provides about the original study and underlying theory. For example, narrow replications, or computational reproductions, allow replicators to determine whether there are any reporting errors in the original paper. If the replication fails, the type

and size of the discrepancy are extremely important. A small discrepancy (e.g., the discovery of a rounding error) may have few to no consequences for the conclusions of the study. However, a large discrepancy (e.g., a regression coefficient found to have the incorrect sign) may substantively change the study's results and substantive meaning. A successful replication, on the other hand, may slightly increase confidence in the veracity of the reported values, given the particular dataset and analysis strategy. However, it cannot speak to the truth or the generalizability of the underlying phenomena measured by the original study.

Due to the fact that narrow replications reanalyze the original study's data, they cannot provide evidence for whether the results found in the original study are real or simply due to sampling error. One way to gather more information about these two competing possibilities is through direct replications – anywhere on the continuum from "exact" to "close." Direct replications keep theoretically relevant variables constant across original and replication studies, and, therefore, a string of failed direct replications may allow for the falsification of the theorized relationship between the predictor and outcome variables. Each direct replication also provides additional information about the likely effect size of the relationship between the particular operationalizations of the predictor and outcome variables. Though individual studies, replication or original, may not provide a particularly precise estimate on their own, multiple replications can be meta-analytically combined to give a more precise estimate.

In the past, some have viewed direct replications in a negative light, perceiving them to divert from other types of replications that could build and extend theories (Crandall & Sherman, 2016). However, "very close" and "close" direct replications can provide robustness and generalizability information within a narrow scope. Researchers have beliefs about what variables in their studies are theoretically relevant and irrelevant, but they may not be correct. Due to the fact that theoretically relevant variables are held constant while theoretically irrelevant variables are allowed to vary, direct replications can provide information about whether or not the irrelevant variables affect the relationship between the predictor and outcome variables. When done in large numbers (e.g., Many Labs style project), the direct replications can be meta-analytically combined and the heterogeneity of the effect sizes can be tested to determine whether there is more variability in the effect size across samples than would be expected due to sampling error alone. If the effect size is homogeneous, then this provides evidence of the robustness of the relationships between the predictor and outcome variable across small deviations in protocols and the generalizability of the finding to different study populations. On the other hand, if there is evidence of heterogeneity, this would suggest that there is an unpredicted variable or hidden moderator, influencing the outcome of the replications. This can lead to the development of auxiliary hypotheses which can be tested in future studies to investigate boundary conditions or moderators of the original effect. Therefore, in aggregate, direct replications can provide information about both the presence and size of the relationship between the original operationalizations of the variables and the extent to which proposed,

theoretically irrelevant variables truly are irrelevant and whether claims about the generalizability of the original finding were warranted.

Though they provide important information, one drawback of direct replications is that they often do not generate information about whether the explanation for a finding is correct. Because the operationalizations of variables do not change across studies in direct replications, any confound in the original design is also included in each replication. Therefore, further direct replications would produce the same results as the original study, but they would do so because of the repeated confound rather than the relationship between the predictor and outcome variables hypothesized by the original research. In order to gain further insights into the mechanisms underlying previously detected relationships, conceptual replications are needed.

Conceptual replications, or "far" replications, purposefully change the way independent and/or dependent variables are operationalized between original and replication studies. Successful replications across different operationalizations increase the evidence of a link between the theoretical constructs of interest, helping to illustrate the generalizability of the findings across different methodologies. This can provide further evidence for the theory under study and help to triangulate the mechanism at work. However, the larger design differences between original and replication studies that occur in conceptual replications can make failures to replicate difficult to interpret. In particular, it is impossible to know if the failure is due to: a) the underlying theory being incorrect, b) a newly introduced confound in the design, or

c) a boundary on the generalizability of the original theory. Conceptual replications may lead to **sophisticated falsificationism** (Lakatos, 1970), which produces auxiliary hypotheses that can be tested. However, this process can easily create a circular system in which more and more auxiliary hypotheses are created to explain away failed replications, leading to a **degenerative** research program (Meehl, 1990). This should engender skepticism of the underlying core theory but often does not because successful conceptual replications are published and given more weight than unsuccessful replications. Thus, in current practice, conceptual replications rarely allow for falsifying theories and often provide only one-sided evidence.

All three broad categories of replications have strengths and weaknesses in terms of the type of information that they can provide. Because of this, replications accumulate the most power in terms of their ability to build knowledge when they are used, in concert, in an iterative loop building from computational reproductions to direct replications to conceptual replications and then computational and direct replications of these conceptual replications, and so forth. Thus, it is important to value and conduct multiple types of replications within a line of research in order to create robust, cumulative science and theory building.

14.3 Past Replication Attempts

Recent efforts to understand the replicability of the published literature have highlighted the diversity of replication approaches. As an example of computationally reproducing published results, McCullough, McGeary, and Harrison (2008) attempted to replicate published results in the *Journal of Money, Credit and Banking* by accessing the originally used data and analysis scripts provided by the authors. They found that sixty-nine of 186 articles they analyzed met the data archive requirement of the journal and, out of those sixty-nine, only fourteen could achieve the same result using the provided information (McCullough et al., 2008). Similarly, Chang and Li (2015) examined sixty-seven empirical articles in macroeconomics published from 2008 to 2013. The identified articles were from journals that required sharing of data and analysis scripts, of which they were able to obtain this information from twenty-nine of the thirty-five papers, as well as journals that did not have a policy, where only eleven of the twenty-six papers provided this information (six had confidential data) (Chang & Li, 2015). Using a definition similar to McCullough et al. (2008) for replication success, they achieved the same result twenty-nine times. These examples expose some challenges when conducting replications – specifically, obtaining access to the original data and analysis scripts as well as having the code execute. Providing the data and code is largely the responsibility of the researchers. However, some journals, such as the *American Journal of Political Science* and the *Quarterly Journal of Political Science* have implemented in-house computational reproducibility checks prior to publication.

These replication examples attempted to use the same data and analysis approach as the original study, thus examining both the success of achieving the same result and if the original data and code were shared. In addition to this computationally focused approach, replication efforts have also extended beyond the original data and code. For example, the Replication Programme of the International Initiative for Impact Evaluation (3ie), which conducted replications of development impact evaluations using the same data and code as the original study, also conducted alternative analytic strategies to test the robustness of the original results (Brown, Cameron, & Wood, 2014).

In many cases, new data will be collected as part of a replication project. While this is a common part of the research process, to repeat the work of others before building on it, it is often not published (Collins, 1992; Schmidt, 2009). Recently, however, there are a growing number of replications being published, either as single replication attempts, large-scale projects conducting multiple replications, or multilab replication efforts. For example, a study to estimate the reproducibility of psychology results conducted a single replication of 100 experimental and correlational studies published in three psychology journals during 2008 (Open Science Collaboration, 2015). Multiple replicability indicators were reported, including statistical significance of the replication, mean relative effect size of the replications, if the original effect size was within the 95% confidence interval of the replication, subjective assessments made by the replication teams, statistical significance of a meta-analysis of the original and replication effect sizes, and prediction market beliefs before the replications were completed (Dreber et al., 2015; Open Science Collaboration, 2015). Of the multiple replicability indicators reported, which ranged from 36% to 68%, no single indicator was able to sufficiently describe replication

success, highlighting the challenge in defining what it means to successfully replicate a previous finding (Dreber et al., 2015; Open Science Collaboration, 2015). A similar investigation was conducted to examine the replicability of eighteen laboratory experiments in economics published in the *American Economic Review* and the *Quarterly Journal of Economics* between 2011 and 2014 (Camerer et al., 2016). Similar to the psychology study, replicability indicators varied (67% to 78%) (Camerer et al., 2016). The focus of these two efforts was to estimate replication rates and the predictors of replication. Importantly, the individual replications within these large-scale efforts cannot, by themselves, explain if an original finding is correct or false, similar to any single published replication attempt. Instead, collective evidence through multiple, diverse lines of investigation are necessary for increasing an understanding of a given phenomenon.

Other recent large-scale replication projects have taken a different approach than the two previous projects. Rather than replicating many studies once, these Many Labs projects replicate a few experiments at many different test sites and meta-analytically combine the replications to gain a more precise estimate of the effect and measures of the heterogeneity of the effects between different collection sites. In the first investigation of this type, the variations of thirteen classic and contemporary effects were independently replicated at thirty-six sites (Klein et al., 2014). The collective data indicated that ten of the effects investigated were consistently observed, one showed weak support for the effect, and two did not show collective evidence of the effect (Klein et al., 2014). In another Many

Labs style project that investigated the ego-depletion effect, twenty-three independent labs conducted replications of the sequential-task paradigm. The meta-analytic results suggested a small effect size with 95% confidence intervals encompassing zero (Hagger et al., 2016). Unlike individual replications of a single effect, which can be influenced by a number of different factors, these projects illustrate ways in which replication can be utilized to understand methodological bases of an area of research.

As more replications have appeared in the scientific literature, debates regarding the approach and interpretation of replications have arisen. These debates have focused largely on the similarities and differences between the methodologies employed during the replication and the original study. If there is some variation between the original study and the replication, such as updating a protocol to adjust for cultural differences, the debate often focuses on the appropriateness of those decisions and how it impacts the conclusions that are drawn (Freese & Peterson, 2017). Another common topic of debate is the definition of replication success and failure. Since the majority of replication studies that have received the most attention have been failures to replicate, concerns have been raised that, if only studies that fail to replicate are being published, a misleading picture of the replicability of results could be created (Zwaan, Etz, Lucas, & Donnellan, 2017). However, published Many Labs style replication efforts have also been successful in observing the same result as the original study (e.g., Alogna et al., 2014) and demonstrate robust findings for future work to build on. Furthermore, the use of Registered Reports, a publishing format where peer review occurs in advance of observing outcomes, with results provisionally accepted for publication before any data are collected or analyzed (Chambers, 2013; Nosek & Lakens, 2014), provides a means of alleviating concerns of bias for or against successful replications. Thus, while there are still likely to be concerns and debates about replications, there are now mechanisms to ensure replications make it into the published literature.

14.4 *How to Conduct Replications*

Given the importance of replications to building scientific knowledge, conducting them should be an ordinary part of the research process. The strength of the design of a replication study, like an original study, dictates the knowledge that can be gained from the results. This section outlines a number of theoretical and practical considerations researchers should keep in mind when planning replications that involve the collection of new data.

14.4.1 What Is the Purpose of the Replication?

Due to the fact that there are many different types of replications, the first step when planning a replication is to decide the intended purpose of the study. The type of replication chosen, in terms of both category and scale, will have consequences for the conclusions the researcher can draw from the findings. For example, if a researcher is interested in gaining a more precise estimate of the effect size, and determining if there is heterogeneity across theoretically unimportant differences (e.g., different study populations), then a direct replication with multiple labs collecting data would be appropriate, and the researcher may try to recruit labs that vary in the populations they will be able to recruit. Other times, a researcher may want to rule out a potential confound in their original operationalizations, and a conceptual replication may be more appropriate.

14.4.2 Operationalizing the Replication

After the general type of replication has been decided, the researchers can begin to design the protocol of the replication. An initial step is to clearly define the effect or effects that will be the focus of the replication. This may be simple if the original study only included one statistical test, such as a t-test, but is more complicated if a regression- or model-based analysis was used. In the latter case, choosing which effect is the focus of the replication can be more complicated. For example, is it a particular regression coefficient, a pattern involving a number of regression coefficients, and/or the fit of the overall model?

The researcher also needs to determine if there are auxiliary tests, such as manipulation checks or positive/negative controls that were included in the original study that should also be included in the replication to allow for more interpretable results. This may include tests that were in the original study itself or pilot testing that was completed before the original study in order to validate the manipulations or dependent variables. For example, imagine a researcher is interested in replicating the finding that priming stupidity leads to lower performance on a general knowledge task (Dijksterhuis & Van Knippenberg, 1998). The original researchers chose soccer hooligans to prime stupidity based on pilot data that showed soccer hooligans were rated as low in intelligence. The replicator wants to perform an "exact" direct replication but is not certain whether the same stereotypes exist in their study population. They run the same pilot study as the original paper and find that soccer hooligans were not rated as low in intelligence. Given the results of the replicated pilot study, a direct replication may not make sense given the replicator's goals to test the link between priming stupidity using stereotypes and performance on general knowledge tasks, as the stupid soccer hooligan stereotype does not exist in their population. A "close" or conceptual replication, using a different social group that is rated as low in intelligence in the replicator's recruitment population, may be more appropriate in this case and provide more interpretable results given the replicator's intent.

Replicators should also be aware of potentially pertinent research that has been published since the original study, especially when replicating older studies. For example, if a researcher is interested in replicating an older original study that found a main effect, but later studies on the same theory found that the main effect was moderated by a personality variable, the researcher may want to test the newer, more thorough understanding of the phenomenon.

14.4.3 Contacting Original Authors

While determining what type of replication to conduct and the exact operationalization of that replication attempt, the researcher has a choice of whether to contact the original author(s) or not. Though some have argued that this is "essential to a valid replication"

(Kahneman, 2014), we do not view this as a required step, but it can be helpful. In theory, methods sections should provide a description of how a study was conducted that is detailed enough for another researcher to recreate the methodology. However, in practice, methods sections are generally under the same word limit constraints as the rest of the paper, and, thus, it is possible some methodological details may have been cut for the sake of the word limit or writing style. Additionally, example scale items or pictorial stimuli may be shown in place of sharing entire measurement scales or manipulation protocols. By contacting the original authors, replicators may get access to exact study materials that were not shared in the original paper and which would make conducting a direct replication easier and more accurate. The replicators may also gain insight into specific details of the methodology that were not included in the paper.

While contact with the original authors can be highly informative, the opinions of the original authors are not binding. For example, original authors and replicators may disagree about whether the replication represents a "direct," "close," or conceptual replication of the original work. Often this is because original papers do not explicitly specify the minimum essential conditions to produce the effect (Lykken, 1968) or theoretically grounded boundary conditions (Simons, Shoda, & Lindsay, 2017). This lack of original specificity can make it difficult to come to a consensus around whether a deviation from the original protocol or study population is theoretically inconsequential enough for the study to still be classified as a "close" replication or whether they mean that the replication is a conceptual

replication. While the original authors' scholarly opinions can be informative, neither the original author nor the replicator has the final word about what type of replication the study is and how the results of the replication speak to the original findings and theory. This will likely be considered in the peer review and post peer review processes.

14.4.4 Choosing a Sample Size

For replications to be highly informative, they, like original studies, need to be adequately powered (80 % or higher). However, choosing a sample size to provide high power can be difficult because published effect sizes can be inflated due to publication bias and underpowered original studies (Simonsohn, 2015). This means that, if replication sample sizes are chosen based on power analyses using the effect sizes from original published studies, the replications will likely be underpowered to find the true population effect size of interest. Additionally, even adequately powered original studies may not have highly precise effect size estimates, making the true population effect size highly uncertain. This combination of uncertainty and upward bias of original effects can make it difficult to properly power replications if only the point estimate from the original study is used.

A number of different procedures have been suggested in order to correct for these influences. One approach is safeguard power (Perugini, Gallucci, & Constantinit, 2014). In this approach, the researcher calculates the 60 % confidence interval, two-tailed, around the original effect size. By using the lower bound of this interval as the point estimate when conducting the power analysis, researchers should achieve 80 % power to find the true population effect size 80% of the time. A conceptually similar procedure for taking publication bias and effect size imprecision into account in power analyses is implemented in the BUCSS R package and accompanying paper (Anderson, Kelley, & Maxwell, 2017). The package allows researchers to model both effect size uncertainty and potential publication bias so these factors can be taken into account directly when performing power calculations. A third alternative is to decide on a theoretical smallest effect size of interest (SESOI) and choose a sample size so that the replication has adequate power to detect this SESOI. While useful, in some areas of research it can be difficult to specify a SESOI. In such cases, the previously mentioned techniques could still be used.

All these procedures will generally require larger sample sizes than the original study – sometimes substantially larger – and may lead to overpowered studies in the case where the true population effect size is larger than the effect size used for the power analysis. If replicators wish to balance concerns of highly powered replications with efficient use of participant resources, the methods described can be combined with a sequential analysis, in which a prespecified number of optional stopping points are planned out with an alpha-spending function. This allows researchers to test their hypotheses repeatedly during data collection, stopping early if they pass a pre-planned significance threshold or continuing data collection to obtain a larger sample if they do not meet the threshold at an early checkpoint. By a priori portioning out the total false positive rate across the different checkpoints, researchers can repeatedly peek at their results and potentially stop data collection early without increasing their overall false positive rate (Lakens, 2014).

When researchers plan to perform a sequential analysis, it is important that they pre-register to avoid inflating the false positive rate. If researchers did not pre-register a sequential design, but wish to collect more data after finding promising results at the end of their originally planned sample size, this will inflate their false positive rate. The amount of inflation will usually be small if researchers only increase their sample size once. Sagarin, Amber, and Lee (2014) describe a method, *p*-augmented, for transparently reporting the true false positive rate associated with a data collection that has been extended in such a post hoc fashion.

If a multi-site replication project is planned, the same issues arise. However, due to the hierarchical nature of the replication design (participants nested within collection sites), the overall power of the replication attempt is determined jointly by the number of participants at each site and the number of collection sites. In general, the number of sites is more important than the number of participants per site, once a reasonable sample size at each site has been reached.

14.4.5 Pre-registering Replications

Replications, like original studies, are individual pieces of evidence; they contribute the most to the understanding of phenomena when they can be accurately combined with additional studies to build a cumulative model of knowledge. For this to occur, researchers need to be able to find the results of replications, no matter their outcome. One way to increase the discoverability of the body of replications that have been conducted is to pre-register replications in public, read-only repositories. A pre-registration includes information about the purpose, design, and planned analyses of a study, and this information is recorded in a repository before a study is conducted. There are several domain-specific registries, including the American Economic Association registry (www.socialscienceregistry.org), and research from any discipline can be pre-registered on OSF (https://osf.io/registries).

Along with increasing discoverability, pre-registration is also important for clearly distinguishing exploratory and confirmatory analyses. Data-dependent analyses, sometimes referred to as *p*-hacking or researcher degrees of freedom (RDF), can severally inflate the false positive rate of studies (Simmons, Nelson, & Simonsohn, 2011). The analysis plan section of a pre-registration is recorded before data have been looked at, thus the planned analyses are independent of how the data turn out. The pre-registration, therefore, allows researchers, reviewers, and readers to determine which analyses were data independent and which were data dependent. This distinction is critical because exploratory analysis, while important, does not provide strong tests of hypotheses.

Pre-registrations can be created for original or replication work. Due to the fact that replications, especially direct, are testing particular hypotheses, and are often using analyses already specified in the original work, they are particularly suited to pre-registration, and the analysis plan may be easier to develop than for an original study. If a large-scale replication project is being conducted, it is important that the analysis plans for the individual studies and the analysis plan for the aggregation of the studies are both pre-registered, as both analyses should be data independent for hypothesis testing.

14.4.6 Reporting Replications

Reproducibility and replicability are as important for replications as they are for original studies. This means that replications should be reported in transparent and replicable ways. To ensure narrow replicability, the data underlying the manuscript and well-commented analysis scripts should be publicly shared and linked to within the replication manuscript. If the data used in the replication are from an extent dataset (e.g., the General Social Survey), then this dataset should be clearly referenced in the manuscript and pertinent analysis scripts shared.

As previously mentioned, replications gain informational value when they can be aggregated with other original studies and replications. Pre-registration helps with discoverability of replication attempts but does not include the results of the replications. For aggregates to be as comprehensive as possible, all results, whether successful, failed, or inconclusive, should be released, either through a published article, a preprint, and/or by posting data, code, and a short write-up on a public, discoverable repository such as OSF (https://osf.io).

14.5 Interpreting Replication Results

Throughout this chapter we have referenced "successful" and "failed" replications but have not yet defined these terms. This is because there is no definitive answer for what determines whether a replication has succeeded or failed. Some metrics that have been used in prior research include: Did the replication result in a statistically significant effect in the same direction as the original study? Are the original and replication effects similar in size? Is the meta-analytic estimate of the original and replication effect sizes statistically significant? Does the original effect fall within the confidence intervals of the replication study or vice versa (Camerer et al., 2016; Open Science Collaboration, 2015; Valentine et al., 2011)? However, whether a single replication succeeded or failed rarely provides definitive evidence for or against a particular phenomenon, just like a single original study is rarely definitive. Individual studies may marginally increase or decrease confidence in particular claims and phenomena, but they are simply individual pieces of evidence. A much richer understanding of a phenomenon can be gained by moving away from viewing replications as dichotomous (succeed/fail) indicators and instead aggregating the evidence from multiple replications using hierarchical or meta-analytic techniques (Tackett & McShane, 2018). For this reason, deep replication projects (e.g., Many Labs style projects) can often provide more information at once than a single replication. However, replications of varying types are still needed to gain a fuller understanding of the size, generalizability, and boundary conditions of a theorized phenomenon and even single replications can, over time, be combined meta-analytically to provide a richer understanding of the phenomena under investigation.

14.6 Conclusion

Replication is an integral part of scientific knowledge building. While neglected in the past, replication of many types are gaining popularity in the social sciences as we move away from a dichotomous view of research and toward a more cumulative model of science. For this reason it is important to understand the different types of replications, the types of knowledge they convey, and how to design replication studies.

KEY TAKEAWAYS

- There are different types of replications that vary along a continuum in terms of how closely they match the procedure of the original study. Different kinds of replications provide different types of knowledge about the phenomena under investigation.
- Well-designed replications and original research share the same characteristics: Clear determination of the research question, adequate power, checks on the construct validity of manipulations and measures, and interpretations of results that are consistent with the evidence produced.
- A single replication, just like a single original study, is not definitive evidence for or against an effect. Instead, over time, similar studies can be combined to provide a richer understanding of the phenomena under investigation.
- New ways to report research have emerged, such as preprints and Registered Reports, that ensure that

research can be discovered by others, regardless of the outcome. Importantly, whether it is a replication or an original research design, reporting in a transparent and replicable way ensures others can understand and build from previous work.

IDEAS FOR FUTURE RESEARCH

- How replicable are findings in the various fields across the social and behavioral sciences?
- What factors improve and hurt the ability to replicate findings in the social and behavioral sciences?
- Over time, how will changes to grants, journals, and the field impact replicability in the social and behavioral sciences?
- Will changes to the training of young scientists improve replicability in the social and behavioral sciences?

SUGGESTED READINGS

Duvendack, M., Palmer-Jones, R., & Reed, W. R. (2017). What is meant by "replication" and why does it encounter resistance in economics? *American Economic Review*, *107*, 46–51. https://doi.org/10.1257/aer.p20171031

Nelson, L. D., Simmons, J., & Simonsohn, U. (2018). Psychology's renaissance. *Annual Review of Psychology*, *69*. https://doi.org/10.1146/annurev-psych-122216-011836

Peng, R. D. (2011). Reproducible research in computational science. *Science*, *334*, 1226–1227. https://doi.org/10.1126/science.1213847

Zwaan, R. A., Etz, A., Lucas, R. E., & Donnellan, M. B. (2017). Making replication mainstream. *Behavioral and Brain Sciences*, 1–50. https://doi.org/10.1017/S0140525X17001972

REFERENCES

Alogna, V. K., Attaya, M. K., Aucoin, P., Bahník, Š., Birch, S., Birt, A. R., … Zwaan, R. A. (2014). Registered replication report: Schooler and Engstler-Schooler (1990). *Perspectives on Psychological Science*, *9*, 556–578. https://doi.org/10.1177/1745691614545653

Anderson, S. F., Kelley, K., & Maxwell, S. E. (2017). Sample-size planning for more accurate statistical power: A method adjusting sample effect sizes for publication bias and uncertainty. *Psychological Science*, *28*, 1547–1562. https://doi.org/10.1177/0956797617723724

Begley, C. G., & Ellis, L. M. (2012). Raise standards for preclinical cancer research: Drug development. *Nature*, *483*, 531–533. https://doi.org/10.1038/483531a

Berry, J., Coffman, L. C., Hanley, D., Gihleb, R., & Wilson, A. J. (2017). Assessing the rate of replication in economics. *American Economic Review*, *107*, 27–31. https://doi.org/10.1257/aer.p20171119

Brown, A. N., Cameron, D. B., & Wood, B. D. K. (2014). Quality evidence for policymaking: I'll believe it when I see the replication. Replication Paper 1, March 2014. Washington, DC: International Initiative for Impact Evaluation (3ie).

Camerer, C. F., Dreber, A., Forsell, E., Ho, T.-H., Huber, J., Johannesson, M., … Wu, H. (2016). Evaluating replicability of laboratory experiments in economics. *Science*, *351*, 1433–1436. https://doi.org/10.1126/science.aaf0918

Chambers, C. D. (2013). Registered Reports: A new publishing initiative at Cortex. *Cortex*, *49*, 609–610. https://doi.org/10.1016/j.cortex.2012.12.016

Chang, A. C., & Li, P. (2015). Is economics research replicable? Sixty published papers from thirteen journals say "usually not." *Finance and Economics Discussion Series*, *2015*, 1–26. https://doi.org/10.17016/FEDS.2015.083

Clemens, M. A. (2017). The meaning of failed replications: A review and proposal. *Journal of Economic Surveys*, *31*, 326–342.

Collins, H. M. (1992). *Changing Order: Replication and Induction in Scientific Practice*. Chicago, IL: University of Chicago Press.

Crandall, C. S., & Sherman, J. W. (2016). On the scientific superiority of conceptual replications for scientific progress. *Journal of Experimental Social Psychology*, *66*, 93–99. https://doi.org/10.1016/j.jesp.2015.10.002

Dijksterhuis, A., & van Knippenberg, A. (1998). The relation between perception and behavior, or how to win a game of trivial pursuit. *Journal of Personality and Social Psychology*, *74*, 865–877.

Dreber, A., Pfeiffer, T., Almenberg, J., Isaksson, S., Wilson, B., Chen, Y., ... & Johannesson, M. (2015). Using prediction markets to estimate the reproducibility of scientific research. *Proceedings of the National Academy of Sciences of the United States of America*, *112*, 15343–15347. https://doi.org/10.1073/pnas.1516179112

Duvendack, M., Palmer-Jones, R., & Reed, W. R. (2017). What is meant by "replication" and why does it encounter resistance in economics? *American Economic Review*, *107*, 46–51. https://doi.org/10.1257/aer.p20171031

Errington, T. M., Iorns, E., Gunn, W., Tan, F. E., Lomax, J., & Nosek, B. A. (2014). An open investigation of the reproducibility of cancer biology research. *Elife*, *3*. https://doi.org/10.7554/eLife.04333

Freese, J., & Peterson, D. (2017). Replication in social science. *Annual Review of Sociology*, *43*, 147–165.

Hagger, M. S., Chatzisarantis, N. L. D., Alberts, H., Anggono, C. O., Batailler, C., Birt, A. R., ... Zwienenberg, M. (2016). A multilab preregistered replication of the ego-depletion effect. *Perspectives on Psychological Science*, *11*, 546–573.

Hamermesh, D. S. (2007). Viewpoint: Replication in economics. *Canadian Journal of Economics/Revue canadienne d'économique*, *40*, 715–733.

Jasny, B. R., Chin, G., Chong, L., & Vignieri, S. (2011). Again, and again, and again ... *Science*, *334*, 1225–1225.

Kahneman, D. (2014). A new etiquette for replication. *Social Psychology*, *45*, 310–311.

Klein, R. A., Ratliff, K. A., Vianello, M., Adams, R. B., Bahník, Š., Bernstein, M. J., ... Nosek, B. A. (2014). Investigating variation in replicability: A "Many Labs" replication project. *Social Psychology*, *45*, 142–152.

Kuhn, T. S., & Hacking, I. (2012). *The Structure of Scientific Revolutions* (4th ed.). Chicago, IL: University of Chicago Press.

Lakatos, I. (1970). Falsification and the methodology of scientific research programmes. In I. Lakatos & A. Musgrave (Eds.), *Criticism and the Growth of Knowledge* (pp. 91–196). Cambridge: Cambridge University Press. https://doi.org/10.1017/CBO9781139171434.009

Lakens, D. (2014). Performing high-powered studies efficiently with sequential analyses: Sequential analyses. *European Journal of Social Psychology*, *44*, 701–710. https://doi.org/10.1002/ejsp.2023

LeBel, E. P., Berger, D., Campbell, L., & Loving, T. L. (2017). Falsifiability is not an option. *Journal of Personality and Social Psychology*, *113*, 254–261.

Lykken, D. T. (1968). Statistical significance in psychological research. *Psychological Bulletin*, *70*, 151–159. https://doi.org/10.1037/h0026141

Makel, M. C., & Plucker, J. A. (2014). Facts are more important than novelty: Replication in the education sciences. *Educational Researcher*, *43*, 304–316. https://doi.org/10.3102/0013189X14545513

McCullough, B. D., McGeary, K. A., & Harrison, T. D. (2008). Do economics journal archives promote replicable research? Economics journal archives. *Canadian Journal of Economics/Revue canadienne d'économique*, *41*, 1406–1420.

Meehl, P. E. (1990). Appraising and amending theories: The strategy of Lakatosian defense and two principles that warrant it. *Psychological Inquiry*, *1*, 108–141. https://doi.org/10.1207/s15327965pli0102_1

Merton, R. K. (1973). *The Sociology of Science: Theoretical and Empirical Investigations*. Chicago, IL: University of Chicago Press.

Mobley, A., Linder, S. K., Braeuer, R., Ellis, L. M., & Zwelling, L. (2013). A survey on data reproducibility in cancer research provides insights into our limited ability to translate findings from the laboratory to the clinic. *PLoS One*, *8*, e63221. https://doi.org/10.1371/journal.pone.0063221

National Academies of Sciences, Engineering, and Medicine. (2016). *Statistical Challenges in Assessing and Fostering the Reproducibility of Scientific Results: Summary of a Workshop*. Washington, DC: National Academies Press. https://doi.org/10.17226/21915

Nelson, L. D., Simmons, J., & Simonsohn, U. (2018). Psychology's renaissance. *Annual Review of Psychology*, *69*, 511–534.

Nosek, B. A., & Errington, T. M. (2017). Making sense of replications. *eLife*, *6*. https://doi.org/10.7554/eLife.23383

Nosek, B. A., & Lakens, D. (2014). Registered Reports: A method to increase the credibility of published results. *Social Psychology*, *45*, 137–141.

Open Science Collaboration. (2015). Estimating the reproducibility of psychological science. *Science*, *349*, aac4716. https://doi.org/10.1126/science.aac4716

Peng, R. D. (2011). Reproducible research in computational science. *Science*, *334*, 1226–1227. https://doi.org/10.1126/science.1213847

Perugini, M., Gallucci, M., & Costantini, G. (2014). Safeguard power as a protection against imprecise power estimates. *Perspectives on Psychological Science*, *9*, 319–332. https://doi.org/10.1177/1745691614528519

Pesaran, H. (2003). Introducing a replication section. *Journal of Applied Econometrics*, *18*, 111–111. https://doi.org/10.1002/jae.709

Popper, K. R. (1959/1934). *The Logic of Scientific Discovery*. New York: Routledge.

Prinz, F., Schlange, T., & Asadullah, K. (2011). Believe it or not: How much can we rely on published data on potential drug targets? *Nature Reviews Drug Discovery*, *10*, 712–712. https://doi.org/10.1038/nrd3439-c1

Sagarin, B. J., Ambler, J. K., & Lee, E. M. (2014). An ethical approach to peeking at data. *Perspectives on Psychological Science*, *9*, 293–304. https://doi.org/10.1177/1745691614528214

Schmidt, S. (2009). Shall we really do it again? The powerful concept of replication is neglected in the social sciences. *Review of General Psychology*, *13*, 90–100. https://doi.org/10.1037/a0015108

Simmons, J. P., Nelson, L. D., & Simonsohn, U. (2011). False-positive psychology: Undisclosed flexibility in data collection and analysis allows presenting anything as significant. *Psychological Science*, *22*, 1359–1366. https://doi.org/10.1177/0956797611417632

Simons, D. J., Shoda, Y., & Lindsay, D. S. (2017). Constraints on Generality (COG): A proposed addition to all empirical papers. *Perspectives on Psychological Science*, *12*, 1123–1128. https://doi.org/10.1177/1745691617708630

Simonsohn, U. (2015). Small telescopes: Detectability and the evaluation of replication results. *Psychological Science*, *26*, 559–569. https://doi.org/10.1177/0956797614567341

Tackett, J. L., & McShane, B. B. (2018). Conceptualizing and evaluating replication across domains of behavioral research. *Behavioral and Brain Sciences*, *41*, e152.

Valentine, J. C., Biglan, A., Boruch, R. F., Castro, F. G., Collins, L. M., Flay, B. R., ... Schinke, S. P. (2011). Replication in prevention science. *Prevention Science*, *12*, 103–117.

Zwaan, R. A., Etz, A., Lucas, R. E., & Donnellan, M. B. (2017). Making replication mainstream. *Behavioral and Brain Sciences*, *41*, 1–50. https://doi.org/10.1017/S0140525X17001972

15 Research Ethics for the Social and Behavioral Sciences

Ignacio Ferrero

Javier Pinto

Social and behavioral science research aims to understand human behavior in society and to produce useful knowledge. But such knowledge can only really have a positive impact on the well-being of society if it is acquired in accordance with the norms of scientific inquiry, assuring the reliability and validity of the indicators and outcomes and respecting the dignity of individuals, groups, and communities (Social Research Association, 2003). As such, the social and behavioral sciences can be a means to a better world if the knowledge that they produce is informed by ethical and responsible research (RRBM, 2017).

However, reports of fraud and malpractice in professional research are often heard and written about in the press. From as early as the 1970s, the social and behavioral sciences seem to have suffered a wave of incidents of scientific misconduct. Scholars have documented the prevalence of questionable research practices (RRBM, 2017). In 1997, Dotterweich and Garrison administered a survey to a sample of professors at institutions accredited by the Association to Advance Collegiate Schools of Business (AACSB) to identify those actions that they felt were unethical and to gauge the state of research ethics among business academics. The survey was centered on eleven substantive issues concerning business research ethics. More than 95% of respondents condemned five of the eleven activities studied, including falsifying data, violating confidentiality of a client, ignoring contrary data, plagiarism, and failing to give credit to coauthors (Dotterweich & Garrison, 1997). Almost fifteen years later, in 2011, the National Academy of Sciences showed a tenfold increase in retractions and related misconduct in the sciences over the past decades (Wible, 2016). In addition, only a few years ago, a study by the Open Science Collaboration (2015), aiming to reproduce the findings of 100 articles published in elite psychology journals, reported that most of these findings failed to replicate (although many of the failures to replicate may have occurred for reasons outside of ethical concerns).

These studies highlight the fact that the science community had started to attract the attention from public authorities who brought into question the reputability of the research that was being done. A number of experiments on human subjects in the United States during the 1960s and 1970s also sparked a public outcry. They were declared both unethical and illegal, since they were performed without the knowledge or informed consent of the test subjects, and also set in motion a national debate that would eventually lead to stricter controls governing medical and social research conducted on humans. The most infamous of these experiments, namely the Tuskegee syphilis experiment, the MK-Ultra project, the Monster Study, the Stanford prison experiment, and the Milgram experiment, offered insight into the extent to which human dignity and rights could be violated in the name of research. Various federal agencies in the United States now require systematic research ethics training programs as a mandatory part of the grant submission process. Similarly, the European Union's Horizon 2020 program encourages applicants to embed research ethics within their proposals.

Moreover, the concern has led to the emergence of professional bodies whose role is to specify the rules and regulations that govern best practices, including codes of conduct and ethical committees in research institutions. Indeed, the proliferation of research codes of practice in use at academic institutions and research centers worldwide is further proof of the commitment to create a culture of best practices in the field of social and behavioral science research. Some examples of the better-known codes and bodies are the Singapore Statement on Research Integrity, the European Code of Conduct for Research Integrity, the World Medical Association's Declaration of Helsinki, the National Institutes of Health (NIH), and the National Science Foundation (NSF). Other influential research ethics policies, such as the Code of Ethics of the American Psychological Association, the Ethical

Principles of Psychologists and Code of Conduct, the Statements on Ethics and Professional Responsibility (American Anthropological Association), and the Statement on Professional Ethics (American Association of University Professors) are also coming into prominence.

In March 2015, after a series of symposia, conferences, and meetings, the Council of the Academy of Social Sciences formally adopted five guiding ethical principles for social and behavioral science research. These are: 1) social science is fundamental to a democratic society and should be inclusive of different interests, values, funders, methods, and perspectives; 2) all social sciences should respect the privacy, autonomy, diversity, values, and dignity of individuals, groups, and communities; 3) all social sciences should be conducted with integrity throughout, employing the most appropriate methods for the research purpose; 4) all social scientists should act with regard to their social responsibilities in conducting and disseminating their research; and 5) all social sciences should aim to maximize benefits and minimize harm (Dingwall, Iphofen, Lewis, Oates, & Emmerich, 2017).

What these codes and principles have in common is that they adopt and promulgate preventative strategies that aim to produce more reliable and actionable knowledge for better policies and practices. In doing so, they contribute to a sense of professional practice integrally linked to compliance with rules and regulations and not just the authentic fulfillment of a professional role per se (OECD, 2007). This is, perhaps, why in recent times researchers have noted a gradual increase in the amount of time they devote to bureaucratic procedures, paperwork, and institutional accreditation processes. Understandably, some scientists are concerned that authorities may have inadvertently set up oversight and reporting systems that are bureaucratic, unnecessarily burdensome, intrusive, or even unfair, and that they risk jeopardizing the research process (Bell & Bryman, 2007). In particular, social and behavioral science researchers are understandably preoccupied about the proposal of applying wholesale the strict requirements of the medical trial programs to social and behavioral science research in that they can disproportionately frustrate and lengthen research projects.

Research involves humans and, hence, ethical research conduct cannot only be about the rules to be followed, but also the application of informed moral reasoning founded on a set of moral principles (The British Psychological Society, 2010). This moral reasoning is absolutely necessary and is the basis of professional practice and meaningful work which impact positively on workplace experience (Kanungo, 1992). Tsoukas and Cummings also explain that a competent action is much more than just an instrumental application of formal-cum-abstract formulae but also involves the use of practical wisdom, which has to do with knowing what is good for human beings in general, as well as having the ability to apply such knowledge to particular situations (1997) and making the right decisions in the workplace (Schwartz & Sharpe, 2010). Moral reasoning and practical wisdom are crucial aspects of professional practice, in general, and social and behavioral science research, in particular.

In social and behavioral science research, the capacity for moral reasoning must be reinforced by ethical principles, how they relate to research practice, and how they contribute to sustaining the ethical research ecosystem (Drenth, 2012). Thus, an ethical culture of social and behavioral science research has to encompass: a) general principles of professional integrity and b) the principles applying to each of the constituencies that form the social ecosystem in which scientists work (Bell & Bryman, 2007). The next section will examine what these ethical principles are and delve into why they are important to the research profession.

15.1 The Ethical Principles of Research

Integrity is *the* defining characteristic of being ethical at work. Integrity implies consistency and coherence in the application of both technical and ethical-professional principles, and the highest standards of professionalism and rigor (Hiney, 2015). Integrity also involves understanding and following the various legal, ethical, professional, and institutional rules and regulations in relation to the activity at hand (Shamoo & Resnik, 2015). Hence, integrity is not only a question of the theoretical acceptance of certain values but also the challenge to put such principles into practice (Resnik, 2011).

Although integrity is considered to be the cornerstone for ethics in research, it is typically accompanied by five broad values that also shape the professional conduct of the researcher.

15.1.1 Honesty

Integrity implies acting with honesty. A greater access to well-grounded knowledge serves society, and, hence, researchers should accept the responsibility to disseminate their acquired knowledge to society, in general. In this regard, the professional principle of honesty encompasses disclosure, transparency, and confidentiality.

First, researchers should disclose information related to the research process: The methods, procedures, techniques, and findings. Although this information may not have been requested, society benefits from its provision in that it stimulates academic debate among interested researchers and the public at large (Drenth, 2012; Social Research Association, 2003).

Second, the publication of this information must be done transparently, without confusion or deception, drawing a clear distinction between their professional statements and any comments made from a personal point of view (Resnik & Shamoo, 2011). This implies impartiality.

Third, the dissemination of this information must respect confidentiality, which has become a pertinent issue as a result of the ongoing digitalization of society. There are now new forms available to disseminate research findings, including online, open-source and open-access publishing. These new channels provide greater opportunities to share the research, but they also threaten the protection of data relating to the privacy of individuals. Research data cannot be made public without authorization, except for cases where withholding the information might be detrimental to a greater good. For instance, if maintaining a confidentiality agreement were to facilitate the continuation of illegal behavior which has come to light during the research process, then such an agreement should be annulled. Imagine an employee of a tobacco company who discovers that her firm deliberately works on illegally increasing the addictiveness of the cigarettes, but, because of the confidentiality agreement, she is bound to secrecy even though this finding should be in the public domain.

15.1.2 Objectivity

A second principle tightly coupled with integrity is that of objectivity, which basically means the use of appropriate methods in research. Researchers must draw conclusions from a critical analysis of the evidence and communicate their findings and interpretations in a complete and objective way (Resnik & Shamoo, 2011). Despite the fact that we cannot be entirely objective, since the selection of topics usually reveals a systematic bias in favor of certain cultural or personal values, we must not engage in methods of selection designed to produce misleading results or in misrepresenting findings by commission or omission.

As an ethical-professional principle, objectivity concerns the need to reach and communicate findings in order to broaden and enhance knowledge but not for one's own personal gain or for that of a third party at the expense of the scientific community or the general public (Drenth, 2012; OECD, 2007; Sutrop & Florea, 2010). Researchers must have the best of intentions regarding the scientific purpose of research, even when they are tempted not to act independently of the interests of funders or donors, who may try to impose certain priorities, obligations, or prohibitions (Social Research Association, 2003).

Objectivity is especially important in aspects of research such as experimental design, data analysis, data interpretation, peer review, personnel decisions, grant writing, expert testimony, and tenure, among others. Interpretations and conclusions on these areas must be underpinned by facts and data that are amenable to proof and secondary review. There should be transparency in the collection, analysis, and interpretation of data, and verifiability in scientific reasoning.

Objectivity must also withstand the threat of possible conflicts of interest on financial, political, social, and religious grounds (Israel, 2014). In such cases, researchers must consult ethics protocols dealing with such concerns (e.g., The British Psychological Society, 2010). If a conflict of interest, of any kind, is present, the researcher has a duty to disclose it immediately.

15.1.3 Accountability

Accountability covers administrative actions that are transparent and verifiable and that disclose the intents and purposes of the professionals who undertake them. The meaning of accountability is not limited to expressions of honesty or good intentions; accountability means being able to justify that one has acted in an honest and well-intentioned way. Hence, researchers ought to keep a complete record of the documentation of the research project in such a way that others may verify and/or replicate the work. Similarly, researchers must take responsibility for their contributions to any publications, funding applications, reports, and other presentations or communications relating to their research (Resnik & Shamoo, 2011).

15.1.4 Authenticity

Our penultimate value is authenticity, which is violated when findings are misrepresented.

The goal of misrepresentation is often to gain an advantage from new scientific discoveries – rewards for publication, workplace promotion, a boost in professional prestige, etc. Misrepresentation runs counter to the spirit of research practice (OECD, 2007) and takes three general forms: Fabrication, falsification, and plagiarism – which we discuss below (Hiney, 2015; Wible, 2016).

Fabrication. Fabrication is the invention of data or results, which are then presented as real, so as to prove a working hypothesis (Drenth, 2012). For instance, claims based on incomplete or assumed results are a form of fabrication. This practice is never merely a matter of negligence since it is almost always intentional and fraudulent and any fabrication of findings and passing them off as true is normally considered a serious offense in the science community and society in general.

Falsification. Falsification is the negligent or fraudulent manipulation of existing data to achieve a result that might be expected from the research process. For example, changing or omitting research results and data to support hypotheses, claims, etc. The seriousness of falsification lies in the fact that it involves false claims about information that may be relevant for scientific research (Drenth, 2012), rather than whether it satisfies special or specific interests.

Plagiarism. Plagiarism is the appropriation of another person's work, results, processes, etc. (Hiney, 2015) without giving credit to the originator, even if done unintentionally. Plagiarism, therefore, is a form of theft, which undermines the integrity of the scientific community and the status of science itself.

The problem of plagiarism is particularly acute in relation to citations and acknowledgements in publications. Researchers ought to acknowledge the names and titles of all those who contributed in a significant way to the research project (the European Code of Conduct for Research Integrity), including editors, assistants, sponsors, and others to whom the criteria of authorship do not apply (Resnik & Shamoo, 2011). Guest authorship and ghost authorship are not acceptable because it means giving undue credit to someone. The criteria for establishing the sequence of authors should be agreed by all, ideally at the start of the project. Respect for all aspects of intellectual property is especially important in this regard: Patents, copyrights, trademarks, trade secrets, data ownership, or any other kind of property in science (Shamoo & Resnik, 2015). Researchers should also cite all sources and make sure to avoid self-plagiarism, duplication of their own work, or the publication of redundant papers.

These cases of misconduct are examples of how the integrity of the research process and self-governance by the research community can be undermined. They breach the trust that allows scientists to build on each other's work and that allows policymakers and others to make decisions based on scientific and objective evidence (Arrison, 2014).

15.1.5 Compliance

Our last value in this section is compliance: The attitude of attentiveness and respect for rules and regulations, be they the laws of a country, the code of conduct in an organization, trade union conditions, or the norms applicable within science associations. Compliance involves ongoing awareness of new rules and regulations governing professional practice (Caruth, 2015; Resnik & Shamoo, 2011; Social Research Association, 2003) in the country or countries where the research is carried out (OECD, 2007). Without such due diligence, unjustifiable lapses may occur through negligence (Sutrop & Florea, 2010).

In situations where others may have engaged in irregular practices, researchers must immediately disassociate themselves from the situation and work toward correcting and redressing the problem. Whistleblowing – the official reporting of such malpractice – may, at times, be the only way of putting a stop to the irregular situation. Researchers ought to keep the relevant authorities informed of any concern they may have about inappropriate research conduct, including fabrication, falsification, plagiarism, or other forms of malpractice that may undermine the reliability of research, such as negligence, the incomplete acknowledgement of authors, a lack of information about contradictory data, and/or the use of dishonest methods of analysis (Resnik & Shamoo, 2011).

15.2 Turning to Context

Proclaiming principles and integrity is necessary but not sufficient to professional moral reasoning. Scholars need to be made aware of the professional ecosystem of research within which they operate. Hence, it is important to reflect on a second criteria for acting in a professional manner during the process of social research, namely taking into consideration the related third parties (constituencies or stakeholders) that constitute the ecosystem that may be affected by research malpractice, or that may influence the way research is done (OECD, 2007). These third parties and actors in the research domain are universities, research centers, government agencies, the community of scholars, journal editors and publishers, the general public, and, in particular, the subjects of the study itself.

15.2.1 Universities, Research Centers, and Government Agencies

These institutions should develop a strategy to encourage researchers to make positive and quality contributions with societal relevance (Social Policy Association Guidelines on Research Ethics, 2009) creating and providing an ethical culture of scientific work (Drenth, 2012). To do so, such institutions must act in consideration of two principles to fulfill their responsibilities toward social research practices. The first is to narrow the gap between research and practice. Since research is primarily evaluated by its placement in elite journals and its impact on subsequent research, its impact on real-world problems is often seen as diminishing. Universities and research centers usually rely on proxy measures of usefulness such as the impact factor of the journal where the work is published, and the system of journals often favors novelty over cumulative insights. This threat could increase considerably in the following years "thanks to the pervasiveness of electronic monitoring inside and outside organizations. Coupled with our current standards of evaluation, this could result in a high volume of novel papers with sophisticated econometrics and no obvious prospect of cumulative knowledge development" (Davis, 2015, p. 180). The second concern is the quality of the research itself. Academic evaluation systems encourage quantity over quality (Gupta, 2013), and novelty over replicability, resulting in little cumulative progress in knowledge (RRBM, 2017).

Moreover, institutions must be cognizant of the fact that responsible research is about both useful and credible knowledge. Therefore, institutions ought to appreciate the obligations that social researchers have to them, society at large, research subjects, professional colleagues, and to other contributors. Promotion and tenure should assess the reliable incremental knowledge as well as the innovativeness and its potential for scholarly and societal impact. For the same reason, funding agencies and governments should broaden the criteria for funding decisions to include potential and societal impact in addition to intellectual merit (RRBM, 2017).

This commitment to integrity should involve clear policies and procedures, training, and mentoring of researchers. Institutions should also demonstrate integrity in their activity and avoid any form of malpractice. Such oversight may be carried out by ethics committees, departments, or teams, or by means of codes of conduct (Social Research Association, 2003). In addition, research centers, journals, and any other research-related professional associations or organizations should have clear procedures on how to deal with accusations of ethical failures or other forms of irresponsible malpractice (Resnik & Shamoo, 2011).

The responsibility of researchers to their employer or institution includes accountability for financial matters, how the project marks progress in knowledge, whether or not it meets the legitimate interests of the institutions, and if it satisfies the relevant economic, cultural, and legal requirements (Iphofen, 2011; Sutrop & Florea, 2010). This overall responsibility may be complemented with an explanation of how the research endeavor contributes to the development of skills, competencies, and knowledge across the institution (Social Research Association, 2003).

15.2.2 The Scientific Community

Research is a collective endeavor, not amenable to the work of isolated individuals. It normally involves cooperation among a number of researchers, as well as the sharing of data and findings with the rest of the scientific community. However, the idea of scientific community is a broad one and extends beyond the immediate research team (Sutrop & Florea, 2010), encompassing all national and international researchers (Drenth, 2012), and even those still in training. Thus, the professional activity of the researcher must consider the overall purpose of the scientific community, including its reputation and prestige (Hansson, 2011). This commitment certainly involves:

Responsible Teamwork. The relationships between members of a research team should not be limited to merely appropriate behavior between professional colleagues. It should also contribute to the professional development of each researcher. This should never come at a cost to other team members nor limit their professional growth and should always envisage progress in the scientific field in which it unfolds (Adams Goertel, Goertel, and Carey, Chapter 16; Social Policy Association, 2009).

Responsible Authorship. Responsible authorship requires the publication of quality scientific research, the enhancement of scientific knowledge, meeting the needs of the funding institution, and ensuring that the findings published are relevant to society as a whole. Another key aspect, in this regard and alluded to previously, is acknowledgement – the right of coauthors to be recognized as such and receive whatever benefits may be due to them as a result (Drenth, 2012). Researchers should ensure that decisions about authorship are made in a fair and transparent way, acknowledging every relevant contributor according to the established conventions of the discipline (Adams Goertel et al., Chapter 16).

Responsible Peer Reviewing and Editing. The responsibility of the anonymous reviewers and the journal editors involves ensuring high scientific standards in publications and guaranteeing that there is an advancement in knowledge. In particular, the reviewers and editors should help researchers whose applications for publication do not meet the standards of the journal or the expectations of the scientific community, providing them with recommendations and further readings that will improve the researchers' work.

Reviewing and editing should be carried out in accordance with objective criteria and be attentive to possible conflicts of interest. For instance, a reviewer should decline to revise a work if he or she knows the author(s) or if that work could compete with his or her own work to be published in the same journal (Social Policy Association, 2009). It would also be highly unethical for reviewers to make use of any materials submitted to them for their own purposes or the purposes of any third parties, without the express permission of the authors.

Responsible Mentoring. Responsible mentoring implies training new researchers, PhD candidates, postgraduate students, or post-doctoral scholars to make them capable of contributing in a significant way to the scientific community (Drenth, 2012) and helping them progress in their academic careers (OECD, 2007). For instance, the mentors should train mentees to present their findings at conferences and congresses and in being professional in conducting and presenting their research.

Responsible Data Management. The goal of data management is to effectively administer any information that may be of use to current or future social and behavioral scientists (Drenth, 2012). Taking into account that much research data – even sensitive data – can be shared ethically and legally, if researchers employ strategies of informed consent, anonymization, and controlling access to data (the UK Data Services, 2012–2017), good data management is nothing less than an ethical duty for the scientific community.

Research Ethics Committees (RECs) should design policies in relation to the availability of data and the criteria governing appropriate use, counseling researchers on how their data may be stored in a secure and protected manner, shared, and made available for reuse, while complying with the requirements of data protection legislation and best practices concerning confidentiality. In this way, the role of RECs is to protect the safety, rights, and well-being of research participants and to promote ethically sound research. In addition,

confidentiality agreements should always be respected, irrespective of whether they were originally established for the purposes of previous research projects, except for cases where withholding the information might be detrimental to a greater good, as we previously mentioned.

15.2.3 Dealing with Human Subjects

The treatment of human subjects may be the most significant ethical concern in social and behavioral science research. Current codes of research conduct largely came about as a result of the findings of the Nuremberg trials concerning the medical experiments performed by the Nazis (Kopp, 2006; Sutrop & Florea, 2010) and other medical, surgical, clinical, and psychological experiments conducted in the United States during the twentieth century. Similarly, the National Commission for the Protection of Human Services of Biomedical and Behavioral Research, in the United States, published the Belmont Report in 1974 in response to the unethical treatment of patients who partook in a medical study. This growing recognition of the dignity of the human person has culminated in a broad consensus that social and behavioral science research must always respect the inalienable principle of the dignity of each and every human being, living or dead (Wilkinson, 2002).

Dignity has many meanings in common usage (Dan-Cohen, 2012; Sulmasy, 2008), but, principally, it refers to the intrinsic worth or value of every human being, which distinguishes him or her from any other being, and, as such, merits respect (Sison, Ferrero, & Guitián, 2016). Such worth or value is often associated with the capacity for reason and autonomy or "self-determination" through free choice. It also implies the need for consensus or mutual recognition among fellow human beings. In short, dignity refers to a preeminent social position, which has come to be attributed universally to all human beings (Dan-Cohen, 2012).

In general, the protection of the personal dignity of human subjects implies social and behavioral science research projects taking the following issues into account:

Avoid Any Kind of Harm. Social and behavioral science research does not expose human subjects to as many harmful effects as other forms of scientific research might do (Bell & Bryman, 2007). Nevertheless, it is vital to ensure that no research project involves serious harm or injury to any human subject in the study (General Assembly of the World Medical Association, 2014), be it physical, psychological, or moral (Barrett, 2006; Social Research Association, 2003; Sutrop & Florea, 2010).

Valid Consent. The dignity of the person precludes any form of coercion obliging an individual or individuals to participate in a research project. Participation must be freely undertaken on the basis of an informed decision (Israel, 2014), having given explicit consent in line with a clear protocol, and in accordance with the law and the culture of the participants (Sutrop & Florea, 2010). The latter condition includes the responsibility to offer a complete description of the project, including all relevant research details required to give truly informed consent (see Festinger et al., Chapter 6).

This informed consent implies that the participant has to understand the relevant information about the process and goals. The participation also has to be voluntary. In other words, the consent is given freely and not as a result of coercive pressure (real or perceived). It must also be competent, meaning the consent has to be given by somebody capable, by virtue of their age, maturity, and mental stability, of making a free, deliberate choice (Houston, 2016; Macfarlane, 2010).

The need for informed consent does not preclude the possibility of addressing those who lack the competence to offer free and informed consent. To ignore the possibility of carrying out research among subjects, such as children younger than the required age to give consent (from fourteen to eighteen years old depending on the country), people with learning or communication difficulties, patients in care, people in custody or on probation, or people engaged in illegal activities (such as drug abuse), may in itself constitute a form of discrimination (Social Research Association, 2003). This kind of research often yields results that may contribute to bettering the situation and quality of life of these groups of people, since the effectiveness of public policies enacted in relation to them depends on previous research of this kind. In these cases, informed consent should be given either by parents or by legal guardians. Such third parties must give valid consent, under the same conditions as a standard expression of consent, and confirm that there are no conflicts of interest at play (Social Research Association, 2003). In addition, depending on the potential risks to the participants, it may be necessary to obtain advice and approval from an independent ethics committee (General Assembly of the World Medical Association, 2014).

Respecting Privacy. The principle of personal dignity also requires that the legitimate privacy and decorum of research subjects be safeguarded, regardless of the value of the research being undertaken and/or the potential of the new technologies now deployed for research purposes (Social Research Association, 2003). This condition sets limits on the information that may be sought about a person for the purposes of any research project, especially if she or he has not consented to disclosing every detail about their personal life (Barrett, 2007).

We can look back to 2017 to find a cogent example of the violation of personal consent and privacy. The Facebook and Cambridge Analytica exposé triggered a public debate on the responsibility of social media and internet companies and their third party partners regarding the use of personal data that they have access to and the degree of consent that users willingly give. Cambridge Analytica, a political consulting firm that combines data mining, data brokerage, and data analysis, alongside strategic communication for electoral campaigning purposes, managed to collect data and build profiles of millions of Facebook users using sources such as demographics, consumer behavior, internet activity, and, most worryingly, by collaborating with others who were given user data by Facebook under the pretext of academic research. In March 2018, *The New York Times* and *The Observer* reported that Cambridge Analytica, however, used this personal information for its commercial service offering to influence the outcomes of the 2016 US elections and the Brexit referendum, without the users' permission or knowledge and without permission from Facebook (Cadwalladr & Graham-Harrison, 2018; Rosenberg, Confessore, & Cadwalladr, 2018).

Another example of this might be research studies involving the everyday behavior of individuals who may not be aware of the full range of situations in which they are to be observed by a third party. The condition of privacy also limits the use researchers may make of the data they collect because such information is provided to a given researcher or research group, not offered to the scientific community or society at large, even though it might be beneficial to them. Therefore, access to research data must be subject to tight control (Bell & Bryman, 2007). Anonymization is a valuable tool that enables the sharing of data while preserving privacy. The process of anonymizing data requires that identifiers are changed in some way, by being removed, substituted, distorted, generalized, or aggregated.

15.3 Conclusion

Contributing to a better world is the ultimate goal of science (RRBM, 2017). Social and behavioral science research can live up to this duty if it continues to hold the values outlined here in the highest esteem. Ethics helps researchers to carry out their research honorably, honestly, objectively, and responsibly toward the ultimate goal of enhancing our understanding of human behavior in social contexts. On the other hand, given that research is a collective endeavor, involving a significant degree of collaboration and cooperation, ethics facilitates the progress of science by underlining key values such as teamwork and trust, responsibility, generosity, respect, fairness, and authorship, among others that we discussed in this chapter.

Moreover, ethics ensures that research meets the aims and needs of funding institutions, respecting both the legitimate interests of those bodies and the broader interests of society as a whole. In this way, ethically grounded research aims to generate knowledge that may be of use and value to society, publishing it in a transparent way that, at the same time, is wholly respectful of the safety, privacy, confidentiality, and dignity of the participants in research projects. It can be said that ethical research contributes to civil society as well as to the funding bodies that finance research studies.

Finally, the service and benefit afforded by rigorous social and behavioral science research means that methods, procedures, techniques, and findings which contribute to refining and enhancing knowledge are made available to the scientific community, thus furthering the academic endeavor in general.

These are some of the reasons that account for the growing importance of the field of research ethics. We believe that it is imperative that ethics be included in the curricula for the pedagogical development of anyone pursuing work in professional research. Education in ethical research will lead to a greater understanding of ethical standards, ethics policies, codes of conduct, and, above all, to fostering the application of ethical judgement and practical wisdom among researchers engaged in decision-making.

KEY TAKEAWAYS

- Ethics helps researchers to carry out their research honorably, honestly, objectively, and responsibly with the goal to improve a comprehensive understanding of human behavior in social contexts.
- Research is a collective endeavor that involves a significant degree of collaboration and cooperation.
- Ethics facilitates the progress of science by underlining such key values as teamwork, trust, responsibility, generosity, respect, fairness, authorship, etc.
- Ethical research aims to generate knowledge that may be of use and value to society, publishing it in a transparent way and wholly respectful of the safety, privacy, confidentiality, and dignity of the participants in research projects.
- Education in research ethics will lead to a greater understanding of ethical standards, ethics policies, codes of conduct, and above all, to fostering the application of practical wisdom in decision-making.

IDEAS FOR FUTURE RESEARCH

- How can researchers remain ethical while adopting many of the newest and most convenient means of data collection?
- What must companies do to ethically handle our data while potentially still making it available for research purposes?

- What incentives can be introduced to encourage researchers to maintain an ethical code of conduct when performing research?

SUGGESTED READINGS

Hiney, M. (2015). *Research Integrity: What It Means, Why It Is So Important and How We Might Protect It*. Brussels: Science Europe.

Israel, M. (2014). *Research Ethics and Integrity for Social Scientists: Beyond Regulatory Compliance*. Thousand Oaks, CA: Sage Ed.

Macfarlane, B. (2010). *Researching with Integrity: The Ethics of Academic Enquiry*. New York: Routledge.

Social Research Association. (2003, November 20). Ethical Guidelines. Retrieved from http://the-sra.org.uk/wp-content/uploads/ethics03.pdf

REFERENCES

Arrison, T. S. (2014). *Responsible Science: Ensuring the Integrity of the Research Process*. Washington, DC: National Academy of Sciences.

Barrett, M. (2006). Practical and ethical issues in planning research. In G. Breakwell, S. Hammond, C. Fife-Schaw, & J. A. Smith (Eds.), *Research Methods in Psychology* (3rd ed., pp. 24–48). London: Sage.

Bell, E., & Bryman, A. (2007). The ethics of management research: An exploratory content analysis. *British Journal of Management*, *18*(1), 63–77.

The British Psychological Society. (2010). *Code of Human Research Ethics*. Leicester: Author.

Cadwalladr, C., & Graham-Harrison, E. (2018, March 17). Revealed: 50 million Facebook profiles harvested for Cambridge Analytica in major data breach. *The Observer*. Retrieved from www.theguardian.com/news/2018/mar/17/cambridge-analytica-facebook-influence-us-election

Caruth, G. D. (2015). Toward a conceptual model of ethics in research. *Journal of Management Research*, *15*(1), 23–33.

Dan-Cohen, M. (2012). Introduction: Dignity and its (dis)content. In J. Waldron (Ed.), *Dignity, Rank, and Rights* (pp. 3–10). Oxford: Oxford University Press.

Davis, G. F. (2015). Editorial essay: What is organizational research for? *Administrative Science Quarterly*, *60*(2), 179–188.

Dingwall, R., Iphofen, R., Lewis, J., Oates J., & Emmerich, N. (2017). Towards common principles for social science research ethics: A discussion document for the Academy of Social Sciences, in Ron Iphofen FAcSS (ed.) *Finding Common Ground: Consensus in Research Ethics Across the Social Sciences* (Advances in Research Ethics and Integrity, Volume 1), Bingley, UK: Emerald Publishing Limited, pp. 111–123.

Dotterweich, D. P., & Garrison, S. (1997). Research ethics of business academic researchers at AACSB institutions. *Teaching Business Ethics*, *1*(4), 431–447.

Drenth, P. J. (2012). A European code of conduct for research integrity. In T. Mayer & N. Steneck (Eds.), *Promoting Research Integrity in a Global Environment* (pp. 161–168). Singapore: World Scientific Publishing.

General Assembly of the World Medical Association. (2014). World Medical Association Declaration of Helsinki: Ethical Principles for Medical Research Involving Human Subjects. *The Journal of the American College of Dentists*, *81*(3), 14.

Gupta, A. (2013). Fraud and misconduct in clinical research: A concern. *Perspectives in Clinical Research*, *4*(2), 144.

Hansson, S. O. (2011). Do we need a special ethics for research? *Science and Engineering Ethics*, *17*(1), 21–29.

Hiney, M. (2015) *Research Integrity: What It Means, Why It Is So Important and How We Might Protect It*. Brussels: Science Europe.

Houston, M. (2016, November 20). *The Ethics of Research in the Social Sciences: An Overview*. Retrieved from www.gla.ac.uk/colleges/

socialsciences/students/ethics/
ethicstrainingresources/

Iphofen, R. (2011). Ethical decision making in qualitative research. *Qualitative Research*, *11*(4), 443–446.

Israel, M. (2014). *Research Ethics and Integrity for Social Scientists: Beyond Regulatory Compliance*. Thousand Oaks, CA: Sage Ed.

Kanungo, R. N. (1992). Alienation and empowerment: Some ethical imperatives in business. *Journal of Business Ethics*, *11*(5–6), 413–422.

Kopp, O. (2006). Historical review of unethical experimentation in humans. In R. McDonald & D. Yells (Eds.), *Ethics Professions* (pp. 2–11). Orem, UT: Utah Valley State College.

Macfarlane, B. (2010). *Researching with Integrity: The Ethics of Academic Enquiry*. New York: Routledge.

OECD. (2007, November 20). Global Science Forum, best practices for ensuring scientific integrity and preventing misconduct. *Report from the World Conference on Research Integrity*. Retrieved from www.oecd.org/sti/sci-tech/40188303.pdf

Open Science Collaboration. (2015). Estimating the reproducibility of psychological science. *Science*, *349*. doi:10.1126/science.aac4716

Resnik, D. B. (2011, May). What is ethics in research & why is it important? *The National*. Retrieved from www.niehs.nih.gov/research/resources/bioethics/whatis/index.cfm

Resnik, D. B., & Shamoo, A. E. (2011). The Singapore Statement on Research Integrity. *Accountability in Research*, *18*(2), 71–75.

Rosenberg, M., Confessore, N., & Cadwalladr, C. (2018, March 17). How Trump exploited the Facebook data of millions. *The New York Times*. Retrieved from www.nytimes.com/2018/03/17/us/politics/cambridge-analytica-trump-campaign.html

RRBM Responsible Research in Business & Management. (2017). A vision of responsible research in business and management: Striving for useful and credible knowledge. Retrieved from www.rrbm.network/position-paper/.

Schwartz, B., & Sharpe, K. (2010). *Practical Wisdom: The Right Way to Do the Right Thing*. New York: Penguin.

Shamoo, A. E., & Resnik, D. B. (2015). *Responsible Conduct of Research*. Oxford: Oxford University Press.

Sison, A., Ferrero, I., & Guitián, G. (2016). Human dignity and the dignity of work: Insights from Catholic social teaching. *Business Ethics Quarterly*, *26*(4), 503–528.

Social Policy Association Guidelines on Research Ethics. (2009, November 20). Retrieved from www.social-policy.org.uk/downloads/SPA_code_ethics_jan09.pdf

Social Research Association. (2003, November 20). Ethical guidelines. Retrieved from http://the-sra.org.uk/wp-content/uploads/ethics03.pdf

Sulmasy, D. (2008). Dignity and bioethics. History, theory, and selected applications. In the President's Council on Bioethics, *Human Dignity and Bioethics* (pp. 469–501). Washington, DC: The President's Council on Bioethics.

Sutrop, M., & Florea, C. (2010). *Guidance Note for Researchers and Evaluators of Social Sciences and Humanities Research*. European Commission.

Tsoukas, H., & Cummings, S. (1997). Marginalization and recovery: The emergence of Aristotelian themes in organization studies. *Organization Studies*, *18*(4), 655–683.

The UK Data Service. (2012–2017, November 20). Legal and ethical issues. Retrieved from www.ukdataservice.ac.uk/manage-data/legal-ethical

Wible, J. R. (2016). Scientific misconduct and the responsible conduct of research in science and economics. *Review of Social Economy*, *74*(1), 7–32.

Wilkinson, T. M. (2002). Last rights: The ethics of research on the dead. *Journal of Applied Philosophy*, *19*(1), 31–41.

16 Interdisciplinary Research

Rachel Adams Goertel

James P. Goertel

Mary G. Carey

16.1 What Does Interdisciplinary Research Mean?

We'd like to say that interdisciplinary research (IDR) is more enjoyable, less stressful, and more rewarding than individual research pursuits. Many hands make light work, right? Well, not always. This chapter of about 6000 words, with no needed data collection, statistical analysis, or relevance of findings, still required for our small collaborative team eighteen hours of face-to-face discussion, eighty-eight texts, seventy-six emails, four Skype conferences, two heated debates, and one extremely tense conversation about shared workload. No, it wasn't easier than working alone. However, we had agreed ahead of time on a conflict resolution strategy (which we'll talk about later). So, our collaboration, albeit a bit bumpy, was successful and led to both a deeper understanding about IDR and a more focused interdisciplinary goal of that shared knowledge in order to complete this chapter.

For some time now, research has been moving toward collaboration, especially in the social and behavioral sciences. The investigation and revelation of knowledge is shifting from individual efforts to group work, from single to multiple institutions, and from national to international. There is a clear expectation from our academic institutions to collaborate, interface, cooperate, join forces, coproduce, partner, or co-act with one another. Whether in education, nursing, business, psychology, biology, or any of the other dozens of disciplines, conducting research collaboratively is strongly encouraged. The interdisciplinary process involves collaboration, framing the right question, information searches, and knowing how to utilize that information, and is, thus, particularly well suited to the contemporary students. Klein (1990) led the early exploration of modern interdisciplinary studies, recognizing both the advantages and barriers. Davis (1995) wrote *Interdisciplinary Courses and Team Teaching: New Arrangements for Learning*, which specifically addressed interdisciplinary, team-teaching best practices. Augsburg (2005), Klein and Newell (1996), Repko (2008), and Szostak (2002) are among others who have made considerable contributions to the understanding and application of interdisciplinary research. Collaboration and integration, they believe, are the keys to scientific progress. Interdisciplinarians are usually academics involved in research that surpasses the typical disciplinary boundaries; while focusing on research within their respective disciplinary boundaries, they utilize concepts and techniques from other disciplines as well.

IDR focuses on particular problems or questions that are too comprehensive to be answered satisfactorily by any one discipline. This interactive process leads to the blurring of disciplinary lines and expertise because of the need to be informed and influenced by other disciplines. A pure IDR process results in the emergence of new knowledge and methodologies being shared across varied disciplinary sciences (COSEPUP, 2004; Nissani, 1997). The IDR process is collaboration beyond a disciplinary focus – it is a procedure that requires high levels of communication and knowledge sharing (Stokols, Harvey, Gress, Fuqua, & Phillips, 2005). In fact, IDR leads to new disciplines. Chemical engineering was birthed from IDR in the 1920s:

> A plant was needed that was large (to meet demand), controllable (for quality and safety), and economically viable (to compete). So experts from peripheral, but relevant, disciplines worked together. Civil engineers designed pressure-safe reaction vessels and plants, including water supply and waste treatment facilities. Mechanical engineers designed pumps, valves, piping, heat exchangers, etc. Materials engineers (actually, metallurgists at the time) designed corrosion-resistant alloys. Electrical engineers designed sensing, monitoring,

> and control systems. And, of course, chemists developed the necessary reactions. (Messler, 2004, p. 46)

Though an integral aspect of the world, chemical engineering no longer seems like a new and innovative discipline nor do other "once" innovative disciplines such as computational finance. As Szostak reminds us, "scientific breakthroughs usually reflect creative integration" (2007, p. 11). Consider the recent disciplines to surface from collaborative projects: Behavioral economics, biotechnology, nanoscience, and graphic communication. What used to be interdisciplinary collaboration has evolved into particular and niche disciplines. Nevertheless, although interdisciplinary collaboration is increasingly recognized as a vital component of research, all interdisciplinary team projects should not be expected to have innovative and drastic implications to humanity. "IDR must involve a *process* of integrating across the insights generated from disciplinary theories and methods" (Szostak, 2007, p. 11). Furthermore, IDR is central to future competitiveness in academia, science, and business because knowledge, creation, and innovation commonly occur at the interface of disciplines.

There are quite a variety of definitions of IDR, but a common characteristic of many of the specific definitions is a focus on both the integrative nature of IDR and its problem-solving orientation. For the sake of the discussion presented in this chapter, we'll first clarify the meaning of IDR. In a report entitled *Facilitating IDR*, the National Academies defined IDR as a mode of research by teams or individuals that integrates information, data, techniques, tools, perspectives, concepts, and/or theories from two or more disciplines or bodies of specialized knowledge to advance fundamental understanding or to solve problems whose solutions are beyond the scope of a single discipline or area of research practice (COSEPUP, 2004). Thus, IDR researchers are a community of scholars who rely on individual knowledge but who share a set of guiding questions, concepts, theories, and methods. The National Institutes of Health (NIH) validates this movement in its statement that IDR has the potential to improve communication and accelerate dis-coveries and their translation to practice (Morgan et al., 2003). In other words, rather than generating new knowledge, IDR efforts are often related to the implementation of research results or recommendations, also called implementation science.

Because IDR involves integration across disciplinary theories and methods, guidance as to how to proceed through this research process must differ from guidance on specific disciplinary specialization research. Szostak (2007) fully details the approach to successful IDR collaboration.

Here, we present an abbreviated outline of his suggestions:

- The IDR researcher must start by asking an interdisciplinary question. This may seem a simple step, yet it cannot be known if a question can be answered well within one discipline unless there is an understanding of the strengths and weaknesses of different disciplines.
- The next step(s) must involve the gathering of relevant disciplinary insights. There are two complementary strategies: One is to reflect on the nature of different

disciplines and identify those that are likely to have something to say about the issue of concern – on the interdisciplinary question posed. The second is to ask what phenomena, theories, or methods are implicated, and then ask which disciplines study each phenomenon identified and/or apply each theory or method.

- The IDR researcher must then critically examine the interdisciplinary question posed, exercising critical thinking skills such as distinguishing assumptions from arguments and evidence from assertions.
- The IDR researcher establishes common ground that would integrate the insights of various disciplines. This calls for creativity, intuition, and inspiration.
- The interdisciplinarians should then reflect on the results of the research into the posed question and contemplate the biases that might have crept into the work. They should seek to identify ways in which their integrative understanding might be tested. Finally, they should disseminate the results in a format that is accessible to multiple audiences.

Combining the knowledge and expertise of multiple disciplines yields outcomes with a synergistic nature. Within the diversity and richness of interdisciplinary exchange, new questions may arise that have not been asked in one's field. IDR can be an extremely productive and rewarding pursuit that can provide both innovation and a platform for conversations that lead to new knowledge. The concept of interdisciplinarity involves the integration of disciplines, as well as a focus on shared problems, topics, or questions. Individ-uals across disciplines work together to find commonalities that lead to new definitions for existing concepts and allow for new applications of existing concepts.

16.2 How Is an Effective Team Assembled and Managed?

IDR is more than the act of working with scientists from other disciplines; it is a process that transcends subject matter expertise.

(Klein, 1990)

So, you have a great idea for an IDR project. Perhaps you are a sociologist interested in the perceptions of rehabilitated addicts in service-industry jobs. Or, you are in ecology and would like to examine the financial feasibility of permanent ice fishing facilities. Or maybe you are a psychologist looking to explore perceptions of immigrant parents who do not share the same language as their child's teachers. All of these queries would certainly involve IDR. Nevertheless, however brilliant your idea may be, finding the right people, with the right qualifications, with the right work ethic, and with the right personality may seem like an almost impossible task. Understanding this is the first step. Assembling a good team means just that: Good. Everyone won't be perfect. However, together, each member's strengths should outweigh any weakness so the team can pull together and produce the desired outcome.

There is undoubtedly a need to continue to inspire researchers to combine a rigorous disciplinary depth with the ability to interface with other disciplines to further understanding. The key component of interdisciplinary collaboration is similar to traditional research:

It is a process to produce new knowledge, deepen the understanding of a topic or issue and implement changes into practice. Importantly, it should be understood that IDR often involves more resources and requires more time, effort, and imagination than single discipline research, but the rewards can be substantial in terms of advancing the knowledge base and helping to solve complex societal problems. Remembering the four components to an effective IDR team is essential – strong leadership, a well-comprised interdisciplinary team, shared goals, and effective communication are the keys to an efficacious endeavor. The conduct of research through interdisciplinary collaboration adds the benefits of intellectual depth and greater likelihood of gaining that sought after knowledge.

16.3 Strong Leadership

Once an interest in an interdisciplinary collaborative project has been identified, strong leadership is the foundation of success. As practitioners, we have observed that some individuals are natural leaders, while others may be effective project executioners. Identifying a strong leader may take insight from all the participants. A principal investigator (PI) is the primary individual responsible for the preparation, conduct, implementation, and dissemination of a research project while complying with applicable laws and regulations and institutional policy governing the conduct of research. Strong leadership can also be in the form of co-principal investigators (Co-PIs). The selected Co-PI should have the traditional skill set of any successful leader: Competence, high integrity, the ability to motivate, effective communication skills, and the ability to mentor others. Furthermore, coordinating efforts of a diverse team requires credibility as a lead researcher, skills in managing personalities, the ability to give and receive criticism, the insight and know-how to draw out individual strengths, and the ability to build and sustain team confidence. IDR leaders need to be especially strong in the areas of interpersonal skills, negotiation, time management, and conflict management. All of these qualities of leadership are completely interdependent and cannot stand alone. The strength of an IDR project ultimately falls on the leader's ability to assemble a strong, collaborative team.

16.4 Selected Team Members

Identifying fellow researchers is often the biggest obstacle in starting an IDR project. We know that personalities do matter. A congenial, interpersonal climate offers the potential to maximize productivity and the quality of the IDR. The PI/Co-PIs must select team members who demonstrate the range of skills and expertise needed to undertake and achieve the shared project goals. For individuals to work effectively together in any setting, respectful relationships need to be fostered and maintained. Koch, Egbert, and Coeling (2005) describe optimal relationships in collaborative teams as respectful, accepting, and trusting. For mutual respect to be earned, fellow researchers must reconcile discrepancies in perspectives, priorities, and processes by actively listening to their colleagues' perspectives, contributions, and anxieties. They must convey empathy and maintain pathways for open communication. Mutually respectful working relationships are essential to an interdisciplinary team's appreciation of each other's contributions and shared enthusiasm to fulfill scientific objectives. In addition, competent, capable skills are equally important in a fellow researcher. In the initial stages of a collaborative IDR project, certain core skills may be required of all participants. However, as the nature of the project evolves, the set of skills to continue success may need to be augmented. Thus, strong leaders will seek team researchers who have sufficient skill to investigate initial issues but sufficiently broad experience to identify strategies and revise applications for successful ongoing research. Clarification of roles and expectations, such as what needs to get done, when, how, and by whom, increases the probability of success. Thus, strong leaders should recruit members who have competent skills in an area that is central to achieving the project's goal but who also have a breadth of knowledge and/or experience that will inform the findings of other team members.

With the above understanding of the qualifications for an IDR partner(s), how do you find your team? Networking: It is not just who you know but who knows those you know. Habitually attending campus faculty activities, such as Lunch & Learn, Grand Rounds, and Brown Bag, provides opportunities to meet other active researchers. Consider presenting your idea at one of these venues to inspire others to join your project. Begin by writing an abstract that outlines your research goal. Keep it succinct, but include a clear purpose. Academic relationships are especially interconnected and dynamic. Begin with researchers and faculty with whom you have worked. Share your idea. Be direct in asking for specific recommendations in regards to a potential IDR team member. You can also go to the departments where your collaborative need would be based. Identify the contact person and bring your idea and your credentials. Great teams often start when tracking down the potential members in person. Remember, if you can successfully penetrate influential networks within your institution, you may be more likely to find success.

16.5 Optimizing Attributes of Interdisciplinary Collaboration: Strong Working Relationships

Once you have identified potential team members, you should initially assess the ability to successfully collaborate. Meet together for a preliminary IDR evaluation. Talk about the project, individual professional goals, shared goals, obligations, funding, etc. Strive to be open with each other in the first meeting because personalities and communication styles drive all matter. These characteristics need to be revealed at the beginning. Also, keep in mind that, although potential team members may be a terrific fit and very upbeat about the collaboration, if they are over-obligated, they will not be able to contribute to your project and, in some cases, may derail your efforts. For example, all NIH grants are reviewed for scientific merit. But the proposal also requires a determination of the feasibility for the PI and Co-PI to conduct the work given their obligations. Those who do not seem to fit well within the team parameters or are over-obligated should not be included in further team discussion. Emphasizing that it is critical that all existing and potential members of the group share the vision and purpose, the National Network for Collaboration (NNC,

1998) has developed a framework for collaboration. NNC posits that several catalysts may initiate collaboration including a problem, a shared vision, or a desired outcome. An IDR team should employ various strategies to cultivate effective ongoing collaboration. Again, strong effective leadership is a key feature to a successful project. The PI keeps the team on task, keeps the communication flowing, keeps the workload fair, and provides clarity in completing and prioritizing ongoing tasks.

Good interdisciplinary collaborators are likely to be open-minded, willing to learn from other disciplines, and have a broad appreciation for the culture of different disciplines. These characteristics should be evident from the preliminary meeting. Be sure that every potential team member gets a chance to speak and to respond. All participants' individual goals should be discussed and revised to fit the scope of the project. An effective interdisciplinary project will be goal oriented but requires a degree of flexibility since research objectives may need to change as the project proceeds. Revisiting goals throughout the project is important. Reflective practice among team members is essential for effective communication. Thus, team members should be responsive and willing to engage in open discussion of the project's evolution, displaying a willingness to give and take both responsibility and credit. Some themes identified as the competencies of a good interdisciplinary team (Nancarrow, Booth, Ariss, Smith, Enderby, & Roots, 2013) are summarized below:

1. Identifies a leader who establishes a clear direction and vision for the team, while listening and providing support and supervision to the team members.

2. Incorporates a set of work values that clearly provide direction for the team's service provision.

3. Demonstrates a team culture and interdisciplinary atmosphere of trust where contributions are valued and consensus is fostered.

4. Ensures appropriate processes and infrastructures are in place (e.g., referral criteria, communications infrastructure).

5. Utilizes communication strategies that promote intrateam communication, collaborative decision-making, and effective team processes.

6. Provides sufficient team staffing to integrate an appropriate mix of skills, competencies, and personalities to enhance smooth functioning.

7. Facilitates recruitment of staff who demonstrate interdisciplinary competencies, including team functioning, collaborative leadership, communication, and sufficient professional knowledge and experience.

8. Promotes role interdependence while respecting individual roles and autonomy.

9. Facilitates personal development through appropriate training, rewards, recognition, and opportunities for career development.

The flexibility, adaptability, and innovation of interdisciplinary work make it ideal for solving complex problems (Harris & Holley, 2008). An IDR team needs care in the form of communication based on reflection and thoughtful revision through the entire project. Keep in mind that building and maintaining an effective IDR team is a complex and challenging process. IDR benefits from a research team who can engage in interdisciplinary synthesis in order to develop more complete pictures. While disciplinary depth is essential for investigating complex issues, IDR also requires what Howard Gardner calls a "synthesizing mind" (2006, p. 3). Gathering people with these qualities will surely yield better-informed interdisciplinary discussion.

16.6 Communication

If you have worked with others, you know that poor communication can create uncertainty that leads to stress and conflict. Conversely, the ability to communicate more effectively across disciplines fosters collaboration and innovation. Being able to communicate the relevance and impact of ideas constructs support for disciplines, promotes understanding of the relevance of interdisciplinary research to society, and encourages more informed decision-making at all levels – from government, to academe, to communities, to industry, to individuals. Ultimately, effective IDR teams will develop effective communication patterns which help teams improve and develop rewarding working relationships. It is important to remember that communication is not just verbal. Although spoken words contain critical content, their meaning is influenced by the style of delivery, which includes the way speakers use nonverbal cues. Furthermore, crucial information is often shared via handwritten notes, emails, or text messages, which can lead to serious consequences if there is misinterpretation. Thus, differences in culture and communication should be expected. Some cultures prefer to talk in depth and discuss all aspects of the project, whereas other cultures may prefer to listen and reflect internally before sharing ideas and responses. Be tolerant of these differences. Time commitments are also cultural. For example, Americans may be frustrated when their French colleague is on the Riviera for "holiday" for the entire month of August. Therefore, it would be important for the French researcher to communicate about his planned absence in August.

Communication is key – from making oneself understood to trusting people we do not know, do not typically work with, or do not easily understand. The PI should keep open communication among all team members and quickly rectify any miscommunication that arises. The PI should regularly and actively solicit feedback to ensure an atmosphere of safe and open communication. Researchers involved in interdisciplinary projects may find barriers in specific areas of communication. The language and culture of one discipline could be much different from that of another. Social cues can vary, and the methods, data analysis, and even the reporting of findings could result in a breakdown of communication as an interdisciplinary team hashes out the final manifestations of a research project. For example, the PI should identify the first author so that author can take the lead, ensuring the writing is cohesive with a consistent voice throughout. Then, the first author can distribute the writing for team review. Team writing, where each person is assigned a section, can lead to confusion and fracture. Ongoing collaborations may consider swapping around first authorship of different projects to help balance workloads and fairly distribute recognition. Regardless of the types of challenges, good communication, both verbal and nonverbal, should revolve around key elements, such as giving undivided

attention, taking time to listen, being mindful of nonverbal cues, and creating an atmosphere of openness.

We all know the expression, "You are hearing but you are not listening." Barriers to effective listening can be external or internal. External distractors can include cellphone ringing/texting, background noise, or an uncomfortable temperature or seating. Be aware of external distractions and avoid them, if possible, by arranging a routine time to meet, well ahead, so team members are prepared and committed to the timeframe. Internal distractions are more complicated to control. Internal distractions include apathy, defensiveness, hearing only facts and not feelings, not seeking clarification, resistance to change, or automatic dismissal of ideas or suggestions. These behaviors pose challenges and may require a level of subtlety to address. Modeling an attentive nature and good listening skills is a start. Encourage listening that keeps an open mind. Maintain eye contact that shows interest. Promote listening for the central themes. Consider the speaker's nonverbal actions, and definitely put away that cellphone during your collaborative meeting.

Inevitably, communication breakdowns occur. These clashes of opinion may simply represent gaps in knowledge, theory, methodology, philosophy, or application. Identifying the root cause of miscommunication can go a long way in quickly getting the interdisciplinary team back on track. Remember that your IDR team members will experience an increase in morale, trust, productivity, and commitment to the project if they are able to communicate and if what they communicate is heard and validated.

16.7 Challenges in Working across Disciplines

The more researchers collaborate, the greater are the chances for success and problem-solving. However, practical challenges will arise. In writing this chapter, our IDR team was no different. Because working with others on a consequential project inevitably invites conflict, the end of this section suggests some strategies you may find useful in resolving issues.

Research on collaborative processes points to several issues that can contribute to a greater failure rate for IDR than for individual research. Latané, Williams, and Harkins (1979) examined social loafing, which is the reduction in motivation and effort when individuals work collectively. This is sometimes seen as a factor when groups are less productive than the combined performance of their members working as individuals. "Although diffusion of responsibility and social loafing may diminish the efforts of the individuals involved in the collaborative research, the total contribution of the group usually is greater than what any one individual could do alone" (p. 831). Researchers considering IDR should be aware of the obstacles they may face. As discussed earlier, lack of appreciation for the numerous components of different disciplines within an IDR project could be a cause of conflict. Varied disciplines have distinct epistemologies and cultures that define the way researchers conduct, disseminate, and evaluate research. Challenges will vary from interpersonal to practical. For example, the difference between a qualitative and a quantitative approach can lead to serious dissention, and a compromise of using a mixed-methods approach, distasteful to all. It

is no less complicated to resolve an issue with a team member in regards to determining an interdisciplinary theoretical framework than it is when faced with the realization that one team member may insist on APA citation format while another insists on MLA. To ignore these differences in opinion invites failure.

Sá (2008) discusses the challenge issued by the National Academies of Sciences and the Association of American Universities to restructure the traditional disciplinary specialization of most institutions of higher learning. Sá contends that separating disciplines into specialties creates departmental silos and results in fragmentation of knowledge because researchers fail to communicate across disciplines. Furthermore, departments within each academic field naturally control curriculum and programming. This decentralization of control impedes attempts to incorporate interdisciplinarity into research. Universities are divided by disciplines, with each area functioning independently, so the focus tends to be on individual achievements (Brewer, 1999). An academic culture that preserves the traditional structures of colleges and universities may discourage faculty from engaging in interdisciplinary collaboration unless administration rewards those who collaborate across disciplines (Sá, 2008). Harris and Holley (2008) found that leaders (deans, chairs, etc.) play a key role in fostering IDR and setting the tone for interdisciplinary work by making interdisciplinary research a priority and harnessing resources from donors and external supporters. However, administrators often struggle to fit interdisciplinary work into existing institutional structures and may not be sincere in promoting such projects over discipline-specific research. The desire to keep these long-time arrangements and to adhere to traditional research can make innovations, like interdisciplinary research, difficult to implement (Gumport & Snydman, 2002). Furthermore, for researchers seeking tenure and promotion, individual authorship may carry more merit than an IDR project. In the humanities, authorship is still decided on contribution, where, in some sciences, authorship is actually alphabetical making everyone on the team equivalent for true IDR! The benefits for promotion-seeking researchers are significant in resume building. Conversations regarding the amount of contribution and authorship should be expected and revisited throughout the project.

Conflict among coworkers is not uncommon. Nevertheless, unresolved, long-running conflicts result in antagonism, breakdown in communications, inefficient teams, stress, and low productivity. Communication breakdowns may periodically occur when team members are not cooperative, lack commitment, or do not meet expectations or deadlines. Furthermore, during meetings, members may be sarcastic, display boredom, blatantly make judgemental comments intended to challenge the integrity and expertise of another teammate, or show a general lack of respect. Additionally, power struggles may emerge and be manifested in disagreements and arguments. These issues are not unique to IDR, but, of course, you want to avoid them. The PI should quickly identify team members who begin to show signs of dissatisfaction. Keep in mind that the tone of communication usually starts at the top. The cause of the problem should be identified and explored before escalation occurs. If the PI communicates openly and clearly, his/her approach tends to filter down throughout the organization. In short, unresolved conflicts make people resentful and unhappy. Avoid exacerbating conflict by keeping communication open. Ultimately, the PI may have to conclude that certain team members are too disruptive or lack significant contribution to merit continued team membership. Such conversations, though uncomfortable and difficult, are the responsibility of the PI and should occur early on to correct the direction of the collaboration.

Finally, it is important to note that, while the unique perspective each individual brings to an IDR project is one of the features that potentially make it distinctive and valuable, research shows that group members spend much more time discussing shared information (information already known by all IDR team members) than they do unshared information (information uniquely held by one team member in relation to his specialty) (Wittenbaum & Park, 2001). Clearly, this syndrome may limit the depth of some aspects of interdisciplinary projects. As an IDR team, discuss this beforehand, recognizing which areas will be focusing on breadth and which areas on depth. At this time, sincerely reflect on the interdisciplinary nature of your project. Sometimes, what disciplines think is IDR is not. The mere addition of researchers from various disciplines or with different academic and professional credentials is not sufficient to make a research effort interdisciplinary. Analysis of the conceptual framework, study design and implementation, data analysis, and conclusions can be used to establish the true degree of interdisciplinarity. Researchers', faculty's, students', and others' decisions about whether and when to participate in IDR may be influenced by various contexts and cultures including the organization/department, the college, the institution, or even the social community. Recognition, formal rewards, and incentive structures, although generally focusing on individual research contributions, are beginning to recognize interdisciplinarity. Some universities have merged disciplinary departments and forged partnerships with government, industry, and business. Nevertheless, many researchers understand the rigidity of disciplinary structures and are reluctant to move to an interdisciplinary mode of working. Combining methods of different disciplines remains a challenge, and the lack of recognition or value for team science can deter faculty members from pursuing IDR activities.

This chapter began by mentioning the one tense conversation among the three of us who collaborated on this chapter. The above section discussed challenges you may encounter with your own IDR project. An agreed-upon conflict resolution strategy may be useful in avoiding escalation of conflicts. In our instance of writing this chapter, we adhered to six guidelines, agreed upon at the start:

- Take a day or two to reflect on the conflict
- Deal with conflict soon after
- Let the each person talk (finish sharing his/her side)
- Listen sincerely (avoid making mental counterpoints)
- Consider seeking a trusted arbitrator
- Find a compromise.

There is no guarantee that the method described here will resolve your conflict. What is important, though, is realizing your team will inevitably face conflict and having some type of protocol in place. Our PI invited suggestions for our conflict resolution strategy at the preliminary meeting, so we all felt committed to the process. In some cases, just referencing the protocol and that initial conversation may be enough to shake sense into team members and get them back on track. Leading a team will require the PI's fortitude and tact in having the ability to maintain members' commitment to the conflict resolution strategy and the commitment to shared goals.

16.8 Promoting IDR

A pattern is beginning to emerge that offers insight into the values of IDR: most successful IDR attracts the very best researchers because they are the individuals who are experienced and capable of taking knowledge from their domain and applying it to problems outside their field. Many other researchers are beginning to conduct IDR because they have recognized the limitations of their disciplinary perspective and are seeing the results of interdisciplinary endeavors. A vision of IDR may begin with basic steps and actions that foster the practice of collaboration. For example, a university could create more opportunities for faculty to work with other faculty, students, and postdoctoral faculty within different disciplines. Also, a university might give IDR high priority in its financial allocations or its fundraising and help to make the case with outside funding sources to support interdisciplinary projects. In 2003, the National Academies launched a program to realize the full potential of IDR. The National Academic Keck Futures Initiative was created to "stimulate new modes of inquiry and breakdown the conceptual and institution barriers to IDR that could yield significant benefits to science and society"

(COSEPUP, 2004). This forty million dollar grant is still providing opportunities to conduct IDR. IDR must increasingly become the standard rather than the exception to continue the public and private funding and the social and academic support for ongoing success.

In 2004, The Committee on Science, Engineering and Public Policy (COSEPUP) reviewed the state of IDR and education. The resulting report is a call to action with recommendations for academic institutions, funding organizations, and professional societies. Below are the major recommendations:

Academic Institutions:

- Develop ways to fund graduate students across departments.
- Develop joint programs and internships with industry.
- Provide training opportunities that involve research, analysis, and interactions across different fields.

Funding Organizations:

- Focus calls for proposals around problems rather than disciplines.
- Support universities to provide shared resources across disciplines.
- Provide seed funds to allow researchers across disciplines to develop research plans.

Professional Societies:

- Host workshops on communication and leadership skills needed to foster interdisciplinary teams.
- Promote networking to establish interdisciplinary partners and collaborations.
- Establish special awards that recognize IDRers or research teams.

Colleges and universities are heeding the call for interdisciplinary options. To bridge the gap between the humanities and sciences, in 2015, Johns Hopkins University's Krieger School of Arts and Sciences launched an interdisciplinary major that gives undergraduates an opportunity to pursue the natural sciences and humanities rather than having to choose between the two (Hub Staff Report, 2015). The Medicine, Science, and Humanities major is offered to students wanting to explore medical and scientific issues through the lens of humanities studies. Whether developed through the support of private funding agencies or the leadership of a university, IDR illustrates the breadth of what can be achieved when disciplines come together to explore issues and solve problems.

Faculty, researchers, and students can be very influential in receiving grants from other sources than the university, such as those mentioned above, and by exploring information about local businesses, product development, and applied research fields. To ensure the future of IDR for solutions to complex problems, continued education, funding, and training are essential to prepare the generations of investigators to tackle these interdisciplinary tasks. Without the cooperation of researchers in several different fields, many of today's important discoveries would not have been possible.

16.9 Funding Opportunities

A commonly held belief is the widespread perception that interdisciplinary research projects are less likely to be funded than single discipline research. However, funding is on the rise and federal funding agencies, such as the NIH and the National Science Foundation (NSF), give high priority to interdisciplinary research that explicitly "bridges" between and among disciplines. NIH has recognized the value of IDR and is encouraging participation: "The goal of the Common Fund's Interdisciplinary Research (IR) program was to change academic research culture such that interdisciplinary approaches and team science spanning various biomedical and behavioral specialties are encouraged and rewarded" (National Institutes of Health, 2017). NIH has identified interdisciplinarity as an essential strategy to discover new knowledge and made it an explicit priority. NSF recognizes that the "integration of research and education through interdisciplinary training prepares a workforce that undertakes scientific challenges in innovative ways" (Bililign, 2013, p. 83). Also, those conducting IDR can, at times, receive priority over single discipline proposals (Holley, 2009). For example, NSF seeks to advance IDR through a grant program known as Fostering IDR in Education (FIRE). The program aims to facilitate the innovative development of theoretical and analytical approaches needed to understand complex issues of national importance in the areas of science, technology, engineering, and math. Furthermore, NSF supports IDR through a number of solicited and unsolicited mechanisms: Solicited Interdisciplinary Programs; Areas of National Importance; Center Competitions; Unsolicited Interdisciplinary Proposals; Education and Training; and Workshops, Conferences, and Symposiums. More information regarding all of these areas can be found on the NFS's website (nsf.gov).

Not only are public monies beginning to fund IDR, but private dollars have also been invested in interdisciplinary endeavors at unprecedented levels. The Janelia Farm Research Campus was launched in late 2002. Speculated to cost $500 million and funded entirely by the Howard Hughes Medical Institute, the focus is on interdisciplinary and collaborative research for the discovery and application of cutting-edge technological tools with creativity and a high degree of scientific risk-taking. As mentioned earlier, the program created to promote IDR, the National Academies Keck Futures Initiative, received a $40 million, fifteen-year grant underwritten from the Keck Foundation. Also in 2003, funded largely by a $90 million grant from a Silicon Valley entrepreneur, the James H. Clark Center opened as the new home of the Stanford University Bio-X Program, which is designed to accelerate IDR for high-tech innovation in the biosciences.

Finally, to allow optimal access of funding dollars, it is important to target inquiries that would most benefit from interdisciplinary approaches. Only after these questions are recognized and goals are acknowledged, should specific funding sources be identified and targeted.

16.10 Conclusion

We pursue IDR with the hopes that new integrative knowledge will be produced, researchers will learn to work together, and interdisciplinary breakthroughs will, in time, contribute to innovation. IDR has led to scientific breakthroughs that would not otherwise have been possible. Only after police collaborated with biologists did a whole new discipline, forensic evidence, emerge. Additionally, the significance of a DNA match was not relevant until a third discipline was introduced – population genetics (statisticians who calculate the odds). Nevertheless, IDR can be challenging and overwhelming. It is important for researchers to strategically determine whether to invest time and energy in IDR projects or to focus on individual investigations. Overall, across disciplines, the practice of IDR is rapidly accelerating because the combination of researchers from different disciplines allows complicated problems to be solved. There is an urgent need for IDR and specific interdisciplinary training to address the pressing social, political, and economic challenges society faces. Additionally, the necessity to prepare students for an increasingly interdisciplinary, collaborative, and global future also calls for interdisciplinary exposure in post-secondary education. Despite the benefits, there are obstacles to working with IDR teams, so research leaders must emerge to provide guidance and successful completion of projects. There has been a national effort by academic institutions, funding organizations, and professional societies to better understand, support, and promote IDR across all disciplines. The move toward IDR will produce diverse interdisciplinary researchers who conduct high-quality, community-engaged, action-oriented research to drive improvements in healthcare, to solve scientific dilemmas, ensure future competitiveness, and to deepen understanding in the humanities and social sciences.

Finally, the success of IDR results from a commitment to work together toward a common goal.

Mixing such different ways of thinking stimulates a generation of new approaches to problem-solving. Consider the fable of the blind men and the elephant. Each of the six men touched a different part of the beast. One touched the elephant's side and described it as a great wall. Another touched the tusk and described the creature as sleek and hard. Another felt the tail and dismissed the elephant as nothing more than a piece of old rope. The Rajah reminded the blind men, "To learn the truth, you must put all the parts together." Only when they worked together, sharing their different ideas and experiences, were they able to discover the truth about the great creature. IDR can be tremendously valuable and satisfying when researchers work together, as it provides more opportunities for discoveries since consolidated knowledge may lead to novel insights. It presents a real opportunity to find solutions to complex problems. Collaborations are a wonderful opportunity to build long-lasting professional relationships, and an IDR experience may be one of the most rewarding and beneficial to your own pursuits of knowledge. With research now becoming increasingly global and collaborative, IDR is deemed to be the future.

KEY TAKEAWAYS

- *Find Your Team:* The ubiquitous nature of social media provides scholars with a wealth of opportunities to identify, locate, and approach prospective collaborators. Assemble an IDR team with a knowledgeable, experienced PI and like-minded, committed researchers.

- *Develop a Common Mission with Your IDR Project:* A common goal and a well-defined division of responsibilities among IDR team researchers will enable a smooth execution of the collaborative project.
- *Communication:* Constant communication and clear explanations of the goals at each stage of the project are a crucial aspect of successful IDR. Problem-solve quickly to alleviate communication breakdowns further down the line.

IDEAS FOR FUTURE RESEARCH

- Will interdisciplinary research prove to be more robust against the replicability crisis?
- What structural factors could be realistically employed to further encourage interdisciplinary research?
- How can we best train young scientists to approach their research in an interdisciplinary fashion?

SUGGESTED READINGS

Augsburg, T. (2005). *Becoming Interdisciplinary: An Introduction to Interdisciplinary Studies.* Dubuque, IA: Kendall Hunt.

Bililign, S. (2013). The need for interdisciplinary research and education for sustainable human development to deal with global challenges. *International Journal of African Development, 1*(1), 82–90.

Bozeman, B., and Boardman, C. (2014). *Research Collaboration and Team Science – a State-of-the-Art Review and Agenda.* New York: Springer.

Katz, J. S., & Martin, B. R. (1997). What is research collaboration? *Research Policy, 26*(1), 1–18.

Ross, S., and Holden, T. (2012). The how and why of academic collaboration: Disciplinary differences and policy implications. *Higher Education, 64*(5), 693–708.

REFERENCES

Augsburg, T. (2005). *Becoming Interdisciplinary: An Introduction to Interdisciplinary Studies.* Dubuque, IA: Kendall Hunt.

Bililign, S. (2013). The need for interdisciplinary research and education for sustainable human development to deal with global challenges. *International Journal of African Development, 1*(1), 82–90.

Brewer, G. D. (1999). The challenges of interdisciplinarity. *Policy Sciences, 32*(4), 327–337.

Committee on Science, Engineering, and Public Policy (COSEPUP). (2004). *Facilitating IDR.* Washington, DC: The National Academies Press.

Davis, J. R. (1995). *Interdisciplinary Courses and Team Teaching: New Arrangements for Learning.* Phoenix, AZ: American Council on Education and the Oryx Press.

Gardner, H. (2006). *Five Minds for the Future.* Boston, MA: Harvard Business School Press.

Gumport, P. J., & Snydman, S. K. (2002). The formal organization of knowledge: An analysis of academic structure. *The Journal of Higher Education, 73*(3), 375–408.

Harris, M. S., & Holley, K. (2008). Constructing the interdisciplinary ivory tower: The planning of interdisciplinary spaces on university campuses. *Planning for Higher Education, 36*(3), 34–43.

Holley, K. A. (2009). Special issue: Understanding interdisciplinary challenges and opportunities in higher education. *ASHE Higher Education Report, 35*(2), 1–131.

Hub Staff Report. (2015, January 22). Johns Hopkins adds new interdisciplinary major: Medicine, science, and humanities. Retrieved from https://hub.jhu.edu/2015/01/22/major-medicine-science-humanities/.

Klein, J. T. (1990). *Interdisciplinarity: History, Theory, and Practice.* Detroit, MI: Wayne State University Press.

Klein, J. T., & Newell, W. H. (1996). Advancing interdisciplinary studies. In J. G. Gaff & J. Ratcliff (and associates) (Eds.), *Handbook of the Undergraduate Curriculum* (pp. 393–408). San Francisco, CA: Jossey-Bass.

Koch, L., Egbert, N., & Coeling, H. (2005). Speaking of research: The Working Alliance as a model for interdisciplinary collaboration. *Work: A Journal of Prevention, Assessment, and Rehabilitation*, *25*, 369–373.

Latané, B., Williams, K., & Harkins, S. (1979). Many hands make light the work: The causes and consequences of social loafing. *Journal of Personality and Social Psychology*, *37*(6), 822–832.

Messler, R. (2004). Growth of a new discipline. *Materials Today*, *7*(3), 44–47.

Morgan, G. D., Kobus, K., Gerlach, K. K., Neighbors, C., Lerman, C., Abrams, D. B., & Rimer, B. K. (2003). Facilitating transdisciplinary research: The experience of the transdisciplinary tobacco use research centers. *Nicotine and Tobacco Research*, S11–S19.

Nancarrow, S. A., Booth, A., Ariss, S., Smith, T., Enderby, P., & Roots, A. (2013). Ten principles of good interdisciplinary team work. *Human Resource Health*, *11*(1), 19.

National Institutes of Health. (2017). Interdisciplinary Research. Retrieved from https://commonfund.nih.gov/interdisciplinary/grants

National Network for Collaboration (NNC). (1998). *Handout 6.5.C Collaboration.* Retrieved from www.uvm.edu/crs/nnco/

Nissani, M. (1997). Ten cheers for interdisciplinarity: The case for interdisciplinary knowledge and research. *The Social Science Journal*, *34*(2), 201–216.

Repko, A. (2008). *IDR: Theory and Methods*. Thousand Oaks, CA: Sage.

Sá, C. M. (2008). Interdisciplinary strategies in U.S. research universities. *Higher Education*, *55*, 537–552.

Stokols, D., Harvey, R., Gress, J., Fuqua, J., & Phillips, K. (2005). In vivo studies of transdisciplinary scientific collaboration: Lessons learned and implications for active living research. *American Journal of Preventive Medicine*, *28*, 202–213.

Szostak, R. (2002). How to do interdisciplinarity: Integrating the debate. *Issues in Integrative Studies*, *20*, 103–122.

Szostak, R. (2007). How and why to teach IDR practice. *Journal of Research Practice*, *3*(2). Retrieved from http://jrp.icaap.org/index.php/jrp/article/view/92/8924

Wittenbaum, G. M., & Park, E. S. (2001). The collective preference for shared information. *Current Directions in Psychological Science*, *10*, 70–73.

17 Cross-Cultural Research

Fons J. R. van de Vijver

17.1 What Is Cross-Cultural Research?

Historically, empirical cross-cultural research in social and behavioral sciences goes back to the 1960s and 1970s when the first attempts were made to compare cultures in a systematic manner. Much of this work finds its roots in ethnographic archival approaches and attempts to build a world map of cultures, notably Murdock's work on the Human Relations Area Files (http://hraf.yale.edu/) (e.g., Murdock, 1950). The comparative movement started off in different sciences in this time period. There was a movement in psychology that focused on empirical data collection in different cultures. An example is work on the now largely forgotten topic of field (in)dependence, shown to be different across hunter-gatherers and agricultural societies (Berry, 1966). Survey research in sociology (Bendix, 1963) and business studies (Wright, 1970) also became increasingly comparative in this period.

The field of comparative studies has changed substantially over the years. Most work nowadays focuses on established empirical methods to collect data such as surveys and tests. Across the comparative fields, convergence has emerged about how to conduct comparative studies. In fact, empirical comparative studies in psychology, sociology, communication, education, political sciences, and business (to mention the largest fields) are often interchangeable in terms of the methods used, even if the various disciplines have their own topics and theoretical models. Thus, comparative studies of social and cultural capital are rare outside sociology and Bourdieu is not well known in psychology; analogously, comparative studies of personality are rare outside psychology. Still, common themes have emerged in studies of the last decades that also point to a few overarching theoretical frameworks that are employed in different fields. The area with probably the most overlap are studies of values, such as Hofstede's (2001) model, which, according to Google Scholar, has received around 150,000 citations (the highest of any model in comparative research). However, what links these comparative studies even more than shared models is their methods. An aspect that clearly links all comparative approaches is an awareness of the need to consider methodological problems of comparative studies, and there is some similarity across disciplines about how to deal with these issues.

One of the most characteristic features of this comparative research is its growth over the last fifty years. According to PsycINFO, in the early 1950s there were on average six publications per year with the word "cross-cultural" in the summary; this number has continually increased and is now well over 1000 per year (this number increases to about 4000 for studies with "cross-cultural" in any field in the database). In sociology (Sociological Abstracts), the number of publications was the same as in psychology in the 1950s but has increased to about 250 per year with "cross-cultural" in the abstract and about 1100 with "cross-cultural" in any search field. The same development can be seen in other disciplines. In Google Scholar, encompassing many disciplines, the number went up from about 150 per year in the 1950s to more than 50,000 in the last years. Such massive and uninterrupted growth, greatly increasing the overall increment in scientific publications over the period (Van de Vijver, 2006), is truly remarkable.

There are, in my view, two related societal developments that help to explain the growth. The first is globalization. Due to globalization, workplaces and customers have become international, and there was an increasing need for managers to deal with diversity within their own teams and among their customers. The second development is migration. Migration streams have diversified. In earlier days, migration mainly involved permanent migrants who were attracted to build up a new existence elsewhere (pull motive, often economic migrants) or who were forced out of their place of residence due to disasters like war or hunger (push motive, often refugees). More recently, this migration stream has been complemented by people going abroad as part of their study or job. There are more international students than ever before. Most university campuses have become truly multicultural. These developments are accompanied by a need to better understand and deal with diversity, to understand acculturation processes, to create psychological instruments with good psychometric properties in multicultural contexts, and to develop marketing strategies for a multicultural group of customers, to mention a few of the implications of globalization and migration for the social and behavioral sciences.

17.2 Specific Topics of Cross-Cultural Studies

In this chapter, the focus is on the quantitative comparative paradigm. The reason is that the methodological issues of comparative studies have been more thoroughly studied and documented in quantitative than in qualitative studies. If we compare the results of two groups, how can we make sure that the differences in scores can be attributed to the experimental treatment and not to uncontrolled other factors (see Wagner & Skowronski, Chapter 2)? If we collect data on a certain psychological construct, say perceived business culture, how can we make sure that cultural differences in scores refer to business culture differences and not to confounding cultural differences such as those in the nature of the companies studied in the two countries? The next section describes challenges to the validity of score inferences in cross-cultural studies and their implications for comparability of scores.

17.2.1 Taxonomy of Equivalence and Bias

Equivalence. Equivalence is a concept denoting the level of comparability of measurement outcomes across cultures (Van de Vijver & Leung, 1997). Different taxonomies of equivalence have been proposed, yet there is quite some convergence about the basics of equivalence. The equivalence hierarchy presented here does not exactly follow Stevens' (1946) influential classification of nominal, ordinal, interval, and ratio levels of measurement, although there is some similarity (see also Edlund & Nichols, Chapter 1). In cross-cultural work, the emphasis is more on comparability of constructs (e.g., Is depression the same in all cultures?), comparability of items and measurement procedures (e.g., Does this item or type of question have the same meaning across cultures?), and comparability of scores (e.g., Do items measuring the culture of an organization have the same meaning within and across cultures?). Below, I describe four types of (in)equivalence that are linked to these levels: Construct, structural (or functional), metric (or measurement unit), and scalar (or full score) equivalence.

Levels of equivalence need to be empirically confirmed; equivalence is the outcome of a study and is based on statistical evidence. It is all too easy to administer an instrument in different cultures and simply assume that scores can be compared. It is, unfortunately, still not a standard practice to test the equivalence in scores in cross-cultural studies. Boer, Hanke, and He (2018) found that, even in the *Journal of Cross-Cultural Psychology*, where authors can be assumed to be well versed in comparative research methods, tests of equivalence are only conducted in a small minority of the published articles. The situation in large-scale surveys, such as the European Social Survey (www.europeansocialsurvey.org), is much better. Equivalence testing is routinely done in such studies. Full comparability of scores is too readily assumed in most published cross-cultural work, and we know from studies that have conducted equivalence tests that simply assuming full comparability can lead to erroneous conclusions.

Construct inequivalence (or lack of construct equivalence) is observed when there is a difference in psychological meaning of a construct across cultures. Historically, construct inequivalence has been claimed on the basis of three different grounds. The first type is often seen as rather ideological and comes from proponents of emic viewpoints who excluded any type of cross-cultural comparisons as individuals and cultures mutually constitute each other, which makes all psychological phenomena context bound, thereby rendering any cross-cultural comparisons futile (e.g., Shweder, 1991). This approach originates in anthropology, where it is known as relativism (Udy, 1973). The second type of construct inequivalence is more empirically based and involves psychological constructs that are associated with specific cultural groups, such as culture-bound syndromes. There has been debate in psychology whether anorexia nervosa and bulimia nervosa are such culture-bound syndromes (e.g., Simons & Hughes, 2012). Although, for anorexia, both universality and cultural specificity (in Western cultures) have been claimed, it is clear that the prevalence is mainly restricted to affluent contexts where food is in abundance, such as Western societies and affluent groups in low- and middle-income countries. The third type of construct inequivalence occurs if tests of equivalence fail to show any cross-cultural similarity. This could be the case when factor analyses of some psychological instrument show different factors across cultural groups. Cui, Mitchell, Schlegelmilch, and Cornwell (2005) administered a scale to measure consumers' ethics, the Ethical Position Questionnaire, to participants in Austria, Britain, Brunei, Hong Kong, and the United States. Examples of items are "A person should make certain that their actions never intentionally harm another even to a small degree" and "What is ethical varies from one situation and society to another." The scale produced different factor structures in various countries. Rather than concluding that the underlying construct was incomparable across countries, the authors used structural equation modeling on the data obtained with the instrument. After the removal of various items, such as those dealing with a relativistic stance (the second example item), the scale showed cross-cultural equivalence.

An instrument shows **structural equivalence** (or construct equivalence) if it measures the same construct in all groups involved in a study. The concept has received much attention in empirical studies – clearly, if an instrument is administered in multiple groups, we want to confirm that we measure the same construct in all groups. The operational translation of the concept has shifted somewhat over the years. In the early days, exploratory factor analysis was often used; finding the same factors in all groups was seen as sufficient evidence for structural equivalence. In recent times, the use of stricter statistical procedures, such as confirmatory factor analysis, has become common. However, studies of structural equivalence could also use other approaches. Calvo, Zheng, Kumar, Olgiati, and Berkman (2012) examined the link between well-being and social capital,

measured with self-reports of access to support from relatives and friends, volunteering to an organization in the past month, and trusting others, in 142 countries. In each country, a positive association was found, which implies that the link between well-being and social capital was universally found. This finding has practical implications in that it suggests that social capital may be a relevant source for intervention and also provides theoretical suggestions by pointing to the critical role of connectedness for well-being.

The next level refers to the identity of the metric of interval or ratio scales – Likert response scales are a good example. This level of equivalence is called **metric (or measurement unit) equivalence**. It may seem unnatural or unnecessary to have a level of equivalence that checks whether the metric of a scale is identical across cultures. After all, there may not seem to be much reason to expect that measurement units may vary greatly across cultures. In practice, however, metric equivalence is an important concept. The main reason is not so much the expectation that the response scale metric will vary across cultures, but metric equivalence is often the result when the final and highest level of equivalence is not observed. It is important to note that metric equivalence means that scores can be compared within cultures but not across cultures. Metric equivalence is often found in large-scale surveys, such as the Teaching and Learning International Survey (OECD, 2014). Despite the careful wording and translation procedures, only metric equivalence is often found. This finding creates a conundrum, as participating countries are typically interested in comparing scores across countries where such comparisons are problematic, from a methodological perspective.

Scalar (or full score) equivalence means that scores can be directly compared within and across cultures and that individual and cultural score differences have the same meaning. Scalar equivalence assumes both an identical interval or ratio scale and an identical scale origin across cultural groups. It is the type of equivalence that is necessary for conducting comparisons of means in analyses of variance or t-tests. In practice, scalar invariance is often hard to obtain, notably in cross-cultural studies involving many cultures. Given this frequent absence of scalar equivalence, metric equivalence is important, as this is the level of equivalence that is found in the rather common situation that factors are identical across cultures – thereby supporting structural equivalence – but that statistical tests of scalar invariance show that the latter level is not supported. Let me give some fictitious, yet experience-based, examples to show challenges of scalar equivalence. Suppose that an item has been translated inadequately and that, in one of the languages of a personality questionnaire, the word "thrifty" has been used in one language and that the equivalent of greedy was used in another language. The items may still measure the same underlying concept, but the endorsement rates are probably influenced by the translation, as individuals may be more likely to endorse that they are thrifty than that they are greedy. Such an item will have a biasing influence on the mean scores of the cultures and will threaten the scalar equivalence. The item may reduce the equivalence from scalar to metric. Another example comes from the literature on

response styles. It is well documented that individuals in East Asian cultures prefer to use responses close to the midpoint of the scale (known as midpoint responding or modesty bias), whereas individuals in Arab and South American countries tend to prefer to use scale extremes (Harzing, 2006; He & Van de Vijver, 2015). These response styles can engender mean score differences between cultures that challenge scalar equivalence and lead to metric equivalence. As documented below, in more detail, it is important to consider the potentially confounding role of response styles when designing cross-cultural studies.

Bias. The previous section referred to challenges of comparability and implications. The second part of the framework refers to the nature of these challenges: What kinds of threats can have an influence on cross-cultural comparisons? These threats are commonly referred to as bias. So, bias refers to cross-cultural differences in scores that do not reflect corresponding differences on the underlying construct (Poortinga, 1989). For example, the translation issue about being thrifty (in Culture A) versus being greedy (in Culture B) would imply that cross-cultural differences on that item are influenced by the translation problem; observed cross-cultural score differences may confound differences in thrift and translation problems. If there is bias, cross-cultural score differences cannot be taken at face value. Below, I describe a taxonomy of three types of bias, depending on whether the bias emanates from the construct that is measured, its methods of administration, or specific items, called construct bias, method bias, and item bias (differential item functioning), respectively (see Van de Vijver & Leung, 1997).

There is **construct bias**, corresponding to construct inequivalence in the equivalence framework, if attitudes, behaviors, or other features that are characteristic for a construct in one culture are not the same across all cultures of a study. Behaviors and attitudes associated with being a (political) liberal are a good example. The concept exists in many countries, but the meaning differs quite a bit. In most countries, being liberal is associated with being left on the political spectrum, being tolerant and open in ethical issues such as abortion, euthanasia, and recreational drug usage (social liberalism); however, in countries like Australia and the Netherlands, liberal parties are on the conservative part of the political spectrum (classical liberalism). So, using the concept in a cross-cultural study should be done with caution to avoid misunderstanding of what is meant and to do justice to the link between the concept and the cultural context.

Method bias refers to problems of cross-cultural research emanating from method-related factors, which can be described as those aspects discussed in the method section of empirical papers. Three types of method bias can be distinguished, depending on the origin of the problems: The sample, administration, or instrument (any test or survey instrument). Sample bias refers to sample differences due to differences in sampling frames or in the composition of the population in the participating countries. For example, comparisons involving sub-Saharan and European countries involve samples that are usually massively different in level of affluence. Another example is the usage of social media

to compare countries. It is becoming more popular to use social media, such as Twitter or Facebook, to compare countries. As an example, Park, Baek, and Cha (2014) were interested in comparing the use of emoticons in tweets. Such cross-country comparisons have to assume that the tweets analyzed are a fair representation of nonverbal emotional expression in all countries involved. However, internet penetration rates are very different across countries, which makes it unlikely that samples from countries with small penetration (and access to Twitter being restricted to a small part of the population) are comparable to samples from Western countries with a high internet penetration. Administration bias comes from differences in test administration characteristics, such as physical conditions (e.g., ambient noise, presence of others) of home interviews, which are difficult to standardize in cross-cultural studies. Finally, instrument bias refers to differential experience with instrument characteristics, such as familiarity with types of cognitive items.

Response styles, such as acquiescent, extremity, and midpoint responding, and social desirability, are sources of method bias in cross-cultural survey research. There is extensive evidence that cross-cultural differences in response styles are systematic. First, individuals from more affluent, more individualistic, and less hierarchical countries reveal less acquiescence and social desirability. Second, extremity responding is lowest in Confucian (East Asian) cultures and highest in Arabic and Latin American groups (Harzing, 2006; He & Van de Vijver, 2015). The interpretation of response styles is controversial in cross-cultural psychology. It is taken for granted that these response styles threaten the validity of intergroup comparisons and, like other validity threats, they should be minimized. However, in cross-cultural psychology, there is also another view according to which response styles refer to preferred and culturally embedded ways of expressing oneself (He & Van de Vijver, 2013; Smith, 2004). The debate is far from settled, yet it is quite clear that conventional attempts to minimize response styles by adapting test designs, statistical analyses, or their combination are not very successful. He et al. (2017) compared common procedures to account for response styles, such as the use of anchoring vignettes, score standardization, and sophisticated statistical modeling procedures (to correct for response biases) in a cross-cultural study involving students from sixteen countries. The authors concluded that no procedure could convincingly show validity improvements or other enhanced psychometric properties. It can be concluded that "the jury is still out" about the cross-cultural meaning of response styles and about the adequacy of design and statistical procedures to correct for response styles.

Item bias, better known as **differential item functioning**, refers to problems at item level. An item is biased if respondents from different cultures with the same standing on the underlying construct (e.g., they are equally conservative) do not have the same mean score on the item (Van de Vijver & Leung, 1997). Item bias can be a consequence of many issues, involving problems in item translation or inapplicability of item content in specific cultures. Multiple statistical procedures have been proposed to identify item bias (Van de Vijver & Leung, 1997). There are procedures for all measurement levels. However,

item bias has turned out to be elusive from a conceptual perspective. The first item bias studies were published fifty years ago (e.g., Cleary & Hilton, 1968), and spawned an amazing development of new procedures and applications. Regrettably, the statistical innovations were not accompanied by new substantive insights in item writing; we do not yet know how to write items that are not susceptible to item bias (Linn, 1993). Ryan, Horvath, Ployhart, Schmitt, and Slade (2000) analyzed item bias in an employee survey, administered in a thirty-six-country study involving more than 50,000 employees. Hypotheses about bias were derived from Hofstede's (2001) dimensions. Hypotheses were better confirmed when countries were more dissimilar whereas predictions of more similar countries were often not confirmed. The positive relation between the size of global cultural differences and item bias has been observed before. Comparisons between more dissimilar countries may show both larger differences in mean scores as well as larger item bias. Meriac, Woehr, and Banister (2010) studied work ethic among Millennials, Generation X, and Baby Boomers. The authors found that many items, including those that asked about working hard (e.g., "By working hard a person can overcome every obstacle that life presents") were not perceived in the same way by the three groups. The oldest generation (Baby Boomers) endorsed most items much more than the younger generations, possibly leading to this bias. It is indeed common to find that differential item functioning tends to be larger when comparing groups with highly different mean scores (Linn, 1993).

It can be concluded that many statistical procedures are available to deal with item bias, but that the psychometric sophistication to identify bias is in no way matched by similar developments in our understanding of item bias. True progress in this field is more likely to come from a better understanding of cultural differences than from further psychometric sophistication. The high level of sophistication of psychometric bias models has led to an overemphasis of the psychometrics and an underrating of the importance of instrument design. Still, a high-quality cross-cultural study does not only depend on analysis but also on the quality of instruments. International surveys involving dozens of countries, such as the World Value Survey (WVS, www.worldvaluessurvey.org/wvs.jsp), the European Social Survey (ESS, www .europeansocialsurvey.org), and educational achievement assessments, such as the Programme for International Student Assessment (PISA, www.oecd.org/pisa) and Trends in Mathematics and Science Study (TIMMS, http://timssandpirls.bc.edu), spend much time on developing items that can be expected to show as little cultural bias as possible. However, all these projects also use state-of-the-art statistical procedures to examine cultural bias once the data have been collected. It is this need "to combine the best of both worlds" (i.e., design and analysis) that methodologically guides these studies. The next part of the chapter discusses the emerging trend to pay more attention to procedures to optimize instrument design in cross-cultural studies.

17.3 Test Adaptations

Test adaptations are an interesting way to bridge the gap between emic and etic approaches and ensure equivalence, mainly by optimizing the suitability of instruments in international comparisons or in studies in which instruments are "imported" for use in a new linguistic and/or cultural context. The concept of translation has now largely been replaced by adaptation (Van de Vijver, 2006). It is no longer obvious that close ("literal") translations are the best way of rendering instruments in other languages, as such translations tend to focus on linguistic aspects, thereby leaving out cultural and quantitative quality considerations such as internal consistencies (Hambleton, Merenda, & Spielberger, 2005; Harkness, Van de Vijver, & Johnson, 2003).

Test adaptations are common when new tests are prepared for surveys in multiple countries. During the test development stage, input from researchers in participating countries is sought so that items are suitable in all locations of the survey. Draft items are evaluated in all countries to avoid inadequacy of the items in specific countries. These administrations are often in the form of cognitive interviews (Willis, 2004), in which potential participants in various countries are asked to reflect on item content, comprehensibility, motivate their answer, and provide other information that is input to a quality analysis of a scale in a study. The above-mentioned large-scale surveys and educational achievement comparisons all use this type of procedure. The quality of items in these surveys is also enhanced by previous experience in earlier waves of these multiwave surveys. For many topics and item formats, information about cultural appropriateness is already available.

The second type of application of test adaptation refers to studies in which an existing instrument is to be used in a new linguistic and cultural context. Examples are intelligence tests, such as the Wechsler scales to assess intelligence, that are translated to new languages (Georgas, Weiss, Van de Vijver, & Saklofske, 2003). The suitability of instructions and/or items is then examined for each subtest.

A classification is proposed here that starts from four types of conceptual, cultural, linguistic, and measurement adaptations (Van de Vijver, 2015). Within each type, there are two subtypes, thereby defining eight kinds of adaptations. Related classifications can be found in Harkness et al. (2003) and Malda, Van de Vijver, Srinivasan, Transler, Sukumar, and Rao (2008).

A concept-driven adaptation is a change of the contents of a question, to accommodate differences in the indicators of culture-specific concepts, such as the 4th of July in the United States, which has equivalents in other countries – although it is not always identical what is commemorated and how it is commemorated. Theory-driven adaptations are instrument changes due to theoretical reasons. Malda et al. (2008) administered an American, short-term digit memory test among Kannada-speaking children in India. The test requires the children to repeat a series of digits, read aloud by the test administrator at a rate of one digit per second. The English version has only one-syllable digits. This choice is based on Baddeley's (1992) phonological loop model, according to which the number of items that can be stored in working memory depends on the number of syllables that can be repeated in about 1.5 seconds – more items can be recalled when these are shorter and the test will be more sensitive when shorter digits are

used. All digits in Kannada from 1 to 10 are bisyllabic, except 2 and 9, which have three syllables. Therefore, the original items were changed so that two-syllabic digits came at the beginning of the test.

The two culture-related adaptations refer to "hard" and "soft" aspects of culture. Terminological/factual-driven adaptations refer to country-specific aspects that are unfamiliar elsewhere. An example is the use of currencies (e.g., dollars or pounds), or non-metric or metric measures (gallons or liters) in PISA numeracy texts in different countries. Norm-driven adaptations are used to accommodate cultural differences in norms, values, and practices. An item about a child helping to do the dishes after dinner requires a specific cultural context that may not apply in all countries.

The distinction between "hard" and "soft" also applies to linguistic applications. "Hard" adaptations refer to specifics of languages that cannot be translated. For example, the English "you" in a questionnaire often needs to be adapted to comply with local courtesy rules, as many languages distinguish between a more informal and a more formal way to address other persons, and there is often only one culturally accepted option.

Measurement-related changes can be familiarity or recognizability driven. Such adaptations are common in cognitive tests and are meant to accommodate differential familiarity of cultures with assessment procedures for specific stimuli. For example, in international comparisons of educational achievement, such as PISA, it is important to compare curricula and topics treated before administering instruments to ensure that students in all countries have been exposed to the topics assessed. When dealing with very different cultures, controlling for differential familiarity can be exceedingly difficult to achieve. The need to make an analysis of the cultural context of the participants is not just important for educational achievement tests. A good knowledge of a culture is needed to understand whether items refer to familiar contents.

Finally, format-driven adaptations are typically used to minimize or avoid unwanted cross-cultural differences that are due to item formats. For example, differences in extreme response style (i.e., the tendency to overuse the endpoints of a Likert scale) may be reduced by using more options in Likert-type response scales (Hui & Triandis, 1989). However, the same caveat is needed as in the previous paragraph – test adaptations can reduce the impact of response styles, and an adequate test design (e.g., with both positively and negatively worded items – see Stassen & Carmack, Chapter 12) can enable an adequate statistical analysis of response style. However, it is unrealistic to expect that response styles can be eliminated by careful test design or adaptation.

17.4 Conclusion

The chapter has described methodological advances by starting from the roots of ethnographic and comparative research. Although the field of methods may look like a stagnant and slowly developing one from an outsider's perspective, the chapter has clarified progress from different perspectives. The most obvious domain of progress is the tremendous increase in number and sophistication in statistical procedures. Structural equation modeling and item bias procedures are examples of such procedures. However, there is also progress, less visible and demarcated, in the way cross-cultural research is conducted. Test adaptation procedures have often been applied. We can build on this experience when designing new studies. Various methodological guidelines for constructing cross-cultural studies have been formulated. It is, nowadays, possible to conduct and report cross-cultural research without having to explain or defend all the methodological choices made. Such established procedures hardly existed some decades ago. This development, suggesting that cross-cultural research often amounts to a certain set of prescribed, well-defined procedures, is important for the field. In psychology alone, on average, more than ten cross-cultural studies are published every day. Google Scholar lists over 50,000 studies published in 2016. With these large numbers, it is unlikely that all cross-cultural research is conducted by specialists. Rather, most current publications are written by researchers who occasionally conduct cross-cultural studies. Having a certain technology or received wisdom regarding how to conduct those studies will facilitate the quality of this research. We are quickly moving toward a situation in which conducting cross-cultural research comprises many standard components; interesting internet sites about how to conduct cross-cultural studies can be found at http://ccsg.isr.umich.edu/index.php, www.census.gov/srd/papers/pdf/rsm2005-06.pdf, and www.intestcom.org/files/guideline_test_adaptation.pdf.

Still, there is considerable room for improvement of cross-cultural research practice. Cultural and methodological sophistication are still too infrequently combined. We are better than before in combining cultural and methodological considerations procedures but still have a long way to go before integrating the two in a balanced manner. It is interesting to look back at the beginning of the cross-cultural empirical movement decades ago. Udy (1973) gave an overview of the issues comparative ethnography (and social and behavioral sciences, by implication) would have to deal with. Some of the issues he described are no longer seen as relevant. For example, he focused on large-scale cultural comparisons, such as the Human Relations Area Files, and pointed to the need to conduct random sampling of the cultures involved in a study. We seem to have become "sadder and wiser" and have given up the idea of randomly sampling cultures and usually have more modest aims (Boehnke, Lietz, Schreier, & Wilhelm, 2011). Comparisons involving small numbers of countries prevail in the literature. The choice of countries in large-scale surveys is governed by availability, which (combined with the high cost of these surveys) usually will lead to an overrepresentation of rich countries and an underrepresentation of low- and middle-income countries.

Urdy also mentioned another important characteristic of cross-cultural studies: The need to start these studies from a theoretical background. There are some popular models of cultural differences, such as individualism–collectivism (Hofstede, 2001) and affluence differences between countries (Georgas, Van de Vijver, & Berry, 2004). However, theoretical progress is nowhere near the statistical progress we made in the last decades. The statistical models and procedures for analyzing cross-cultural data are very sophisticated. It is unfortunate that the link between theory and data, so much emphasized in the comparative ethnographic models of half a century ago, has become much weaker. Still, who would have expected then that comparative studies would have become so prominent in the social and behavioral sciences? Cross-cultural encounters have become everyday experiences for many of us, and it can be expected that the interest in cross-cultural studies will continue in the near future.

KEY TAKEAWAYS

- What you see is not always what you get in cross-cultural studies. It is important to be aware of potential problems when using psychological instruments in other cultures.
- A profound knowledge of the new context will help to adapt instruments, when necessary.
- Many statistical procedures are available to test whether instruments measure consistently across cultures.

IDEAS FOR FUTURE RESEARCH

- How can we better integrate quantitative (comparative) and qualitative (ethnographic) approaches so that we can better distinguish universal and culture-specific components of constructs and their measures?
- Can statistical procedures be developed that are useful in large-scale studies and that overcome shortcomings of existing procedures (such as a poor fit in multigroup confirmatory factor analyses)?

- How can "Big Data" (such as large-scale analyses of social or other media) inform the quality of our measures (e.g., identifying good operationalizations of constructs)?
- How can we more effectively incorporate cultural factors into mainstream psychology?
- How can we better integrate methods and theory in cross-cultural research?

SUGGESTED READINGS

Iliescu, D. (2017). *Adapting Tests in Linguistic and Cultural Situations*. New York: Cambridge University Press.

Johnson, T., Pennell, B.-E., Stoop, I., & Dorer, B. (Eds.) (2018). *Advances in Comparative Survey Methods: Multicultural, Multinational and Multiregional (3MC) Contexts*. New York: Wiley.

There is a free online encyclopedia of cross-cultural psychology, *The Online Readings in Psychology and Culture*, which has various contributions on cross-cultural methods: https://scholarworks .gvsu.edu/orpc/vol2/iss2/

REFERENCES

Baddeley, A. (1992). Working memory. *Science*, *255*(5044), 556–559.

Bendix, R. (1963). Concepts and generalizations in comparative sociological studies. *American Sociological Review*, *28*, 532–539.

Berry, J. W. (1966). Temne and Eskimo perceptual skills. *International Journal of Psychology*, *1*, 207–229.

Boehnke, K., Lietz, P., Schreier, M., & Wilhelm, A. (2011). Sampling: The selection of cases for culturally comparative psychological research. In D. Matsumoto & F. J. R. van de Vijver (Eds.), *Cross-Cultural Research Methods in Psychology* (pp. 101–129). New York: Cambridge University Press.

Boer, D., Hanke, K., & He, J. (2018). On detecting systematic measurement error in cross-cultural research: A review and critical reflection on equivalence and invariance tests. *Journal of Cross-Cultural Psychology*, *49*(5), 713–734. https://doi.org/10.1177/0022022117749042

Calvo, R., Zheng, Y., Kumar, S., Olgiati, A., & Berkman, L. (2012). Well-being and social capital on planet earth: Cross-national evidence from 142 countries. *PLoS One*, *7*(8), e42793.

Cleary, T. A., & Hilton, T. L. (1968). An investigation of item bias. *Educational and Psychological Measurement*, *28*, 61–75.

Cui, C. C., Mitchell, V., Schlegelmilch, B. B., & Cornwell, B. (2005). Measuring consumers' ethical position in Austria, Britain, Brunei, Hong Kong, and USA. *Journal of Business Ethics*, *62*, 57–71.

Georgas, J., van de Vijver, F. J. R., & Berry, J. W. (2004). The ecocultural framework, ecosocial indices and psychological variables in cross-cultural research. *Journal of Cross-Cultural Psychology*, *35*, 74–96.

Georgas, J., Weiss, L. G., Van de Vijver, F. J., & Saklofske, D. H. (Eds.) (2003). *Culture and Children's Intelligence: Cross-Cultural Analysis of the WISC-III*. New York: Academic Press.

Hambleton, R. K., Merenda, P. F., & Spielberger, C. D. (Eds.) (2005). *Adapting Educational Tests and Psychological Tests for Cross-Cultural Assessment*. Mahwah, NJ: Lawrence Erlbaum Associates.

Harkness, J. A., van de Vijver, F. J. R., & Johnson, T. P. (2003). Questionnaire design in comparative research. In J. A. Harkness, F. J. R. van de Vijver, & P. Ph. Mohler (Eds.), *Cross-Cultural Survey Methods* (pp. 19–34). New York: Wiley.

Harzing, A. W. K. (2006). Response styles in cross-national survey research: A 26-country study. *International Journal of Cross Cultural Management*, *6*, 243–266.

He, J., & van de Vijver, F. J. R. (2013). A general response style factor: Evidence from a multi-ethnic study in the Netherlands. *Personality and Individual Differences*, *55*, 794–800.

He, J., & van de Vijver, F. J. R. (2015). Effects of a General Response Style on cross-cultural comparisons: Evidence from the Teaching and Learning International Survey. *Public Opinion Quarterly, 79*, 267–290.

He, J., van de Vijver, F. J. R., Fetvadjiev, V. H., Dominguez-Espinosa, A., Adams, B. G., Alonso-Arbiol, I., ... Zhang, R. (2017). On enhancing the cross-cultural comparability of Likert-scale personality and value measures: A comparison of common procedures. *European Journal of Personality, 31*, 642–657.

Hofstede, G. (2001). *Culture's Consequences: Comparing Values, Behaviors, Institutions and Organizations across Nations*. Thousand Oaks, CA: Sage.

Hui, C. H., & Triandis, H. C. (1989). Effects of culture and response format on extreme response style. *Journal of Cross-Cultural Psychology, 20*, 296–309.

Linn, R. L. (1993). The use of differential item functioning statistics: A discussion of current practice and future implications. In P. W. Holland & H. Wainer (Eds.), *Differential Item Functioning* (pp. 349–364). Hillsdale, NJ: Lawrence Erlbaum Associates.

Malda, M., van de Vijver, F. J. R., Srinivasan, K., Transler, C., Sukumar, P., & Rao, K. (2008). Adapting a cognitive test for a different culture: An illustration of qualitative procedures. *Psychology Science Quarterly, 50*, 451–468.

Meriac, J. P., Woehr, D. J., & Banister, C. (2010). Generational differences in work ethic: An examination of measurement equivalence across three cohorts. *Journal of Business and Psychology, 25*, 315–324.

Murdock, G. P. (1950). Feasibility and implementation of comparative community research: With special reference to the Human Relations Area Files. *American Sociological Review, 15*, 713–720.

OECD. (2014). *TALIS 2013 Results: An International Perspective on Teaching and Learning*. Paris: OECD Publishing.

Park, J., Baek, Y. M., & Cha, M. (2014). Cross-cultural comparison of nonverbal cues in emoticons on twitter: Evidence from big data analysis. *Journal of Communication, 64*, 333–354.

Poortinga, Y. H. (1989). Equivalence of cross-cultural data: An overview of basic issues. *International Journal of Psychology, 24*, 737–756.

Ryan, A. M., Horvath, M., Ployhart, R. E., Schmitt, N., & Slade, L. A. (2000). Hypothesizing differential item functioning in global employee opinion surveys. *Personnel Psychology, 53*, 541–562.

Shweder, R. A. (1991). *Thinking Through Cultures: Expeditions in Cultural Psychology*. Cambridge, MA: Harvard University Press.

Simons, R. C., & Hughes, C. C. (2012). *The Culture-Bound Syndromes: Folk Illnesses of Psychiatric and Anthropological Interest*. Dordrecht: Springer.

Smith, P. B. (2004). Acquiescent response bias as an aspect of cultural communication style. *Journal of Cross-Cultural Psychology, 35*, 50–61.

Stevens, S. S. (1946). On the theory of scales of measurement. *Science, 103*, 677–680.

Udy, S. H., Jr. (1973). Cross-cultural analysis: Methods and scope. *Annual Review of Anthropology, 2*, 253–270.

Van de Vijver, F. J. R. (2006). Culture and psychology: A SWOT analysis of cross-cultural psychology. In Q. Jing, H. Zhang, & K. Zhang (Eds.), *Psychological Science Around the World* (Vol. 2, pp. 279–298). London: Psychology Press.

Van de Vijver, F. J. R. (2015). Methodological aspects of cross-cultural research. In M. Gelfand, Y. Hong, & C. Y. Chiu (Eds.), *Handbook of Advances in Culture & Psychology* (Vol. 5, pp. 101–160). New York: Oxford University Press.

Van de Vijver, F. J. R., & Leung, K. (1997). *Methods and Data Analysis for Cross-Cultural Research*. Newbury Park, CA: Sage Publications.

Willis, G. B. (2004). *Cognitive Interviewing: A Tool for Improving Questionnaire Design*. Thousand Oaks, CA: Sage.

Wright, R. W. (1970). Trends in international business research. *Journal of International Business Studies, 1*, 109–123.

Part Five: New Statistical Trends in the Social and Behavioral Sciences

18 A Gentle Introduction to Bayesian Statistics

Milica Miočević

Rens van de Schoot

18.1 Overview

Bayesian statistics are usually not part of the common core of statistics courses taken in social and behavioral science graduate programs and are, consequently, not used as often as classical (frequentist) statistics. However, this trend is changing, and, nowadays, there is an increased interest in Bayesian analysis in a number of scientific disciplines, including educational science (König & Van de Schoot, 2018), epidemiology (Rietbergen, Debray, Klugkist, Janssen, & Moons, 2017), health technology (Spiegelhalter, Myles, Jones, & Abrams, 2000), medicine (Ashby, 2006), psychology (Van de Schoot, Winter, Ryan, Zondervan-Zwijnenburg, & Depaoli, 2017), and psychotraumatology (Van de Schoot, Schalken, & Olff, 2017). There are many reasons for using Bayesian instead of frequentist statistics. Researchers may be "forced into" this choice because their model cannot be estimated using other approaches, or their sample size is too small for adequate power. Alternatively, researchers might prefer a framework that allows them to incorporate relevant prior knowledge into the estimation process and allows for sequential updating of knowledge, which is in line with the cumulative nature of science. Furthermore, Bayesian methods provide an easier way to compute interval estimates of parameters that have unknown and/or asymmetrical distributions (Yuan & MacKinnon, 2009).

Null hypothesis significance testing has been criticized for decades (Bakan, 1966; Ioannidis, 2005; Rozeboom, 1960). Some researchers criticize binary decisions based on the p-value and propose using effect size measures and confidence intervals instead (Cumming, 2014). Other researchers recommend abandoning the frequentist paradigm altogether because the p-value does not quantify the probability of the hypothesis given the data (Wagenmakers, Wetzels, Borsboom, & Van der Maas, 2011), gives no measure of whether the finding is replicable (Cohen, 1994), and the $(1-\alpha)\%$ confidence intervals only tell us that, upon repeated sampling $(1-\alpha)\%$ of the confidence intervals will

This research was supported by a grant from the Netherlands Organisation for Scientific Research: NWO-VIDI-452-14-006.

contain the true value of the parameter (Jackman, 2009).

The alternative to frequentist statistics, which provides more intuitive interpretations of findings than the *p*-values and confidence intervals, are Bayesian statistics. In the Bayesian framework, it is possible to quantify the probability of a hypothesis given the data and to compute intervals for an effect for which there is (1-α)% probability that the parameter lies within the interval. Bayesian statistics implement Bayes' theorem to update prior beliefs with new data. This produces updated (posterior) beliefs about the research hypothesis and/or the parameters in a statistical model. Bayes' theorem follows from the laws of conditional probabilities, and it is not controversial by itself. The element of Bayesian statistical analysis that is often debated is how to specify prior beliefs in a way that does not bias results, which are the posterior beliefs about a hypothesis or model parameters.

The current chapter provides a gentle introduction to Bayesian statistics and its applications in the social and behavioral sciences. The reader is guided through the steps in a Bayesian analysis of the single mediator model, using data from the study of PhD delays by Van de Schoot, Yerkes, Mouw, and Sonneveld (2013). We highlight differences between frequentist and Bayesian analyses, seeking to answer the same questions, and point out the new possibilities for evaluating and reporting results offered by Bayesian statistics. The chapter is structured in a series of steps, from the development of the research question to the reporting of the results and sensitivity analysis. All syntax files for reproducing the steps in the Bayesian analysis can be found on the Open Science

Framework: https://osf.io/fud8v/?view_only = 0243cf4f4bd84af6ab73bad571e11d82.

18.2 Data and Research Question

Designing a study, from research question to collecting data, is the same for frequentist and Bayesian analyses, and is described in more detail in Chapters 1–5 of this volume. The differences between frequentist and Bayesian frameworks become apparent in the analysis, when the researcher is faced with the choice between ignoring prior information (i.e., doing a frequentist analysis) and updating prior information with the observed data (i.e., doing a Bayesian analysis).

18.2.1 Data

The data for our illustration of Bayesian analyses come from a study of PhD completion in the Netherlands. We use survey data about Dutch doctoral recipients, gathered between February 2008 and June 2009, at four universities in the Netherlands. For details concerning the full sample, see Sonneveld, Yerkes, and Van de Schoot (2010). All PhD candidates who applied for permission to defend their theses were invited to participate in the survey, and the response rate was 50.7% ($N = 565$). We focus solely on those respondents who reported their start and end date and who reported their status as being an employee ($N = 308$) or scholarship recipient ($N = 25$).

18.2.2 Research Question and Hypothesis

The Dutch system has PhD trajectories with primarily fixed durations. Consequently, a PhD project includes predetermined start and end dates, which makes it possible to compute an exact duration for the PhD, both actual and expected. All respondents indicated the length of their contract (planned PhD duration) as well as how long it took them to complete their thesis (actual PhD duration). The gap (delay) between actual and planned duration was computed using this information.

In the example analysis, the independent variable is employment status (0 = external funding; 1 = employed by the university). The mediator was the mean-centered self-reported level of the student's understanding of the societal relevance of their research. The outcome was the delay in the completion of the PhD, measured in months and operationalized as the difference between the expected completion date and the actual completion date (Figure 18.1).

For illustrative purposes, it was hypothesized that students employed by the university would report higher levels of understanding of the relevance of their research compared to scholarship recipients (*a* path, Figure 18.1). Greater understanding of the relevance of one's research was expected to reduce delay in PhD completion (*b* path). The indirect effect of being employed by the university on delay, through understanding the relevance of

one's research, was expected to be negative and different from zero (i.e., that university employees would complete their PhD with less delay than scholarship holders, by having a clearer understanding of the relevance of their research). Finally, we expected that not all of the effect of type of funding support was mediated by understanding of the relevance of one's research, so we also included a direct effect (*c'* path). It was hypothesized that the direct effect would also be negative, indicating that, for a fixed level of understanding of the societal relevance of one's research, university employees would have a lower delay than scholarship holders.

18.2.3 Parameters to Be Estimated

Mediation analysis allows for the decomposition of a total effect of an independent variable (X) on a dependent variable (Y) into the indirect effect, transmitted through the mediator (M), and the direct effect. When the mediator and the outcome are continuous, the single mediator model is described using three linear regression equations (MacKinnon, 2008):

$$Y = i_1 + cX + e_1, \tag{1}$$

$$M = i_2 + aX + e_2, \tag{2}$$

$$Y = i_3 + c'X + bM + e_3. \tag{3}$$

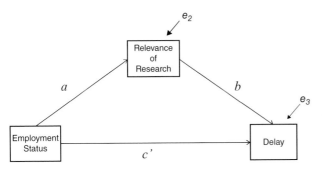

FIGURE 18.1 Proposed mediation model for predicting the delay in PhD completion

The indirect (mediated) effect can be computed as the product of coefficients ab from Equations 2 and 3. The direct effect is represented by the coefficient c' in Equation 3.

18.3 Frequentist Options

The direct effect is obtained by estimating Equation 3 using ordinary least squares (OLS) regression, and, in this example, $c' = -7.721$, $p = .01$. The point estimate of the indirect effect in this example is $ab = 0.17$. There are several frequentist methods for evaluating the significance of the indirect effect. We do not cover frequentist options in depth, but we show results obtained using three popular frequentist methods. We also provide references for readers interested in learning more about these methods. An important consideration in mediation analysis, in both the frequentist and the Bayesian frameworks, is that the indirect effect is computed as the product of two coefficients (both assumed to follow normal distributions), and the distribution of the product of two regression coefficients is not normal (Craig, 1936). We only report results based on methods that had superior statistical properties in simulation studies, meaning that, despite their continuing presence in the literature, we did not use the causal steps approach and normal theory confidence limits (MacKinnon, Lockwood, Hoffman, West, & Sheets, 2002).

In this example, $ab = 0.17$ indicates that the indirect effect of being employed by the university on the delay in dissertation completion, through the understanding of the relevance of one's own research, is 0.17 months. The distribution of the product confidence limits are -0.391 and 0.962, the percentile bootstrap confidence limits are -0.389 and 0.939, and the accelerated bias-corrected bootstrap limits are -0.242 and 1.309. Confidence intervals for the mediated effect, constructed using all these methods, contain 0, thus, the mediated effect is not statistically significant.

In the absence of relevant prior information, interval estimates of the mediated effect obtained using Bayesian methods with uninformative priors have comparable statistical properties to intervals obtained using the percentile bootstrap and the distribution of the product (Miočević, MacKinnon, & Levy, 2017). However, when there is relevant prior information about the a path, the b path, or both, using Bayesian methods with informative priors yields intervals with higher power than those obtained using frequentist methods. Bayesian methods also offer an easy way to approximate functions of parameters, such as effect size measures (Miočević, O'Rourke, MacKinnon, & Brown, 2018) and causal estimators for mediation models (Miočević, Gonzalez, Valente, & MacKinnon, 2018).

18.4 Bayesian Options

In the frequentist framework, estimates of effect size from previous studies are often used for a priori power analyses or to decide which variables to include in the model. However, prior information is not used in the statistical analysis. In the Bayesian framework, previous information is the basis of a prior distribution that represents the researcher's knowledge about the parameters in the model (e.g., a regression coefficient) before data collection. Priors for the analysis can come from any source – a meta-analysis, a previous study, or even expert consensus. If there is no prior information, or the researcher wishes to ignore available prior information, it is possible to conduct a Bayesian analysis with uninformative (diffuse) prior distributions. Numerical results of such analyses often resemble results from frequentist analyses, but the interpretations differ. We illustrate Bayesian mediation analyses with both diffuse and informative priors following the procedure described by Yuan and MacKinnon (2009) and relying on steps in the WAMBS checklist (When to Worry and how to Avoid the Misuse of Bayesian Statistics; Depaoli & Van de Schoot, 2017).

18.4.1 Parameter Estimation with Diffuse Priors

The parameters of the single mediator model are two intercepts (i_2 and i_3), three regression coefficients (a, b, and c'), and two residual variances (σ^2_{e2} and σ^2_{e3}) from Equations 2 and 3. In order to draw inferences about the mediated effect, we approximate the posterior of the product of coefficients ab. The first step in a Bayesian analysis is to select a prior distribution that quantifies our knowledge about the parameters before data collection. When constructing prior distributions, it is helpful to think about **impossible** parameter space and **improbable** parameter space for parameters of interest (see Van de Schoot, Sibrandji, Depaoli, Winter, Olff, & Van Loey, 2018).

In the first analysis, we assigned univariate priors for individual parameters that communicate very little knowledge. Parameters of a prior distribution are called hyperparameters. The hyperparameters of a normal prior distribution are the mean and the variance. The mean hyperparameter encodes the best guess

about the parameter. The variance hyperparameter encodes our confidence about this guess (i.e., the higher the variance hyperparameter, the more spread there is in the prior, and the less confident we are about our best guess). We can also conceptualize the spread of the normal prior (i.e., our confidence about the best guess) in terms of standard deviation or precision – the inverse of the variance.

Following common practice, intercepts and regression coefficients were assigned normal priors with mean hyperparameters of 0 and precision hyperparameters of .001 (i.e., variance hyperparameter of 1000, which translates to standard deviations of 31.62; Miočević, MacKinnon, & Levy, 2017). These normal priors are constructed so negative and positive values close to 0 (e.g., 10 and −10) are almost as probable as the value of 0 (Figure 18.2). We did not indicate any values as *impossible*, but values, such as −50 and 50, are assumed to be *improbable* relative to the value of 0.

Thus, our prior for the intercept in Equation 2 (i_2) encodes the assumption that it is just as probable that a student with external funding will have higher than average understanding of the relevance of their research as it is that they will have lower than average understanding. Our prior for the intercept in Equation 3 (i_3) encodes the assumption that it is just as probable that a student with external funding, with an average level of understanding of the relevance of their research, will finish before their anticipated completion date as it is that they will have a delay in completing their PhD. The same normal prior for the a path indicates that we assume that being employed by the university, instead of having external funding, could lead to either a

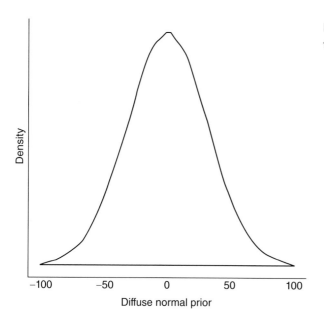

FIGURE 18.2 Diffuse normal prior assigned to intercepts and regression coefficients

reduction or an increase in understanding of the relevance of one's research and that our best guess is that having university versus external funding leads to no change in understanding of the relevance of one's research. Similarly, our prior for the *b* path encodes the assumption that our best guess for the change in predicted delay for a one-unit increase in the understanding of the relevance of one's research for students with external funding is 0. Our prior for the *c'* path encodes the assumption that our best guess for the change in predicted delay for students with average understanding of the relevance of their research due to having university instead of external funding is also 0.

The remaining parameters in Equations 2 and 3 that require priors are the residual variances (σ_{e2}^2 and σ_{e3}^2). In the Bayesian framework, we often think in terms of precision, which is the inverse of a variance. The priors for residual precisions in Equations 2 and 3 encode the assumption that the best guess for the residual precision is 1, and that this best

guess carries the weight of 1 observation. Residual precisions σ_{e2}^{-2} and σ_{e3}^{-2} were assigned gamma priors with both α and β hyperparameters equal to .5. For more gamma distributions, see Appendix A in Gelman, Carlin, Stern, Dunson, Vehtari, and Rubin (2013).

Posterior distributions can sometimes be computed analytically. However, in most situations, analytical solutions are not available, and the posterior is approximated using Markov Chain Monte Carlo (MCMC). The model in this example was estimated using MCMC in the software WinBUGS (Lunn, Thomas, Best, & Spiegelhalter, 2000), run from R (R Core Team, 2014), using packages R2WinBUGS (Sturtz, Ligges, & Gelman, 2005) and coda (Plummer, Best, Cowles, & Vines, 2006). We used three chains with different starting values. In MCMC, it is important to find evidence that the chains have reached the target (posterior) distribution so that draws before the posterior has been reached can be discarded – only draws from the posterior distribution are used for inference. The process of evaluating whether chains have started sampling from the posterior is called "diagnosing convergence." In this example, we used trace plots and values of the Potential Scale Reduction (PSR) factor for diagnosing convergence. Values of PSR factor slightly above 1 (preferably below 1.1 according to Gelman and Shirley, 2011) are considered evidence of convergence. In this analysis, PSR factor values were equal to 1 for all parameters. Trace plots in Figure 18.3 display draws from the three chains for each parameter, and we see that the chains are mixing well before the 1000th iterations. We chose to discard the first 1000 iterations and retain the remaining 4000 iterations from each chain, meaning that the posterior was approximated using 12,000 iterations. For more on convergence diagnostics, see Sinharay (2004).

Recall that, in a Bayesian analysis, the answer is a distribution and not a point estimate with a standard error (Figure 18.4). To interpret the results, we summarized the posteriors of each parameter using point (posterior mean and median) and interval (equal-tail and highest posterior density credibility intervals) summaries. The 95% equal-tail credibility intervals (middle of Figure 18.4) are obtained by taking the 2.5th and the 97.5th percentiles of the posterior distribution. The highest posterior density (HPD) intervals have the property that no value outside of the interval has higher probability than any value within the interval (Gelman et al., 2013) and are obtained by selecting the values with the highest posterior probability (right of Figure 18.4). For parameters that are expected to have asymmetrical posteriors (e.g., the indirect effect), HPD credibility intervals are preferred over equal-tail credibility intervals.

For the sake of brevity, we only interpret posterior summaries for the indirect and direct effects in Table 18.1. Results of the Bayesian analysis, with uninformative priors, indicate that the mean of the posterior for the indirect effect ab is equal to 0.169 and the median is equal to 0.108; this difference between the posterior mean and median indicates asymmetry in the posterior for the indirect effect. Furthermore, the posterior mean is numerically similar to the point estimates obtained using frequentist methods (ab = 0.17). The 95% equal-tail credibility interval for the mediated effect ranges from −0.384 to 0.955 and is numerically similar to confidence limits obtained using the distribution of the product and percentile bootstrap confidence limits. The 95% HPD credibility interval ranges between −0.424 and 0.893. The credibility intervals for the mediated effect contain 0, thus indicating that it is among the most probable values. We conclude that there is no indirect effect of being employed by the university versus having external funding on the delay to complete a PhD, through understanding of the relevance of one's research.

The mean of the posterior for the direct effect (c') is equal to −6.849, and the posterior median is equal to −6.840. The small discrepancy between the mean and median suggests a relatively symmetric posterior distribution. The 95% equal-tail credibility interval ranges between −12.750 and −1.076; the 95% HPD interval ranges between −12.730 and −1.059. These findings suggest that, by being employed by the university instead of having external funding, a PhD student with an average understanding of the relevance of their research is expected to complete their degree

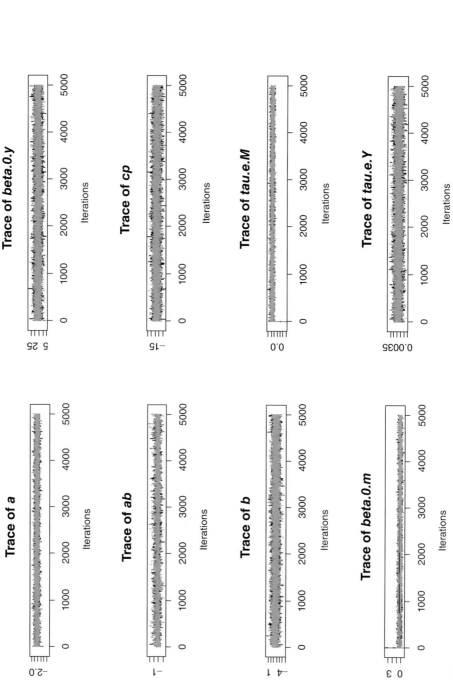

FIGURE 18.3 Trace plots for parameters in the single mediator model

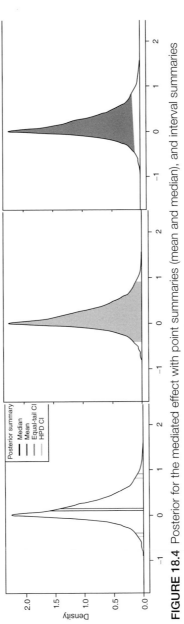

FIGURE 18.4 Posterior for the mediated effect with point summaries (mean and median), and interval summaries (equal-tail and highest posterior density credibility intervals)

Table 18.1 Summaries of marginal posteriors from the analysis with uninformative priors

Parameter	Mean	Median	Equal-tail CI	HPD CI
i_2	0.300	0.301	[−0.127, 0.725]	[−0.129, 0.723]
i_3	16.287	16.270	[10.690, 21.960]	[10.510, 21.740]
a	−0.322	−0.323	[−0.770, 0.125]	[−0.747, 0.141]
b	−0.520	−0.516	[−2.034, 0.965]	[−1.996, 0.996]
c'	−6.849	−6.840	[−12.750, −1.076]	[−12.730, −1.059]
σ_{e2}^{-2}	0.926	0.925	[0.790, 1.074]	[0.784, 1.067]
σ_{e3}^{-2}	0.005	0.005	[0.004, 0.005]	[0.004, 0.006]
ab	0.169	0.108	[−0.384, 0.955]	[−0.424, 0.893]

between 1.06 and 12.73 months earlier than anticipated by their contract.

18.4.2 Reporting Non-Significant Findings

Results of frequentist analyses and Bayesian analyses with diffuse priors indicate that the indirect effect of being employed by the university on delay in PhD completion, through the understanding of the relevance of one's own research, is not significantly different from zero. Confidence intervals tell us that, upon repeated sampling, 95% of the intervals constructed using a given frequentist procedure will contain the true value. Note that this is often phrased as 95% confidence, but we cannot interpret this to mean that there is 95% probability that the interval contains the true value, as mentioned in Edlund and Nichols, Chapter 1.

In the Bayesian framework, we have several options for reporting results. With diffuse priors, we found that there is 95% probability that the mediated effect lies between −0.424 and 0.893. We could also summarize the posterior of ab to find out what proportion of the posterior lies between −0.1 and 0.1 (35%) and report that there is 35% probability that the indirect effect of being employed by the university on delay, through understanding of relevance of one's research, ranges from −0.1 to 0.1 months. Let's suppose that a delay of at least 6 months is considered relevant. Using posterior draws, we can compute that there is 0% probability that the indirect effect of being employed by the university on delay, through understanding of relevance, is greater than or equal to 6 months.

18.4.3 Parameter Estimation with Informative Priors

Prior studies in the United States (Muszynski & Akamatsu, 1991) and Australia (Bourke, Holbrook, Lovat, & Farley, 2004; Pitchforth et al., 2012) also evaluated predictors of delay in PhD completion. Although quite different from our study, the study by Bourke and colleagues (2004) was the most similar to our study in the choice of predictors and because the sample consists of PhD students from a variety of disciplines. Thus, informative priors will be based solely on findings from Bourke et al. (2004).

Bourke and colleagues found that holding a scholarship was a significant negative predictor of time between the start of the degree and degree completion (not delay, as in our example) – standardized β = −0.136, $p <$ .05 – suggesting that scholarship holders are expected to finish sooner than university employees. Completing a PhD in engineering (a discipline with clear applications) was not a significant predictor of time between the start of the degree and degree completion, and the magnitude of the regression coefficient was not reported in the article. There was no information about whether being employed by the university versus holding a scholarship is predictive of the level of understanding of the relevance of one's research. Data in the previous study were collected and modeled in different ways than in our study. Therefore, one ought to be cautious when using the numerical values of coefficients from the previous study for specifying the hyperparameters for the priors in the current model. In light of this ambiguity, suppose two researchers reading the same article by Bourke and colleagues (2004) came up with different priors (see Table 18.2).

Suppose one researcher is more familiar with the differences in how PhD students are funded in the Netherlands versus Australia and decides to construct weakly informative priors based only on the signs and significance of the original results. After converting the reported standardized coefficient from Bourke et al. into an unstandardized regression coefficient, the approximate best guess for c' is 7.434. This quantity was obtained by multiplying the standardized regression coefficient

−.136, reported by Bourke et al., by the standard deviation of the dependent variable in this study (SD = 14.431) and by dividing this quantity by the standard deviation of the independent variable in this study (SD = 0.264). This computation assumes that the independent and dependent variables from the previous study had comparable standard deviations as their counterparts in this study, and that other covariates in the regression, where the independent variable predicts the dependent variable, did not correlate with the independent variable. These assumptions are not ideal. However, given the absence of information about the standard deviations and correlation matrix from the study by Bourke et al., they were necessary.

The first researcher decides that, given this finding, positive values of c' ought to be twice as likely as negative values and encodes this assumption by selecting a prior with a mean hyperparameter of 7.434 and a precision hyperparameter of 0.003 – which places 66% of its weight on values greater than or equal to 0. The precision hyperparameter was chosen empirically to reflect the desired percentage of prior weight on values ≥ 0 using the syntax available at https://osf.io/fud8v/?view_only = 0243cf4f4bd84af6ab73bad571e11d82. The magnitude of the coefficient to be used as a proxy for the b path was not reported in Bourke et al. (2004), so the researcher selects 0 as the mean hyperparameter of the normal prior for b and a precision hyperparameter of 0.003 to encode the same level of informativeness that was encoded in the prior for c'. There was no information about the relationship between being employed by the university and understanding the relevance of one's research, so the researcher selected a prior with a mean hyperparameter of 0 and a precision of 0.001 for a to make the prior for a less informative than the priors for b and c' because it is not based on previous findings.

The second researcher is less familiar with the differences between the Dutch and Australian PhD systems and relies more on the results of Bourke et al. Consequently, the second researcher chooses not to downweigh the prior information as much as the first researcher. This researcher selects a normal prior, with a mean of 7.434 and a precision of 1 for c', to represent the belief that scholarship holders complete their PhD faster than university employees. This prior places almost 100% of its weight on values higher than or equal to 0, thus indicating the belief that c' is almost certainly positive. A normal prior with a mean of 0 and precision of 1 was assigned to regression coefficient b; this prior communicates the expectation that there is no effect of having a clear understanding of the relevance of one's own research on the delay in PhD completion but carries more weight than the prior chosen by the first researcher. The second researcher assigned the same normal prior as the first researcher, with a mean of 0 and a precision of 0.001 for the a path. See Figure 18.5 for visuals of priors in the analyses with weakly informative and informative priors. Priors for the a path are the same in the analyses with weakly informative and informative priors, thus the two lines overlap.

Both researchers specified diffuse univariate priors for intercepts and residual precisions, identical to those in the analysis with diffuse priors. The number of burn-in iterations and criteria for diagnosing convergence

were the same as in the analysis with diffuse priors.

Results of the analysis with weakly informative priors, conducted by the first researcher, indicate that the mean of the posterior for the indirect effect ab is equal to 0.166, and the median is equal to 0.106. The 95% equal-tail credibility interval for the indirect effect ranges from -0.386 to 0.951. The 95% HPD interval ranges from -0.425 to 0.889, thus indicating that 0 is among the 95% most probable values for the indirect effect. The mean of the posterior for c' is equal to -6.536, and the median is equal to -6.527. The equal-tail credibility interval ranges from -12.39 to -0.814, and the HPD interval ranges from -12.4 to -0.832. From the credibility intervals, we conclude that there is 95% probability that a PhD student with an average understanding of the relevance of their own research will finish between a year and a month early. The numerical values of the point and interval summaries from the analysis with weakly informative priors are different than those from the analysis with diffuse priors. However, the conclusions have not changed – there is no indirect effect of being employed by the university on delay, through understanding the relevance of one's research, but there is a direct effect indicating that those employed by the university and who have an average understanding of the relevance of their own research are expected to complete their PhD earlier than scholarship holders with an average understanding of the relevance of their research.

Results of the analysis with informative priors, conducted by the second researcher, who gave more weight to findings from the study in Australia, are slightly different. The

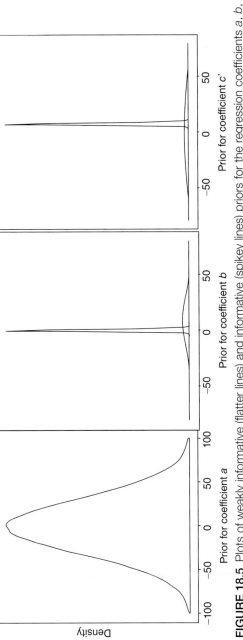

FIGURE 18.5 Plots of weakly informative (flatter lines) and informative (spikey lines) priors for the regression coefficients *a*, *b*, and *c'*

Table 18.2 Summaries of marginal posteriors from analyses with weakly informative and informative priors

Results with weakly informative priors				
Parameter	Mean	Median	Equal-tail CI	HPD CI
i_2	0.300	0.301	[−0.128, 0.725]	[−0.129, 0.723]
i_3	15.997	15.980	[10.440, 21.610]	[10.300, 21.420]
a	−0.322	−0.323	[−0.770, 0.125]	[−0.747, 0.141]
b	−0.513	−0.509	[−2.026, 0.971]	[−2.003, 0.985]
c'	−6.536	−6.527	[−12.390, −0.814]	[−12.400, −0.832]
σ_{e2}^{-2}	0.926	0.925	[0.790, 1.074]	[0.784, 1.067]
σ_{e3}^{-2}	0.005	0.005	[0.004, 0.006]	[0.004, 0.006]
ab	0.166	0.106	[−0.386, 0.951]	[−0.425, 0.889]
Results with informative priors				
Parameter	Mean	Median	Equal-tail CI	HPD CI
i_2	0.297	0.299	[−0.130, 0.722]	[−0.130, 0.722]
i_3	4.362	4.376	[2.010, 6.672]	[2.010, 6.673]
a	−0.320	−0.321	[−0.767, 0.126]	[−0.777, 0.113]
b	−0.156	−0.155	[−1.377, 1.037]	[−1.348, 1.059]
c'	6.050	6.047	[4.149, 7.904]	[4.160, 7.911]
σ_{e2}^{-2}	0.926	0.924	[0.790, 1.074]	[0.790, 1.073]
σ_{e3}^{-2}	0.005	0.005	[0.004, 0.005]	[0.004, 0.005]
ab	0.049	0.024	[−0.438, 0.606]	[−0.447, 0.591]

mean of the posterior for the indirect effect is 0.049, and the median is 0.024. The equal-tail and HPD intervals both include 0 and range from −0.438 to 0.606 and from −0.447 to 0.591, respectively. Notice that the credibility intervals are narrower with informative priors. However, the conclusion about the indirect effect not being different from zero does not change. The numerical values of point and interval summaries of the direct effect changed as did the conclusion! The

mean and median of the posterior for the direct effect are equal to 6.05, and the equal-tail and HPD credibility intervals range from 4.2 to 7.9. Thus, with informative priors selected by the second researcher, the conclusion is that a PhD student with an average understanding of the relevance of their own research is expected to complete their PhD with a delay of 4–8 months, if they're employed by the university instead of on a scholarship.

These analyses exemplify the risks and inherent controversy that come with selecting prior distributions. Two researchers, using not only the same dataset, but also the same source of prior information, can still reach different conclusions.

18.4.4 Sensitivity Analysis

It is important to follow up on the results of a Bayesian analysis with a sensitivity analysis in order to understand the influence that the chosen prior and other potential priors have on the conclusions drawn from the Bayesian analysis. Sensitivity analysis is an evaluation of the influence of the choice of prior on the posterior distribution, and it is generally done after the model has been estimated and the posterior has been summarized. Rindskopf (2012) cautions against doing sensitivity analysis without reporting it and compares such an omission to doing multiple post hoc procedures after ANOVA and only reporting the significant findings.

In this example, the two researchers used the same mean hyperparameters for the normal priors for regression coefficients a, b, and c', and the only differences between the weakly informative and informative priors were in the precision hyperparameters. The posteriors obtained using weakly informative and informative priors for the a and b paths led to the same conclusions about the indirect effect as the posteriors obtained using diffuse priors. However, there was a 4.6% change in the magnitude of the posterior mean of the direct effect with weakly informative priors (relative to the posterior mean with diffuse priors), and a change of 188.3% with informative priors. Any change in posterior summary greater than 10% is said to signal a large

influence of the subjective prior (Depaoli & Van de Schoot, 2017). Recall that the weakly informative prior placed 2 times more weight on the positive values of c' than on negative values (i.e., approximately 66% of the prior was greater than or equal to 0), and almost 100% of the density of the informative prior was positive. The sensitivity analysis evaluates the percent change in the posterior mean and the conclusion about c' from the HPD intervals as a function of the percentage of the prior for c' that is greater than or equal to 0. Percentages considered are 66%, 75%, 80%, and 90%, corresponding to positive values of the c' path being considered 2, 3, 4, and 9 times more likely than negative values. It was empirically determined that these percentages correspond to precision hyperparameters of 0.003, 0.007, 0.015, and 0.028. The priors for the intercepts, residual variances, and the a and b paths remain the same as in the analysis with informative priors (see Table 18.3).

The sensitivity analysis shows that the subjective prior for the c' coefficient starts having a large influence when we assume positive values of c' are at least 3 times more likely, and the conclusion about the significance of the c' path changes when we assume that positive values are at least 4 times more likely than negative values. Note that the influence of these precision hyperparameters is partly due to the choice of mean hyperparameter, and that a mean hyperparameter that is lower in absolute value may not produce priors that are as influential with the above precision hyperparameters. The posteriors for c', obtained using the four precision hyperparameters in the sensitivity analysis, are shown in Figure 18.6.

Table 18.3 Results of the sensitivity analysis

% prior density ≥ 0	Posterior mean c'	% change relative to posterior mean with diffuse prior	HPD CI	Conclusion change
66	−6.477	5.4	[−12.290, −0.747]	No
75	−6.011	12.2	[−11.660, −0.316]	No
80	−5.167	24.6	[−10.750, 0.250]	Yes
90	−3.998	41.6	[−9.372, 1.132]	Yes
→100	6.050	188.3	[4.160, 7.911]	Yes

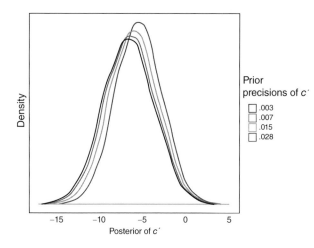

FIGURE 18.6 Posteriors for the coefficient c' with different prior precisions tested in the sensitivity analysis

A change of 10% or more in the point summary of the posterior (here the mean, but sensitivity analysis could also be done on the median) and a change in conclusion relative to the results of the analyses using diffuse priors are not enough to dispute a prior distribution. In some cases, updating information with a prior that carries a lot of weight brings us closer to the truth. However, we have no way of knowing whether the influence of the prior is in a desired direction. In this case, where the prior dataset was collected in a different country with a different doctoral system and where the variables were conceptualized and measured differently, the informative priors selected by the second researcher are not defensible. Therefore, a thorough literature search (maybe even in combination with expert elicitation) and providing full transparency about the process of translating prior information into prior distributions is essential (see the example in Van de Schoot et al., 2018).

18.5 Conclusion

This chapter provided an introduction to Bayesian analyses through the example of a

Table 18.4 Abbreviated form of the WAMBS checklist by Depaoli and Van de Schoot (2017): When to Worry and how to Avoid the Misuse of Bayesian Statistics

TO BE CHECKED BEFORE ESTIMATING THE MODEL
Point 1: Do you understand the priors?
TO BE CHECKED AFTER ESTIMATION BUT BEFORE INSPECTING MODEL RESULTS
Point 2: Does the trace plot exhibit convergence?
Point 3: Does convergence remain after doubling the number of iterations?
Point 4: Does the histogram have enough information?
Point 5: Do the chains exhibit a strong degree of autocorrelation?
Point 6: Does the posterior distribution make substantive sense?
UNDERSTANDING THE EXACT INFLUENCE OF THE PRIORS
Point 7: Do different specifications of the multivariate variance priors influence the results?
Point 8: Is there a notable effect of the prior when compared with non-informative priors?
Point 9: Are the results stable from a sensitivity analysis?
AFTER INTERPRETATION OF MODEL RESULTS
Point 10: Is the Bayesian way of interpreting and reporting model results used?
(a) Also report on: missing data, model fit and comparison, non-response, generalizability, ability to replicate, etc.

mediation analysis where it was of interest to compute the indirect and direct effects. There are many details about Bayesian statistics that had to be omitted for the sake of brevity, and for the interested readers, we recommend textbooks listed under suggested readings. As we have demonstrated in the example in this chapter, Bayesian methods give results with intuitive interpretations, but selecting priors is not an easy task – highly informative priors can sometimes have a dramatic influence on the conclusions drawn from the Bayesian analysis. For this reason, we suggest that researchers using Bayesian analyses also follow the steps provided in the WAMBS checklist (Depaoli & Van de Schoot, 2017), which will at least guarantee full transparency about the choice of prior and its influence on the posterior and conclusions drawn from the analysis. An abbreviated form of the WAMBS checklist is made available in Table 18.4. Future methodological research should focus on providing guidelines for constructing informative prior distributions that reflect the current state of knowledge and feature an appropriate level of confidence about the relevance of the previous findings for the analysis of the current study.

KEY TAKEAWAYS

- The Bayesian framework offers a more straightforward way of computing functions of parameters (e.g., the mediated effect) and evaluating whether these are different from zero.
- The most controversial aspect of a Bayesian analysis is the selection of a prior distribution. It is possible to specify uninformative prior distributions that lead to similar numerical results as those obtained using frequentist methods. However, the results of Bayesian analyses have probabilistic interpretations while the results from frequentist analyses do not.
- Currently, the optimal solution for mitigating the risk associated with specifying informative priors in a Bayesian analysis is to be as transparent as possible about the choice of prior distributions and to follow up each Bayesian analysis with a sensitivity analysis to evaluate the results that would have been obtained using other plausible prior distributions.

IDEAS FOR FUTURE RESEARCH

- What barriers are there to the implementation of Bayesian analyses across the social and behavioral sciences?
- Will the selection of priors be a significant challenge in the review process of manuscripts as Bayesian analyses become more common in the social and behavioral sciences?
- Will more attention to Bayesian approaches help researchers become more comfortable with Bayesian analyses?

SUGGESTED READINGS

Gill, J. (2008). *Bayesian Methods: A Social and Behavioral Sciences Approach* (2nd ed.). Boca Raton, FL: CRC Press.

Jackman, S. (2009). *Bayesian Analysis for the Social Sciences*. Chichester, West Sussex, UK: Wiley.

Kruschke, J. K. (2015). *Doing Bayesian Data Analysis, Second Edition: A Tutorial with R, JAGS, and Stan*. New York: Academic Press.

REFERENCES

Ashby, D. (2006). Bayesian statistics in medicine: A 25 year review. *Statistics in Medicine, 25*(21), 3589–3631. doi:10.1002/sim.2672

Bakan, D. (1966). The test of significance in psychological research. *Psychological Bulletin, 66*(6), 423–437.

Bourke, S., Holbrook, A., Lovat, T., & Farley, P. (2004, November). Attrition, completion and completion times of PhD candidates. Paper presented at the AARE Annual Conference, Melbourne November 28–December 2.

Cohen, J. (1994). The Earth is round (p<.05). *American Psychologist, 49*(12), 997–1003.

Craig, C. C. (1936). On the frequency function of xy. *Annals of Mathematical Statistics, 7*, 1–15.

Cumming, G. (2014). The new statistics: Why and how. *Psychological Science, 25*(1), 7–29.

Depaoli, S., & van de Schoot, R. (2017). Improving transparency and replication in Bayesian statistics: The WAMBS-Checklist. *Psychological Methods, 22*(2), 240–261. https://doi.org/10.1037/met0000065

Gelman, A., Carlin, J. B., Stern, H. S., Dunson, D. B., Vehtari, A., & Rubin, D. B. (2013). *Bayesian Data Analysis* (3rd ed.). Boca Raton, FL: CRC Press.

Gelman, A., & Shirley, K. (2011). Inference from simulations and monitoring convergence. In S. P. Brooks, A. Gelman, G. Jones, & X.-L. Meng (Eds.), *Handbook of Markov Chain Monte Carlo* (pp. 163–174). Boca Raton, FL: Chapman Hall.

Ioannidis, J. P. A. (2005). Why most published research findings are false. *PLoS Med, 2*(8), e124. https://doi.org/10.1371/journal.pmed.0020124

König, C., & van de Schoot, R. (2018). Bayesian statistics in educational research: A look at the current state of affairs. *Educational Review, 70*(4), 486–509. doi:10.1080/00131911.2017.1350636

Jackman, S. (2009). *Bayesian Analysis for the Social Sciences*. Chichester, West Sussex, UK: Wiley.

Lunn, D. J., Thomas, A., Best, N., & Spiegelhalter, D. (2000). WinBUGS – a Bayesian modelling framework: Concepts, structure, and extensibility. *Statistics and Computing, 10*, 325–337.

MacKinnon, D. P. (2008). *Introduction to Statistical Mediation Analysis*. New York: Routledge.

MacKinnon, D. P., Lockwood, C. M., Hoffman, J. M., West, S. G., & Sheets, V. (2002). A comparison of methods to test mediation and other intervening variable effects. *Psychological Methods, 7*(1), 83–104.

Miočević, M., Gonzalez, O., Valente, M. J., & MacKinnon, D. P. (2018). A tutorial in Bayesian potential outcomes mediation analysis. *Structural Equation Modeling: A Multidisciplinary Journal, 25*(1), 121–136.

Miočević, M., MacKinnon, D., & Levy, R. (2017). Power in Bayesian mediation analysis for small sample research. *Structural Equation Modeling, 24*(5), 666–683.

Miočević, M., O'Rourke, H. P., MacKinnon, D. P., & Brown, H. C. (2018). Statistical properties of four effect-size measures for mediation models. *Behavior Research Methods, 50*(1), 285–301.

Muszynski, S. Y., & Akamatsu, T. J. (1991). Delay in completion of doctoral dissertations in clinical psychology. *Professional Psychology: Research and Practice, 22*(2), 119–123.

Pitchforth, J., Beames, S., Thomas, A., Falk, M., Farr, C., Gasson, S., … Mengersen, K. (2012). Factors affecting timely completion of a PhD: A complex systems approach. *Journal of the Scholarship of Teaching and Learning, 12*(4), 124–135.

Plummer, M., Best, N., Cowles, K., & Vines, K. (2006). CODA: Convergence diagnosis and output analysis for MCMC. *R News, 6*(1), 7–11.

R Core Team. (2014). R: A language and environment for statistical computing. R Foundation for Statistical Computing, Vienna, Austria. Retrieved from www.R-project.org/

Rietbergen, C., Debray, T. P., Klugkist, I., Janssen, K. J., & Moons, K. G. (2017). Reporting of Bayesian analysis in epidemiologic research should become more transparent. *Journal of Clinical Epidemiology, 86*, 51–58. doi:10.1016/j.jclinepi.2017.04.008

Rindskopf, D. (2012). Next steps in Bayesian structural equation models: Comments on, variations of, and extensions to Muthén and Asparouhov (2012). *Psychological Methods, 17*(3), 336–339.

Rozeboom, W. W. (1960). The fallacy of the null-hypothesis significance test. *Psychological Bulletin, 57*(5), 416–428.

Sinharay, S. (2004). Experiences with Markov Chain Monte Carlo convergence assessment in two psychometric examples. *Journal of Educational and Behavioral Statistics, 29*(4), 461–488.

Sonneveld, H., Yerkes, M. A., & van de Schoot, R. (2010). PhD trajectories and labour market mobility: A survey of recent doctoral recipients at four universities in the Netherlands. Utrecht: Nederlands Centrum voor de Promotieopleiding/IVLOS.

Spiegelhalter, D. J., Myles, J., Jones, D., & Abrams, K. (2000). Bayesian methods in health technology assessment: A review. *Health Technology Assessment, 4*(38), 130. doi:10.3310/hta4380

Sturtz, S., Ligges, U., and Gelman, A. (2005). R2WinBUGS: A package for running WinBUGS from R. *Journal of Statistical Software, 12*(3), 1–16.

Van de Schoot, R., Schalken, N., & Olff, M. (2017). Systematic search of Bayesian statistics in the field of psychotraumatology.

European Journal of Psychotraumatology,
8(sup1), 1375339. doi:10.1080/
20008198.2017.1375339

Van de Schoot, R., Sijbrandij, M., Depaoli, S.,
Winter, S. D., Olff, M., & van Loey, N. E.
(2018). Bayesian PTSD-trajectory analysis with
informed priors based on a systematic
literature search and expert elicitation.
Multivariate Behavioral Research, 53,
267–291.

Van de Schoot, R., Winter, S. D., Ryan, O.,
Zondervan-Zwijnenburg, M., & Depaoli, S.
(2017). A systematic review of Bayesian papers
in psychology: The last 25 years. *Psychological
Methods, 22*(2), 217. doi:10.1037/met0000100

Van de Schoot, R., Yerkes, M. A., Mouw, J. M., &
Sonneveld, H. (2013). What took them so long?
Explaining PhD delays among doctoral
candidates. *PLoS One, 8*(7): e68839.
http://dx.doi.org/10.1371/journal.pone
.0068839

Wagenmakers, E. J., Wetzels, R., Borsboom, D.,
& van der Maas, H. L. (2011). Why
psychologists must change the way they analyze
their data: The case of psi: Comment on Bem
(2011). *Journal of Personality and Social
Psychology, 100*(3), 426–432.

Yuan, Y., & MacKinnon, D. P. (2009). Bayesian
mediation analysis. *Psychological Methods,
14*(4), 301–322.

19 Development and Applications of Item Response Theory

Clifford E. Hauenstein

Susan E. Embretson

Fundamental to all scientific endeavors is the process of **measurement**, or, broadly defined, the re-representation of properties and relations among real phenomena onto an abstract, numeric scale. Although the philosophical investigations of the inherent nature of measurement remain relatively abstruse, measurement of physical properties is a process we are quite comfortable and familiar with on a day-to-day basis. We measure body weight with a scale to track fitness progress, height with a tape measure to find the best-fitting suit, and the volume of sugar needed for a cake recipe. In each of these situations, we have carried out the fundamental measurement process by mapping a set of relations among physical properties onto a more abstract, numeric scale.

Measurement in the social and behavioral sciences, however, provides its own unique challenges since most of the person variables we are interested in quantifying are not directly observable. That is, there is interest in distinguishing and differentiating persons on abstract characteristics – constructs such as extraversion, agreeableness, spatial reasoning, or auditory processing. Measurement of abstract constructs requires the specification of behavioral indicators that are assumed to

be related to the construct. For example, one might consider accuracy in blending phonemes, deconstructing words into individual phonemes, and counting syllables as all being related to a more general construct of auditory processing. Performance differences across these tasks would be assumed to indicate individual differences in the general trait of auditory processing. Additionally, any effort to measure abstract constructs through a set of observable, behavioral indicators requires some declaration of *how* these indicators relate to the construct of interest.

The field of psychometrics is devoted to developing and refining different approaches to evaluate ability constructs through observed behavior. This is principally done through the development of different *measurement models*, each of which formally define the assumed relationships between tasks, observed behavioral indicators, impact of error, and the construct of interest. These theoretical measurement models are applied to observed response data in order to estimate particular parameters of the model. It is ultimately these parameters that define the individual differences in ability. For example, classical test theory (CTT), which was discussed in a previous chapter in this text, defines the following measurement model:

$$X = T + error$$

Here, the true score () is a parameter of the model, taken as the expected value of the observed raw score (X) on a test or set of tasks for any individual (assuming a random and normally distributed error term with mean of 0). In the CTT model, individuals are scaled according to the ability of interest by their differing true score estimates. Note that CTT represents only one psychometric approach to measuring person abilities.

In this chapter, we will be discussing a psychometric alternative to CTT – item response theory (IRT). IRT represents a family of logistic measurement models that define the probability of a particular observed response, given a set of individual item or task characteristics and the **latent trait** of the individual. The latent trait is defined as an unobservable, underlying faculty of the individual that drives item responses. The logistic function invoked in the IRT family of models offers a number of unique differences and advantages relative to

the simpler, linear model in CTT. These are summarized in Table 19.1.

In the following sections of the chapter, we will introduce and elucidate the basic theoretical and applied principles of IRT – from the most commonly employed models and estimation procedures to evaluating the various assumptions of the models and assessing item fit. Throughout, we will periodically refer back to, and expand upon, the distinct properties of IRT measures listed in Table 19.1.

19.1 The Rasch Model and Basic Principles of IRT

The most significant groundwork for IRT was established by Danish mathematician Georg Rasch. Rasch endeavored to develop a scoring procedure which fulfilled, what he considered, an essential property of measurement – **specific objectivity**. Generally, Rasch considered a measurement model to be ideal if it allowed for **invariant scaling** of objects (along the dimension of interest) across different measurement tools and contexts. To use an example from physical measurement, the

Table 19.1 Comparisons of measurement properties between CTT and IRT

CTT measurement properties	IRT measurement properties
• Interval level scaling of scores only justified when scores are normally distributed	• Interval level scaling of scores justified, especially for the Rasch IRT model
• Same standard error of measurement assumed for all response patterns	• Standard error of measurement estimated differently for different response patterns
• Person ability estimates do not necessarily generalize beyond a specific item set	• Person ability estimates have meaning across different item sets
• Item difficulty estimates do not necessarily generalize beyond a specific examinee pool	• Item difficulty estimates can be estimated and compared across different examinee pools
• Produces difference scores with questionable reliability and validity properties	• Measured differences in performance across time or intervention are meaningful

measured relationships among the heights of three individuals should remain preserved across any set of tape measures if specific objectivity holds. Extending this notion to psychological or educational measurement, specific objectivity means that the relationships among person ability estimates should not vary across different tests with different patterns of item difficulties. Under specific objectivity, the converse should also be true – the relationships among item difficulty estimates should not be impacted by different samples of examinees with different ability levels.

This would provide an enormous benefit over CTT approaches, which model a person's trait level simply as the expected raw score over repeated testings with parallel test forms. A glaring deficit of the CTT approach is the dependency of the expected raw score metric with the distribution of item difficulties. In other words, with CTT, person trait estimates will not necessarily generalize from one item set to another. This limitation of CTT disallows score comparisons among subgroups of examinees taking different sets of test items. For example, under the CTT approach, SAT scores from two different examinees on two different tests would not necessarily be comparable. This is a critical disadvantage, as many measurement efforts necessitate comparisons of scores from different subjects across different test forms.

In 1960, Georg Rasch presented a measurement model that was based upon his ideal of specific objectivity, which would become one of the most significant contributions in the field of educational and psychological measurement. To meet the ideal of specific objectivity, Rasch specified a logistic function that related person ability and item difficulty to response probabilities as follows:

$$P\left(X_{ij} = 1 | \theta_j, \beta_i\right) = \frac{e^{\left(\theta_j - \beta_i\right)}}{1 + e^{\left(\theta_j - \beta_i\right)}}$$

In the equation, θ_j is defined as the location of person j's ability on the latent scale, β_i is defined as the location of item i's position on the latent scale, and X_{ij} is defined as the response to item i by person j ($X_{ij} = 1$ indicates a correct response to item i).

Ultimately, Rasch's model directly defines the probability of a correct response by the relative distance between person ability and item difficulty on a latent scale. A graphical representation of his model is provided in Figure 19.1.

In Rasch's approach, ability was modeled as an unobservable person-level trait to be estimated (represented on the horizontal axis in Figure 19.1, and traditionally referred to by the Greek symbol, θ). Latent ability was defined as an abstract, psychological construct that drove observed responses. Individual differences in the psychological construct were assumed to produce individual differences in response patterns. Note that it is common practice to arbitrarily fix the scale of latent ability to have a mean of 0 and standard deviation of 1, and this scaling is represented on this plot. The vertical axis of Figure 19.1 represents the probability of observing a correct response, and the plotted S-shaped curves (commonly referred to as "item characteristic curves" in the IRT literature) represent the expected probability of a correct response for a given item, conditional on a particular trait level.

A few aspects of the item characteristic curves are worth noting. The inflection points

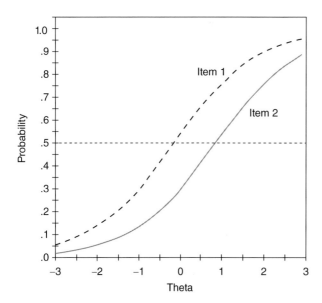

FIGURE 19.1 Example item characteristic curves derived from the Rasch model

for each of these item characteristic curves correspond to the location on the latent scale where an individual has a .5 probability of solving the item (marked by the dotted line). The value of the latent scale at the inflection is defined as item difficulty, parameterized as β_i in the Rasch model. For example, in Figure 19.1, item 1 would have an estimated difficulty of approximately 0, since the item inflection point corresponds to a value of 0 on the latent scale. An individual with an ability estimate of 0 on the latent scale would be expected to have a .5 probability of solving item 1 correctly. Also note that the item characteristic curve for item 2 is shifted to the right, relative to item 1. Thus, the inflection point for item 2 is shifted to the right, and item 2 is defined with a higher difficulty value. Practically speaking, the expected probability of a correct response will always be lower for item 2, relative to item 1, for a given trait level. It is also important to point out that, in Rasch's model, the item characteristic curves never cross; all item characteristic curves display

the same slope and each item has equal discriminating power. When this assumption holds, justifications of specific objectivity and interval level scaling are simplified.

The principle of the joint scaling of item and person parameters on a common latent scale is one of the defining characteristics of IRT and is of particular importance, since it allows for a more discrete level of measurement. In contrast to the more blunt metric of total expected raw score in CTT, IRT considers an individual's position along the latent trait relative to item location. Thus, unique item characteristics are established as central components in the measurement model. This micro-level approach to measurement is advantageous, as it allows measurement errors to be uniquely defined for different persons and response patterns.

Essentially, items provide more information at the point on the latent trait where they have more discriminating power (i.e., where the slope of the item curve is steepest, at the inflection point). The information provided

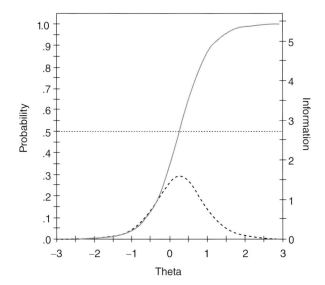

FIGURE 19.2 Item information function for a Rasch item plotted alongside the item characteristic curve

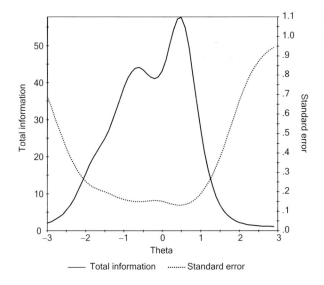

FIGURE 19.3 Test information function and standard error of measure

for a given item can be plotted as a function of ability level as in Figure 19.2; this particular item provides the most information for subjects with an approximate $\theta = 0.25$.

Then, for any given test, the item information functions for all items can simply be summed to create a **test information curve** (see Figure 19.3).

This test information curve displays the location along the latent ability scale where a test has maximum precision. In the example plot in Figure 19.3, it can be seen that the test as a whole provides most information for subjects with $\theta = 0.5$. Note also that the plot shows how the standard error of measure changes with latent ability via the dotted line.

The standard error of measure can be calculated as:

$$SE(\theta_j) = \frac{1}{\sqrt{TI(\theta_j)}}$$

where $TI(\theta_j)$ is the test information value for a given θ_j.

The practical implication is that items can be selected according to the measurement precision objectives of the test. In the example above, for instance, it was noted that this set of items provided maximal precision for students with $\theta = 0.5$. If, instead, the goal was to achieve maximal measurement precision for lower-performing students (those students with $\theta = -1.5$, for example), then a different test with an easier set of items should be developed. This is, in fact, the foundation for modern approaches in computer adaptive testing. Computer adaptive testing has the potential to lessen testing time while decreasing the standard errors of measure. This is done by adaptively selecting the most informative items for a given examinee during the testing period. First, an algorithm is developed which continually updates an examinee's estimated ability – as he or she is responding to items in the testing session. Assuming a large pool of items is available, for which difficulty values are known, the algorithm then selects the most informative items (i.e., those with difficulty values nearest the examinee's current estimated ability level). Because these tests are effectively custom-tailored for each examinee, testing procedures are much more efficient; examinees are only administered the items most relevant to their current performance and ability levels. Furthermore, because of the property of specific objectivity, ability scores for examinees with different sets of custom selected items can be validly compared (a critical limitation of CTT).

19.2 Additional Unidimensional IRT Models

19.2.1 The Two Parameter Logistic Model

The two parameter logistic model (2PL model) follows from the same motivations and framework as the Rasch model, but specifies additional item parameters which might affect the response process (Lord & Novick, 1968). In the 2PL extension, the slopes (or "discriminations") of the item characteristic curves are allowed to vary between items. In this formulation, two item parameters are now considered and estimated for each item: Difficulty and discrimination. The function of the 2PL model is defined below:

$$P(X_{ij} = 1|\theta_j, \beta_i) = \frac{e^{\alpha_i(\theta_j - \beta_i)}}{1 + e^{\alpha_i(\theta_j - \beta_i)}}$$

All terms are defined identically, as in the Rasch model; the only additional element is the scaling factor, α_i, which is estimated separately for each item. Note that, in the traditional Rasch model, this scaling factor was constrained to 1 for all items. Larger values of α_i, for a particular item, result in a steeper item characteristic curve, creating a more discriminating item. That is, the change in probability of success around the inflection point is much more rapid for high α_i. Figure 19.4 shows a plot of item characteristic curves from a 2PL model. Note that, in the 2PL model, the item difficulty parameter still

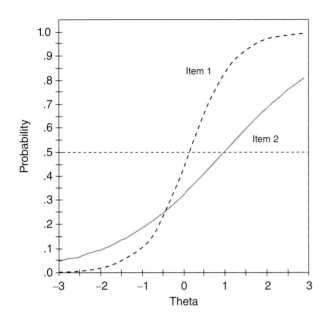

corresponds to the point on the latent trait where the inflection point occurs.

The effect of modeling different slopes for each item is immediately apparent. Item 1, represented with the dashed line, is the most discriminating item; very small increases in ability, near the inflection point, lead to rapid increases in the probability of a successful response. Conversely, item 2 exhibits a more gradual slope. For this item, ability gains near the inflection point do not impact response probability nearly as much. Thus, item 1 can be thought of as being more efficient in distinguishing examinees who are capable of providing a correct response from those who are not. Thus, the discrimination parameter can be considered the IRT analogue of the item-total correlation statistic from classical test theory. Both statistics establish how closely related an item is to the trait of interest. Put another way, the discrimination parameter is an indicator for the amount of information supplied by an item in estimating an examinee's ability.

Note that the desirable measurement properties of the Rasch model become somewhat complicated with the 2PL model. Thus, we generally prefer applications of the Rasch model, when justified. However, if, through either empirical or theoretical grounds, we strongly believe that items exhibit different discrimination parameters, then we welcome estimation with a 2PL model. A 2PL model will always exhibit equal or greater fit to the data, relative to the Rasch model, because of the additional parameters being estimated. Our goal is ultimately to apply a measurement model that reflects the true nature in which item characteristics interact with the latent ability to affect response outcomes. When items truly offer different discriminating power, then we should capture these differences with a 2PL model. In the section on model fit, later in this chapter, we discuss methods to empirically test the fit of various models that can be used to guide model selection.

19.2.2 The Three Parameter Logistic Model

The three parameter logistic model (3PL model) extends the 2PL model by one additional parameter to account for guessing (Lord & Novick, 1968). Both the Rasch and 2PL models assume that, as a person's ability falls far below an item's location on the latent scale, the probability of a correct response approaches zero. That is, if an item is very difficult, relative to one's ability, one can assume a zero probability of correct response. However, it is reasonable to assume that, for multiple choice items, an examinee's probability of success never truly falls to zero. Even by random guessing, one would expect a probability of success of .25 for an item with four response options. Additionally, other attributes of an item may provide some direction to the correct answer, even for examinees with very low ability levels. The 3PL model accounts for guessing by applying a lower asymptote to the item characteristic curves. With this correction to the item characteristic curves, the expected probability never falls to zero, even for the lowest ability examinees. The model takes the following form:

$$P(X_{ij} = 1 | \theta_j, \gamma_i, \alpha_i, \beta_i) = \gamma_i$$
$$+ (1 - \gamma_i) \frac{e^{\alpha_i(\theta_j - \beta_i)}}{1 + e^{\alpha_i(\theta_j - \beta_i)}}$$

The difficulty and discrimination parameters still retain their same meaning from the Rasch and 2PL models. The addition of the guessing parameter (γ_i) essentially distinguishes the probability of a correct response, due to guessing (estimated as γ_i), and the probability of a correct response due to ability, independent of guessing. The sum is then taken as the total expected probability of a correct response. Figure 19.5 comprises a plot of item characteristic curves from a 3PL model.

Here, item 2 is slightly more affected by guessing, though both items display an asymptote > 0. It is important to note that, just as with the 2PL model, demonstrations of specific objectivity are not straightforward with the 3PL model. However, if there is

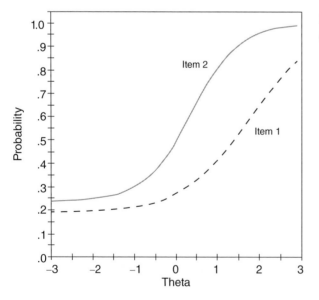

FIGURE 19.5 Example item characteristic curves derived from the 3PL model

sufficient evidence to suggest that guessing may be impacting the response process, then a 3PL model should be preferred over a Rasch or 2PL model. When guessing truly impacts the response process, a 3PL model will provide significantly better fit over the Rasch and 2PL models.

19.3 Unidimensional Models for Polytomous Responses

The models described previously can only be applied to situations with dichotomous outcomes (e.g., multiple choice exams, where a response is scored as either correct or incorrect). However, surveys and attitudinal measures often offer more than two response categories. A common example is items which ask subjects to indicate their level of agreement with a particular attitudinal statement. One distinct advantage of IRT over classical measurement approaches is the flexibility to include items with different numbers of response categories, *in a single test*, without compromising the integrity of the trait estimates. Classical measurement approaches may produce biased scores for tests with such mixed item formats.

To accommodate items with more than two response categories, a number of polytomous IRT models have been developed. Generally speaking, these models fall into two categories with respect to the estimation process: Indirect methods and direct methods. Indirect methods of modeling polytomous data include the graded response model (GRM; Samejima, 1969) and modified graded response model (M-GRM; Muraki, 1990). The direct methods include the partial credit model (PCM; Masters, 1982), generalized partial credit

model (G-PCM; Muraki, 1992), and nominal response model (NRM; Bock, 1972). All of these approaches adopt principles from dichotomous models and extend them to cases with more than two response categories.

The graded response model, for example, essentially applies a 2PL model to each category threshold. For example, consider the following item for a survey of political affiliation:

The Federal Government places too many restrictions on modern corporations			
Strongly Disagree	Disagree	Agree	Strongly Agree
0	1	2	3

There are four response categories and three category **thresholds** (i.e., three transition points between categories). The graded response model estimates a separate 2PL model for each of the category transitions. In other words, one 2PL model is defined for the probability of responding in or above category 1. Another 2PL model is defined for the probability of responding in or above category 2, and so on. Thus, one can estimate the probability of responding in or above any category as follows:

$$P\left(X_{ijt} = 1 | \theta_j, \beta_{it}\right) = \frac{e^{\alpha_i\left(\theta_j - \beta_{it}\right)}}{1 + e^{\alpha_i\left(\theta_j - \beta_{it}\right)}}$$

As before, θ_j represents the latent trait that is driving the item response. For attitudinal surveys, θ_j may specifically represent some underlying degree of conviction related to survey content. In the example above, θ_j may represent the degree of conservative values embraced for an individual. Stronger conservative values would be indexed by a higher score for θ_j and lead to an increased probability of selecting a higher response category ("agree" or "strongly agree"). β_{it} is a "threshold parameter" and refers to the location on the latent scale where there is a .5 probability of responding above threshold t. As an example, one could estimate the probability of subject j responding in category 2 or higher with this model. One could then run a second model to determine the probability of responding in category 3 or higher. To determine the probability of selecting just category 2, one simply needs to *subtract* these probabilities. This defines the indirect, two-step process to deriving category response probabilities for the GRM. If this process is repeated for all thresholds, a plot of **category response curves** can be created (see Figure 19.6).

The category response curves describe how the expected probability of selecting any particular category changes with the latent trait. Note that, in its use of the two-step estimation method, the GRM makes the assumption that the threshold parameters are ordered along the latent trait. The effect of this is to guarantee that each category will be the most likely selected at some location along the latent trait. In Figure 19.6, one can observe a general trend, such that, when conservative values increase, the most probable response category selected sequentially changes from 0 to 1, 1 to 2, 2 to 3, and 3 to 4. The M-GRM follows an identical approach, except that the category threshold values are not allowed to differ between items. Additionally, an item location parameter is estimated that defines the position along the latent trait that the item, as a whole, occupies.

The PCM does not employ a two-step process. Rather, the PCM estimates category response probabilities in a single, direct step via a **divide by total model**:

$$P\left(X_{ijc} = 1 | \theta_j, \delta_{ik}\right) = \frac{\sum_{k=0}^{c} e^{\left(\theta_j - \delta_{ik}\right)}}{\sum_{r=0}^{m} e^{\sum_{k=0}^{r} e^{\left(\theta_j - \delta_{ik}\right)}}}$$

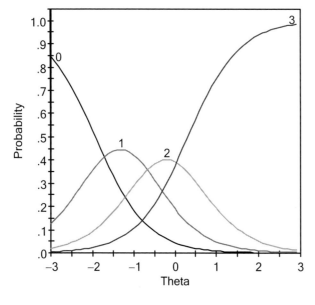

FIGURE 19.6 Example category response curves derived from the GRM

In this model, the subscript k refers to a particular threshold, m refers to the total number of thresholds, and c refers to a particular threshold of interest. Practically, the PCM differs from the GRM and M-GRM by directly modeling the set of category response curves. In the PCM model, δ_{ik} is referred to as a step difficulty parameter and indicates the location on the latent scale where the category response curves intersect for successive categories. Additionally, the PCM is a more flexible approach to modeling polytomous data since it makes fewer assumptions regarding the ordering of response category probabilities. To clarify, consider the plot of category response curves in Figure 19.7.

Notice that, as one moves from left to right along the ability scale, the most probable category selected moves from 0 to 1, then to 3. At no point along the scale is 2 the most likely selected category. Thus, as subjects increase in the latent trait, they effectively "skip" category 2 and move straight from category 1 to category 3. Thus, selection of the GRM or PCM may be guided by one's assumptions regarding the sequential orderings of category intersections. The G-PCM represents a 2PL analogue of the PCM, whereby a separate discrimination value is estimated for each item as a whole.

The NRM is, itself, a generalization of the G-PCM, and estimates a separate discrimination value *for every threshold in every item*. Thus, the NRM makes no assumptions regarding the ordering of response categories relative to the latent trait. In the political survey above, for example, perhaps an item solicits information regarding type of profession. If we have no pre-existing assumption regarding how type of profession relates to conservatism, we may choose to model this item with the NRM. Traditionally, the NRM has been used to evaluate the relative attractiveness of distractors in multiple choice tests or to find the relationship between various incorrect responses and latent ability.

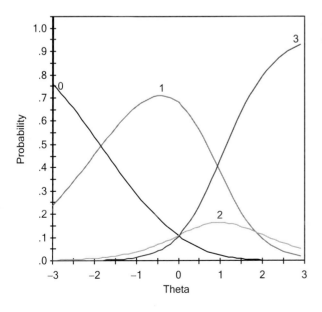

FIGURE 19.7 Example category response curves derived from the PCM model

19.4 Estimating Item and Person Parameters

19.4.1 Maximum Likelihood (ML)

Once a measurement model is selected, item parameters can then be estimated. Although different approaches exist for estimating model parameters, most estimation efforts in IRT are based on deriving a **likelihood value**. The likelihood value indicates the probability of the observed response patterns, given a particular measurement model (e.g., Rasch model) and the set of corresponding parameter values (e.g., values for each item's difficulty value). In this way, a likelihood value can be thought of as a measure of model fit; larger values indicate that the individual observed response patterns are more probable under the applied model. Thus, the likelihood value can serve as a criterion for determining the most optimal parameter values. If we apply a Rasch model to a particular dataset, for example, we want to identify the set of ability and item difficulty values that result in the largest likelihood value possible. To accomplish this, a **likelihood function** is derived that is an equation that relates different likelihood values to different sets of parameter values.

The likelihood function represents the joint probability of observed responses to all items. For example, consider the following response pattern from a single student to a five-item multiple choice test (where 1 indicates a correct response and 0 indicates an incorrect response): 1,1,0,1,0. The joint probability of this response pattern indicates the probability of observing a correct response to item 1 *and* a correct response to item 2 *and* an incorrect response to item 3 *and* a correct

response to item 4 *and* an incorrect response to item 5.

A basic theorem of probability establishes that the joint probability for any set of *independent* events is equal to the product of the individual event probabilities:

$$P(A \cap B) = P(A) * P(B)$$

That is, the probability of events A and B co-occurring is equal to the probability of event A occurring multiplied by the probability of event B occurring. In our example above, the probability of the response pattern for subject *j* would be calculated as:

$$P(X_{1j} = 1) * P(X_{2j} = 1) * P(X_{3j} = 0)$$
$$* P(X_{4j} = 1) * P(X_{5j} = 1)$$

The question that remains is, how are the probabilities for each individual response then computed? Since the Rasch model is applied, in this example, the probabilities for individual item responses are taken directly from the logistic function of the Rasch model. Recall that, for the Rasch model, estimates for the probability of a correct response to item *i* for person *j* are as follows:

$$P(X_{ij} = 1 | \theta_j, \beta_i) = \frac{e^{(\theta_j - \beta_i)}}{1 + e^{(\theta_j - \beta_i)}}$$

To obtain the probability of an incorrect response to item *i* for person *j*, one simply subtracts the probability of a correct response from 1, or:

$$1 - P(X_{ij} = 1 | \theta_j, \beta_i) = 1 - \frac{e^{(\theta_j - \beta_i)}}{1 + e^{(\theta_j - \beta_i)}}$$

Thus, for this particular response pattern, one can substitute in any set of provisional values for the ability and item difficulty parameters

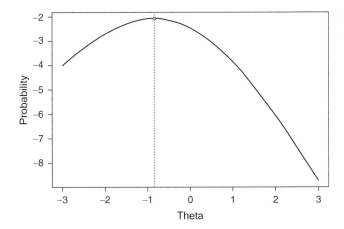

FIGURE 19.8 Log-likelihood function (with respect to θ)

(θ_j and β_i) for the Rasch function and determine the likelihood value. The same principles apply to applications of other IRT models. Note that, if there are a large number of items, the likelihood value can shrink unmanageably small. In the effort to work with values that are larger and more practical, it is common practice to take the natural log of the likelihood function to arrive at the **log-likelihood function**. Critically important in this process is the **assumption of independence** in calculating the likelihood value. The equation for joint probability, which was invoked, assumes independence of events. Thus, a primary assumption of IRT is that all item responses are independent of one another, after conditioning on the latent trait. This point will be elaborated in our section regarding assumptions of IRT.

To further elucidate the utility of the log-likelihood function, consider the situation where item parameter values were known a priori, for the previous example, and only the ability parameter needed to be estimated. With the log-likelihood function constructed, one could specify a provisional θ value for the subject and calculate the log-likelihood value.

Then, one could try a different provisional θ value and calculate a new log-likelihood value. This could be carried out iteratively until an entire plot is constructed, as shown in Figure 19.8.

Here, the calculated log-likelihood values are plotted against different θ values. One simply needs to locate the θ value that corresponds to the maximum observed log-likelihood value. With this example, the log-likelihood function was maximized with a $\theta \sim -.85$. *This θ value represents the latent trait estimate for the subject that would maximize the probability of observing the data, given the specified model.* Note that a brute force method, such as this, is not reasonable to apply in practice. Instead, many software packages use the Newton–Raphson scoring algorithm to locate the maximum. Maximum likelihood estimates for parameters are desirable for several reasons. Specifically, maximum likelihood estimates: a) are consistent, b) asymptotically exhibit the least variability of any estimate, and c) are asymptotically normally distributed.

However, it is commonly the case that neither item parameter estimates nor person ability estimates are known. With both sets of parameters unknown, the estimation process is greatly complicated, since there are no fixed values in the likelihood function. If at least one set of parameters is known, then the other set can be estimated in a relatively straightforward fashion with the likelihood maximization steps illustrated previously. Conditional maximum likelihood (CML) and joint maximum likelihood (JML) are two ML approaches that attempt to resolve this indeterminacy problem. However, both present a rather strict set of limitations. CML can only be applied when a common discrimination value is estimated for all items (e.g., the Rasch model), and JML cannot derive parameter estimates for persons or items with extreme scores (all correct or all incorrect). Additionally, estimates from JML are not consistent (i.e., estimates do not necessarily increase in accuracy with increases in sample size).

19.4.2 Marginal Maximum Likelihood (MML)

Circumventing the issues noted in CML and JML estimation procedures, MML considers person ability as a random variable drawn from an assumed (a priori) distribution. To derive a likelihood value, MML defines the probability of any response pattern as resulting from two probabilities: The probability of the response pattern given a particular ability level and the probability of observing that particular ability level in the population. Ultimately, MML considers the probability of any given response pattern *over the entire ability distribution* (called the marginal probability). Evaluating the marginal probability of the full set of response patterns effectively *removes* the ability parameter from the

likelihood function, allowing for maximization of the likelihood function with respect to just item parameters. In the following section, we elaborate on issues regarding specification of prior distributions.

With this approach, MML rectifies the most severe limitations of CML and JML; it can be applied to non-Rasch models, it produces consistent estimates, and is able to derive parameter estimates for items with extreme score patterns. The lack of information in an extreme response pattern is compensated for by information from the prior regarding the ability distribution.

Although MML is more computationally burdensome than other estimation methods, we recommend it for most procedures because of the aforementioned benefits. In fact, MML is one of the most commonly utilized methods of item parameter estimation and is incorporated into several popular IRT software packages, including BILOG-MG, PARSCALE, MULTILOG, TESTFACT, and IRTPRO.

19.4.3 Maximum A Posteriori (MAP) and Expected A Posteriori (EAP)

Once item parameters have been estimated, the next step is to estimate ability parameters for all persons in the dataset. If the item parameter estimates are treated as fixed and known in the likelihood function, a traditional maximum likelihood approach could be implemented to derive person ability estimates. However, the traditional maximum likelihood approach will have limited utility when a response pattern offers little information regarding latent ability. This occurs when: a) an extreme response pattern is observed, whereby an individual gets all (or nearly all) items wrong or right or b) there are few items located near the subject's ability level. A traditional maximum likelihood approach will be unable to derive ability estimates for all correct or all incorrect response patterns and will produce large standard errors when few items are located near the person's ability level. For example, for an individual who exhibits a perfect score, all that is known is that the ability level of the examinee far exceeds the item difficulty levels. However, it is unknown exactly by *how much* the ability exceeds item difficulties. MAP and EAP circumvent these issues through the specification of a prior distribution.

Generally, both MAP and EAP can be regarded as Bayesian approaches in that both approaches utilize information from a prior distribution of abilities and the set of response patterns to develop a **posterior distribution** (see Miočević & Van de Schoot, Chapter 18 for more on Bayesian approaches). Similar to MML, the prior distribution represents an assumption regarding the distribution of abilities in the population. For example, it is common practice to assume ability levels are distributed normally, with a mean of 0 and standard deviation of 1. In both MAP and EAP, the lack of information provided by extreme response patterns is compensated for by using information from the specified prior distribution. Effectively, a prior distribution functions to "fill in" information for ability estimates when the response pattern offers limited information. The general effect of the MAP and EAP approaches is to pull ability estimates toward the mean of the prior, reduce the variance of ability estimates, and reduce the standard error of the estimates. Both MAP and EAP are similar in their advantages, though EAP is less computationally burdensome so is often preferred.

The one primary challenge in these approaches is specifying a prior distribution of abilities that is valid. Although the standard normal is commonly specified, the true form of the distribution of abilities is often unknown. Thus, the choice of the prior distribution is frequently open to debate. If the prior distribution is grossly misspecified, then parameter estimates will be biased. Fortunately, estimates tend to be robust for mild to moderate misspecifications of the prior.

19.5 Assumptions of IRT

While IRT approaches offer a host of advantages over classical measurement approaches, they also establish a more stringent set of demands. In addition to the increased data and computational demands of IRT models, IRT approaches require a fairly robust set of assumptions. Principally, IRT approaches require that the dimensionality of the model not be underspecified and that local independence holds.

The models described previously are all unidimensional in the sense that a single, underlying trait is assumed to drive item responses. Multidimensional extensions of these models exist for situations in which two or more latent traits underlie the response process. Note that overspecifying the true dimensionality will not inherently bias results.

Local independence essentially states that item responses are expected to be independent, after controlling for the latent trait. Put another way, nothing beyond the trait being modeled should account for item performances; only the underlying trait level should be controlling response patterns. Recall that, during estimation, the probability of a response pattern was taken as the product of expected response probabilities for the individual items. Generally speaking, the joint probability between any two *independent* events can be calculated as the product of their individual probabilities. Thus, ensuring independence of responses, controlling on the latent trait, is fundamental to the development of an appropriate likelihood function. Violations of this assumption may occur when, for example, one item offers a clue to solving another item. Violations may also occur when the model fails to account for particular traits or abilities that affect the response process. In this way, local independence is related to correct specification of dimensionality; if local independence holds, then the dimensionality has not been underspecified.

Evaluation of residual item correlations is a traditional method of evaluating whether local independence holds for a given model. After applying a particular IRT model of interest, it is possible to calculate residuals by subtracting model expected responses from observed responses, for all items. These residual values should be uncorrelated between items if local independence holds. Yen (1984) and Chen and Thissen (1997) have described principled approaches to evaluating these residual correlations to detect violations of local independence.

If local independence is violated, there are a few modifications that can be made to the test structure to account for item dependencies. A popular procedure is to combine dependent items into a single set, or "testlet" (Thissen, Steinberg, & Mooney, 1989). This testlet is then scored as a single item

with a polytamous model, whereby different response patterns in the testlet are treated as different response categories.

Because of the relationship between correct specification of dimensionality and local independence, it is additionally useful to examine the underlying dimensionality of the data. Traditional approaches include evaluation of Eigenvalues from a matrix of tetrachoric correlations and classical internal consistency statistics. However, a few contemporary approaches may offer more powerful means of evaluating the essential dimensionality of a response matrix. Those interested are referred to Stout (1990), the corresponding DIMTEST software package, and the DETECT software package (Stout, Habing, Douglas, Kim, Roussos, & Zhang, 1996).

19.6 Model Fit

19.6.1 Overall Model Fit

Recall that the likelihood value represents the probability of observing the full set of item responses, conditional on the parameters from the specified model. In this way, the likelihood value represents a metric of model fit; the larger the likelihood value, the more probable the pattern of observations, given the model parameters. The likelihood value also forms the basis for a statistical test of comparative fit between two *nested* models, for a given dataset.

Essentially, a set of models is nested if one model can be converted to the other simply by constraining or freeing a set of parameters. For example, the Rasch, 2PL, and 3PL models are all nested models. The 3PL can be converted to the 2PL model simply by constraining all the guessing parameters to zero. Additionally, the 3PL could be converted to the Rasch model by constraining all the guessing parameters to 0 and all the discrimination parameters to 1. When a set of models is nested in this way, a likelihood ratio test can be carried out to determine whether or not the more elaborate model provides significant predictive benefit over the simpler model. It is calculated as follows:

$$Likelihood\ ratio = -2\big[\ln\big(likelihood_{simple}\big) - \ln\big(likelihood_{elaborate}\big)\big]$$

In this equation, the number of parameters estimated for the elaborate model is always less than the number of parameters estimated for the simple model. The likelihood ratio statistic follows a chi-square distribution, with df = the difference in the number of parameter estimates between the two models. This creates the opportunity for deriving statistical tests of relative fit. Many IRT software packages also provide Akaike's Information Criterion (AIC) and the Bayesian Information Criterion (BIC) values in the output file for model fit. These indices provide a sense of model misfit; larger values indicate larger residuals. As with the likelihood value, these values can be used to evaluate the relative fit between two models. However, neither AIC nor the BIC require nested models for the comparison. Unlike the likelihood statistic, though, no test of significance exists for these metrics.

19.6.2 Item Fit

There are a number of different statistics offered in the literature for evaluating item fit, though many are essentially based on the same approach. First, a given measurement model of interest is defined and item and person parameters are estimated according to the model. Once person parameters have been estimated, examinees are clustered into ten to fifteen different groups according to their estimated abilities. Ideally, these groups should be of equal size and evenly spaced by median ability. Next, the proportion of correct responses for a given item can be determined for each group. If the item fits well within the model, these observed proportions should reflect the expected response probabilities from the measurement model at each ability level. If the observed proportions differ substantially from the expected probabilities, the item may need to be eliminated or modeled differently.

Comparison of observed proportions versus expected probabilities can be done visually by plotting the observed proportions alongside the item characteristic curve. Significance tests are also available for this comparison. A traditional approach is to calculate a chi-square statistic for any item i:

$$\chi_i^2 = \sum_{k=1}^{K} N_k \frac{O_{ki} - E_{ki}}{E_{ki}}$$

In this equation, O_{ki} is the observed proportion correct for cluster k, E_{ki} is the expected probability of correct responses from the model for cluster k, and N_k is the number of examinees in each cluster. The chi-square statistic is distributed with df = $K-1-p$, where K is the number of clusters and p is the number of item parameters estimated.

However, the inflated error rate for this approach has been cited as a great concern. Others have noted that the statistic is biased due to the fact that examinees are sorted on a value that is derived from the model itself (the ability estimate). Thus, several modified and updated chi-square tests have been proposed. For example, Orlando and Thissen (2000) developed a method that sorts examinees by **test total score** (instead of estimated ability). This circumvents the bias previously noted in using estimated ability to sort subjects.

19.7 Conclusion

In this chapter, we discussed some of the basic aspects of IRT measurement, with frequent reference to the advantages afforded over classical measurement approaches. In particular, we focused the chapter on the most foundational IRT models. Since the development of these foundational IRT models, there has been an explosion of novel ways to characterize and re-parameterize these models. These distinct advantages, and the unique approach IRT takes in jointly scaling both item characteristics and person ability on a common latent scale, create the opportunity for many different approaches to modeling the response process. Different models have been able to ask different questions about the conditions and abilities that jointly impact item responses. It is our goal that readers use this chapter as a foundation, and continue to explore the wealth of advantages and types of research questions offered by IRT, when selecting a measurement model for future endeavors.

KEY TAKEAWAYS

- IRT is a psychometric, latent trait alternative to CTT that measures latent ability from observed item responses. Unique from CTT, IRT jointly scales both item difficulty and person ability on a common scale.
- The probability of a correct response for an individual on a given item is modeled with a logistic function that relates both item and person characteristics. For example, the more a person's ability falls above item difficulty on the common scale, the higher the probability of a correct response.
- Person scores can be meaningfully compared even when the scores are derived from different item sets.

Conversely, item difficulty estimates can be meaningfully compared across different subject pools (an advantage over CTT approaches).
- IRT approaches have been able to expand the field of measurement beyond the limits of CTT. For example, the IRT approach has led to the development of full information factor analysis, computer adaptive testing, and automatic item generation.

IDEAS FOR FUTURE RESEARCH

- Will advanced computational techniques allow for automatic creation of items?

- Will the field be able to develop proper algorithms for the selection of items from different content domains?
- How will multidimensional models be incorporated into IRT?

SUGGESTED READINGS

Embretson, S. E., & Reise, S. P. (2000). *Item Response Theory for Psychologists*. Mahwah, NJ: Lawrence Erlbaum Associates.

Lord, F. M., & Novick, M. R. (1968). *Statistical Theories of Mental Test Scores*. Reading, PA: Addison-Wesley.

Rasch, G. (1960). *Probabilistic Models For Some Intelligence and Achievement Tests*. Copenhagen: Danish Institute for Educational Research.

Reckase, M. (2009). *Multidimensional Item Response Theory* (Vol. 150). New York: Springer.

REFERENCES

Bock, R. D. (1972). Estimating item parameters and latent ability when responses are scored in two or more nominal categories. *Psychometrika*, *37*(1), 29–51.

Chen, W. H., & Thissen, D. (1997). Local dependence indexes for item pairs using item response theory. *Journal of Educational and Behavioral Statistics*, *22*(3), 265–289.

Lord, F. M., & Novick, M. R. (1968), *Statistical Theories of Mental Test Scores*. Reading, PA: Addison-Wesley.

Masters, G. N. (1982). A Rasch model for partial credit scoring. *Psychometrika*, *47*(2), 149–174.

Muraki, E. (1990). Fitting a polytomous item response model to Likert-type data. *Applied Psychological Measurement*, *14*(1), 59–71.

Muraki, E. (1992). A generalized partial credit model: Application of an EM algorithm. *ETS Research Report Series*, *16* (June), 159–172.

Orlando, M., & Thissen, D. (2000). Likelihood-based item-fit indices for dichotomous item response theory models. *Applied Psychological Measurement*, *24*(1), 50–64.

Rasch, G. (1960). *Probabilistic Models for Some Intelligence and Achievement Tests*. Copenhagen: Danish Institute for Educational Research.

Samejima, F. (1969). Estimation of latent ability using a response pattern of graded scores. *Psychometrika*, Monograph Supplement, No. 17.

Stout, W., Habing, B., Douglas, J., Kim, H. R., Roussos, L. A., & Zhang, J. (1996). Conditional covariance-based nonparametric multidimensionality assessment. *Applied Psychological Measurement*, *20*, 331–354.

Stout, W. F. (1990). A new item response theory modeling approach with applications to unidimensionality assessment and ability estimation. *Psychometrika*, *55*(2), 293–325.

Thissen, D., Steinberg, L., & Mooney, J. A. (1989). Trace lines for testlets: A use of multiple-categorical-response models. *Journal of Educational Measurement*, *26*(3), 247–260.

Yen, W. M. (1984). Effects of local item dependence on the fit and equating performance of the three-parameter logistic model. *Applied Psychological Measurement*, *8*(2), 125–145.

20 Social Network Analysis

Sebastian Leon Schorch

Eric Quintane

20.1 What Is Social Network Analysis?

Social Network Analysis (SNA) is a theoretical perspective and a research method used in the social and behavioral sciences. It distinguishes itself from other research methods by focusing on social relationships and is based on the idea that social actors (e.g., individuals, groups, organizations, countries) are influenced by the patterns of social relations surrounding them. SNA enables researchers to investigate these patterns as well as their antecedents and consequences. Thereby, researchers can move beyond individualistic explanations to develop a more contextual understanding of social phenomena (Wasserman & Faust, 1994).

This chapter provides an introduction to SNA for social and behavioral scientists that are new to the method. We focus on articulating the logic of key network concepts and measures rather than writing an exhaustive review. Accompanying the chapter, Figure 20.1 serves as a roadmap to help the reader navigate through concepts and measures at different levels of analysis. The numbers in the figure correspond to sections in the chapter, in which the terms and

concepts will be further explained, generally from simpler to more complex.

20.1.1 Foundations and Key Ideas

A diverse set of scientific disciplines has contributed to the establishment of the networks research paradigm in the social and behavioral sciences, including psychology, sociology, anthropology, geography, mathematical biology, and political science. Two forms of inquiry are common in network analysis. First, there is a "top down" approach (starting with a large social network), pioneered by early sociologists (such as Émile Durkheim or Georg Simmel), that aims to understand social subdivisions in villages, communities, and organizations (e.g., Davis, Gardner, & Gardner, 1941; Mayo, 1949). Second, a "bottom up" approach (starting with pairs of individuals) was advanced by researchers in social psychology, who explored how individuals form relationships in small groups and how these relationships influence the structuring of larger social aggregates (e.g., Cartwright & Harary, 1956; Heider, 1946; Moreno, 1934). These streams of inquiry were synthesized by George Homans and then by Harrison C. White, who made significant advances in the analytical decomposition and mathematical modeling of

328

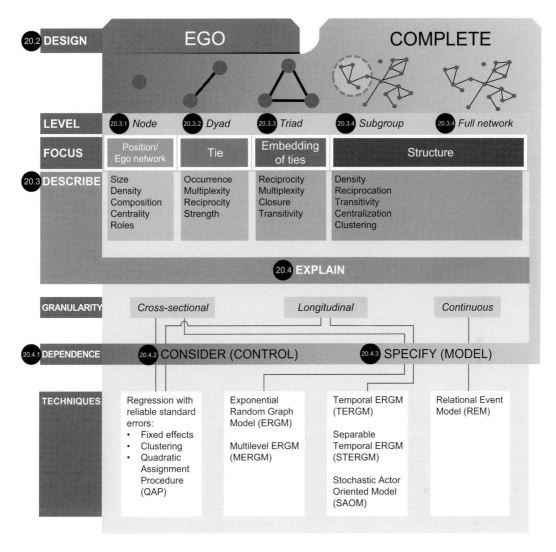

20.2 DESIGN	EGO			COMPLETE	
LEVEL	20.3.1 *Node*	20.3.2 *Dyad*	20.3.3 *Triad*	20.3.4 *Subgroup*	20.3.4 *Full network*
FOCUS	Position/ Ego network	Tie	Embedding of ties	Structure	
20.3 **DESCRIBE**	Size Density Composition Centrality Roles	Occurrence Multiplexity Reciprocity Strength	Reciprocity Multiplexity Closure Transitivity	Density Reciprocation Transitivity Centralization Clustering	
			20.4 **EXPLAIN**		
GRANULARITY	*Cross-sectional*		*Longitudinal*		*Continuous*
20.4.1 **DEPENDENCE**		20.4.2 CONSIDER (CONTROL)		20.4.3 SPECIFY (MODEL)	
TECHNIQUES	Regression with reliable standard errors: • Fixed effects • Clustering • Quadratic Assignment Procedure (QAP)	Exponential Random Graph Model (ERGM) Multilevel ERGM (MERGM)	Temporal ERGM (TERGM) Separable Temporal ERGM (STERGM) Stochastic Actor Oriented Model (SAOM)	Relational Event Model (REM)	

FIGURE 20.1 Social network research map

social structure. The last four decades have seen significant advances in methods and theories that led to an explosion of applications of SNA in a multitude of fields (see Borgatti, Mehra, Brass, & Labianca, 2009). While social network analysis was, historically, strongly influenced by anthropology and qualitative accounts of social structure, we explicitly focus this chapter on the quantitative approaches that have represented the mainstream in network research over the last decades.

20.2 Understanding, Collecting, and Visualizing Network Data

20.2.1 Terminology

A typical social network contains one set of actors (e.g., students in a classroom) which are linked by some type of relationship (e.g., friendship). More generally, networks are composed of a finite set of identifiable entities and of a finite set of identifiable ties that connect these entities (Wasserman & Faust, 1994). Ties have been used to represent social similarities (location, common membership, common attributes, etc.), social relations (kinship, friendship, etc.), interactions (conversations, electronic communications, collaborations, etc.), or flows (information, resources, etc.; Borgatti et al., 2009). Ties can be **directional** (e.g., one actor asks another actor for advice) or **non-directional** (e.g., two actors are married). Ties can also be **binary** (e.g., two actors are married or not) or **valued** (e.g., one actor can seek advice more or less frequently from another actor). When actors have multiple relationships with each other, researchers speak of **multiplex ties**. Any focal actor in a network is typically referred to as **ego** and the actors that are linked to ego via ties are referred to as **alters**.

20.2.2 Network Data

Network data can be cross-sectional, longitudinal, or continuous. While cross-sectional data contains relationships between actors at a given point in time, longitudinal data typically contains the same types of ties, between the same set of actors, at multiple points in time. For continuous network data, the temporal information pertains to interactions between social actors that occur continuously over time (e.g., email communication). Furthermore, network data can be collected to represent relationships between one type of actor (e.g., friendship between individuals), or between two different types of actors (e.g., people's participation in events). The latter yields so-called two-mode networks (see Breiger, 1974) which can span multiple levels of analysis (multi-level networks, see Lazega, Jourda, Mounier, & Stofer, 2008).

20.2.3 Sources of Network Data

The most frequently used method to collect social network data is sociometric questioning, where researchers administer questionnaires to a group of respondents, asking them about relationships of interest. However, increasingly important sources of network data are archival records of interaction/collaboration artifacts or of affiliations with some subordinate entity or outcome. Some examples are:

- logs of professional technology-mediated communication (e.g., Goldberg, Srivastava, Manian, Monroe, & Potts, 2016)
- social media (e.g., Golder, Wilkinson, & Huberman, 2007)
- publication records (e.g., Lazega et al., 2008)
- historical records (e.g., Padgett & Ansell, 1993)
- patents (e.g., Goetze, 2010)
- trade data between countries (e.g., Cranmer, Heinrich, & Desmarais, 2014).

Further data sources include observation of individuals (e.g., Wyatt, Choudhury, Bilmes, & Kitts, 2011) and diary research (e.g., Fu, 2008).

20.2.4 Data Collection

There are two main ways to design the collection of network data. First, in an ego network data collection design, researchers collect information about selected actors' direct relationships to alters and possibly the relationships between these alters. This method resembles traditional data collection methods in the social and behavioral sciences and is particularly suitable for research that focuses on individuals and their immediate relationships (e.g., in studies of social support; Barrera, 1980). Second, in a full network data collection design, researchers obtain ego network data from all actors within a bounded population, which enables them to investigate the relational structure of the population.

Ego Network Design. Ego network data collection commonly begins with the use of a name generator question when interviewing or surveying subjects. Subsequently, researchers ask subjects about their relations with each alter they nominated (e.g., "How frequently do you exchange information with this person?"). While clearly having advantages with regards to the ease of data collection, the method relies on a single informant's recall of relationships, which can be problematic. The recall of one's own interactions is typically biased toward longer-term interactions (Freeman, Romney, & Freeman, 1987) and inaccurate when respondents report their perceptions of relationships between third parties (McEvily, 2014). Extensions for the name generator approach, to measure access to individuals in particular social positions (Lin, Fu, & Hsung, 2001) and to specific resources (Van Der Gaag & Snijders, 2005), are frequently used.

Complete Network Design. The limitations of collecting ego network data can be mitigated, to some extent, by collecting complete network data. First, respondents can be presented with a roster (i.e., a complete list of individuals that are part of a population of interest; Marsden, 1990). This reduces the bias stemming from unreliably recalling interaction partners' names. Second, inaccuracies in perceiving relationships to others can be

cross-checked, since all members of the population answer questions about their relationships toward each other. Third, in a complete network survey, subjects are usually not asked about their perception of relationships between third parties, as this information is obtained directly from the parties involved.

Complete network data can be collected through sociometric questioning for relatively small populations (e.g., departments in organizations or school classes) or extracted from electronic repositories. For larger populations, the use of sociometric questioning may become unreliable due to the length of the resulting questionnaire (Pustejovsky & Spillane, 2009).

Network Boundaries and Sampling. Defining the boundaries of a network is a crucial step in designing a network data collection (Lauman, Marsden, & Prensky, 1983). There are two principal ways to define the boundaries of a network. First, researchers can rely on formal boundaries (e.g., work groups, departments, organizations, or industries). A disadvantage of this approach is that it discards all ties that members within the boundary maintain with actors outside the boundary – which may ultimately have important, yet systematically ignored, consequences. This bias can be reduced by allowing respondents to include alters from outside the boundary (e.g., Reagans & McEvily, 2003). Second, researchers can use boundaries set by the participants themselves (e.g., a collectively shared identity or a common experience). This approach has advantages when the boundaries of the network are difficult to ascertain (e.g., when studying networks among criminals).

When populations are too large to be surveyed completely, representative sampling would normally be an obvious approach. However, drawing samples is problematic when collecting full network data because it is very difficult to ascertain that the relational structure among members of the sample is similar to the relational structure among all members of the population (Burt, 1981). The sampling issue in collecting full network data can be "dealt with" in two different ways: 1) by relying on a (representative) sample of ego networks (thereby disregarding global properties of the network; see Burt, 2004); 2) by using a snowballing or respondent-driven sampling technique (moving from an initial sample deeper into the social structure by questioning the individuals that were nominated by prior respondents; Heckathorn & Cameron, 2017; Milgram, 1967).

Network Survey Questions. Network surveys typically rely on single items for specific relationships. While the use of multiple items would be preferable, it significantly increases the length of questionnaires and may consequently affect reliability. In ego network designs, researchers have mitigated potential misinterpretations of single-item questions by using multiple name generators to identify a social network – for example, "Please list the six persons that you feel closest to" or "Please list all persons with whom you have discussed important matters during the last six months" (Campbell & Lee, 1991). When a complete roster of the population is provided, a common strategy is the use of elaborate prompts that provide an example or a definition of the relationships that are assessed. For example, asking "I consider this person as a friend, as someone that I care about and that cares about me and with whom I like to spend time" is more reliable than asking "this is my friend" when recording a friendship network (Marsden, 1990).

Ethical Considerations. There are several ethical issues to consider when collecting network data (Robins, 2015). First, complete network surveys require the possibility to identify respondents in order to be able to match them across surveys. This implies that researchers cannot offer complete anonymity. However, de-identification methods can (and should) be used in order to protect the identity of participants. Researchers need to emphasize such steps during institutional ethics reviews, when negotiating access to an organization, and when contacting respondents. Second, respondents are often asked to provide information about other people who have not given consent for this information to be shared. Third, the information obtained using network analysis (e.g., an individual's network position) can be very revealing, and care must be taken when informing others about the results to protect respondents. Group-level visualization or layout changes might be required to ensure anonymity (see 20.2.6). For a general discussion of ethical issues in the social and behavioral sciences, see Ferrero and Pinto, Chapter 15.

20.2.5 Data Organization

Network data is organized in matrices or lists. In adjacency matrices, rows indicate the sender of the tie while columns indicate the receiver of the tie. When a network dataset consists of directed ties, its matrix is asymmetrical, and when a network dataset consists of undirected ties, its matrix is symmetrical. For example, the left-hand section of Figure 20.2 shows a data matrix for a binary network in which "1" could indicate that ego considers alter a friend and "0" could indicate that no such friendship relationship exists. The right-hand side of Figure 20.2 shows a data matrix for a valued network, representing the frequency with which ego seeks information from alter with values ranging from "0" (never) to "5" (multiple times per day). When network data is based on affiliations (two-mode networks), the names in the rows and in the columns are distinct.

Network data is also frequently stored in edge lists. These are multi-column lists that contain only present connections between actors, which make them particularly useful for very large or very sparse networks.

20.2.6 Data Visualization

Network visualization is a powerful tool that allows getting a quick overview of various aspects of relational structures (Freeman, 2000). A large number of software packages and plugins that facilitate the visualization of networks exist. We consider NetDraw (a component of UCInet; Borgatti, Everett, & Freeman, 2002), Gephi (Bastian, Heymann, & Jacomy, 2009), nodeXL (Smith, Milic-Frayling, Shneiderman, Mendes Rodrigues, Leskovec, & Dunne, 2010), and igraph (Csárdi & Nepusz, 2006) as most user friendly.

To make a graph easily interpretable, researchers have multiple options. One frequently applied technique is the graphical adjustment of nodes (actors) and ties (relationships). The size of nodes is often proportional to the number of ties a node has. As a result, nodes with many ties appear larger,

and nodes with fewer ties appear smaller. Similarly, ties can be visually distinguished by adjusting their width. For instance, ties that represent a higher interaction frequency can appear thicker and ties that represent a lower interaction frequency can appear thinner. Additionally, nodes and ties can be colored according to their attributes.

A more sophisticated operation is the layout of the graph, which can be determined with the help of algorithms. While differing in detail, many frequently used algorithms arrange the nodes according to three basic principles: First, nodes that share a lot of mutual ties are grouped together. Second, actors or subgroups that bridge the gaps between other actors or subgroups appear between them and are placed more to the center of the graph. Third, the distance between nodes or subgroups is not arbitrary but meaningful in relation to other distances in the graph.

The above adjustments can be observed in Figure 20.3, which shows the graph of a non-directed network with weighted ties (produced in Gephi).

20.3 Describing Network Data

Descriptive statistics of network data can be derived at five levels of analysis: Node (actor) level, dyad level (ties between two actors), triad level (ties between three actors), subgroup level, and network level. The following provides a summary of the most common measures at each level. For mathematical details and a more comprehensive collection of available measures, please refer to Borgatti, Everett, and Johnson (2013) or Wasserman and Faust (1994).

	Paul	Julia	Brian	Max	Rose
Paul	–	1	1	1	1
Julia	1	–	0	0	1
Brian	1	0	–	0	0
Max	0	1	1	–	0
Rose	1	0	1	1	–

	Paul	Julia	Brian	Max	Rose
Paul	–	2	2	4	3
Julia	5	–	0	0	5
Brian	2	0	–	3	0
Max	4	4	5	–	0
Rose	1	2	1	1	–

FIGURE 20.2 Network data in matrix format

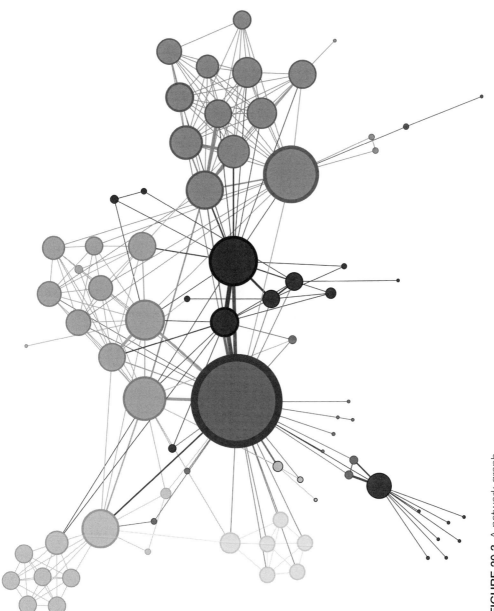

FIGURE 20.3 A network graph

20.3.1 Node Level

Measures at the node level typically assess characteristics of ego networks (based on either ego network data or complete network data) and actors' positions in the wider social structure (most adequate for complete network data). Important actor-related characteristics of ego networks include size, density, and composition. Important positional measures are centrality and roles.

Ego network size is the number of alters that ego is directly connected to. It is often related to actors' access to resources and their influence on others in a social group (Powell, Koput, & Smith-Doerr, 1996). Network density represents the number of ties in a network compared to the number of possible ties. Dense networks are typically considered beneficial for the spread of information and coordinated action (Fleming, Mingo, & Chen, 2007), yet they may contain less diverse information than sparser networks (Lazer & Friedman, 2007). The composition of an ego network refers to the distribution of attributes of alters that ego is directly connected to. Composition can be used to approximate the diversity of resources that an actor is able to access.

Centrality refines the concept of network size and can have strong implications for actors' power and possibilities for action (Bonacich, 1987). While degree centrality is equivalent to ego network size, distinguishing between incoming and outgoing ties provides additional information. In-degree (the number of incoming ties) can be relevant to understand popularity and status in networks, where higher-status actors are likely to have a higher in-degree than lower-status actors (Sauder, Lynn, & Podolny, 2012). Out-degree centrality (the number of outgoing ties) is often considered a proxy for ego's activity. A plethora of additional centrality measures that take into account aspects of the wider network have been developed (see Schoch, 2017 for an overview). Among the most important ones are closeness centrality and betweenness centrality and eigenvector centrality. Closeness centrality represents the average path length (i.e., the number of steps) that is required to reach all other actors in the network. Betweenness centrality approximates the extent to which an actor stands "in-between" other actors in a given network (i.e., the extent to which other actors have to go through the focal actor to reach one another). Lastly, eigenvector centrality provides information about the extent to which an actor is connected to other central actors in the network.

Identifying the roles of actors has been a key area of methodological development in SNA. Actors who are connected to similar alters are called structurally equivalent and are often assumed to fulfill similar roles. There are multiple forms of structural equivalence, which can be calculated using a variety of algorithms (Wasserman & Faust, 1994). A specific actor role, which has generated much research interest, is that of a broker. Brokers connect otherwise disconnected actors in a network, which can be advantageous with regards to work performance and creativity (Burt, 2004). Brokerage is commonly assessed through **constraint**, a measure for the redundancy in actors' networks (Burt, 1992) or through the identification of triadic brokerage relations (Gould & Fernandez, 1989).

20.3.2 Dyad/Tie Level

Networks are structures of interconnected relationships for which dyads are the smallest building blocks. Three types of analyses are typically conducted around dyads. First, where in the social structure do ties exist (i.e., between which types of actors)? Research has shown that specific combinations of actor attributes are conducive for the emergence of ties. Two examples are similarity (homophily) and colocation (propinquity; McPherson, Smith-Lovin, & Cook, 2001). Moreover, the presence of one type of tie often coincides with another (multiplexity). For instance, frequent interaction can be an antecedent for trust (Rivera, Soderstrom, & Uzzi, 2010).

A second type of analysis is related to the reciprocation of ties between actors. Ties in directed networks can be asymmetric (i.e., a tie goes from ego to alter, but not from alter to ego), or reciprocated (i.e., ties go from ego to alter and from alter to ego). Asymmetric affective and asymmetric exchange-oriented ties can often be expected to represent some temporary state on their way toward reciprocation or dissolution (Rivera, Soderstrom, & Uzzi, 2010). Contrarily, advice ties in organizations are often characterized by an absence of reciprocation and imply hierarchy (Lazega, 2001).

A third angle of analysis relates to the strength of ties. The distinction between "strong" ties and "weak" ties (e.g., based on interaction frequency or interpersonal affect) is empirically and theoretically meaningful. The resources that flow along strong ties may be qualitatively different from those flowing along weak ties, and weak ties may provide access to more distant areas of an actor's social network than strong ties (Granovetter, 1973).

20.3.3 Triad Level: Cohesion and Embedding of Ties

The next larger building blocks of networks are triads, which are combinations of ties between three actors. The avenues for exploration outlined for dyads remain, to a large extent, valid, yet, triadic analysis offers additional opportunities. While dyads offer up to four distinguishable connection states (absent, asymmetrical [2], and reciprocated), triads offer sixteen such states (grouped into triad isomorphism classes; see Figure 20.4). Each connection between the three actors can be absent, point from one actor to the other, or be reciprocal.

Triads in networks can be counted (triad census) to assess the non-random occurrence of theoretically meaningful triadic constellations (Wasserman & Faust, 1994). Thereby, closed triads generally point toward cohesion, which has been associated with effective knowledge transfer (Fleming et al., 2007) and the existence of strong behavioral norms and expectations (Portes, 1998). Open triads point toward low cohesion in (ego) networks and brokerage opportunities. Investigating the embedding of a focal dyad within triadic structures can be empirically relevant. For instance, dyads that are embedded in a triad of reciprocal strong ties (so-called Simmelian ties) have been identified as particularly strong, stable, and conducive for knowledge transfer (Tortoriello & Krackhardt, 2010).

The evolution of triadic structures often follows the principles of balance and transitivity. Individuals strive for balance in their attitudes (Cartwright & Harary, 1956). Therefore, a balanced state of mutual liking or mutual disliking in triads is expected to develop over time. The logic of transitivity posits that a triad of A, B, and C is transitive whenever it contains a directed tie from A to B, a directed tie from B to C, and a directed tie from A to C. Transitive triads play an important role in stabilizing social structure, and intransitive

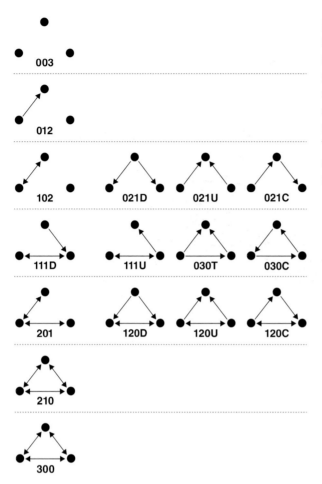

FIGURE 20.4 Triad isomorphism classes. Figure adapted from Wasserman and Faust (1994). The first number represents the number of reciprocal dyads, the second number represents the number of asymmetric dyads, and the third number represents the number of empty dyads. The letter distinguishes between different configurations of asymmetric dyads: U for Up, D for Down, T for Transitive, C for Cyclic

triads tend to develop toward a transitive state (Wasserman & Faust, 1994).

20.3.4 Subgroup and Full Network

A variety of measures and analytical approaches to characterize subgroups and full networks have been developed. Three types of measures are particularly important. First, network-level measures of density, tie reciprocation, and overall transitivity represent general characteristics of a network and can provide information about cohesiveness, hierarchy, or stability in a network (see 20.3.2 and 20.3.3). Second, network centralization (i.e., the extent to which a network resembles a star with few actors in the center and many in the periphery) has been found to influence information and knowledge transfer (Leavitt, 1951). Third, as ties in networks often form more or less cohesive subgroups (clusters), a measure that evaluates the degree of clustering, such as the clustering coefficient, can be insightful when comparing networks. Various approaches have been proposed to identify actors that belong to subgroups. On the one hand, one can identify fully interconnected subgroups (i.e., groups in which all actors have all possible relationships to one another) and gradually relax this requirement to include actors that are less tightly connected but still share significant overlap in their relational structure with other actors in a subgroup (see Wasserman & Faust, 1994). On the other hand, one can identify critical connections without which the network would disassociate in multiple separate components (bi-components or lambda sets). The systematic analysis of the relationships between subgroups (or between groups of structurally equivalent actors) is achieved through blockmodeling (Wasserman & Faust, 1994).

20.4 Explaining Network Structure

The bottom half of Figure 20.1 is a decision tree that helps researchers choose the appropriate inferential network analysis technique, depending on the type of network data (ego network or full network), the granularity (cross-sectional, longitudinal, or continuous), and on the desired treatment of the dependence of observations.

20.4.1 Dependence of Observations

Relationships in social networks are inherently interdependent. In complete network data, each dyad may be included in the calculation of measures for multiple actors. For example, Actor A may have eight advice ties to other actors .These ties are not only counted for Actor A, but also toward the number of advice ties that those eight actors receive, or toward the amount of closure around Actor C, from whom Actor A and others ask advice. Consequently the assumption of independence of observations in regression analysis is systematically violated. Without accounting for this non-independence, the standard errors of (regression) estimates would be unreliable.

There are two general approaches to dealing with this problem. First, researchers can control for the dependence of observations to make it less of a methodological issue. Second, researchers can recognize the theoretical importance of this interdependence and make it an explicit part of the analysis.

20.4.2 Controlling for Dependence of Observations

When using regression-like analyses (linear regression, logistic regression, t-test, etc.), the dependence of observations can be controlled for in three ways. First, you can add a fixed effect (dummy variable) for every actor (sender and/or receiver) that captures network-related autocorrelation as well as other unobserved heterogeneity (e.g., Reagans & McEvily, 2003). Second, standard errors can be clustered on both members of a dyad (Cameron, Gelbach, & Miller, 2011). Third, a permutation algorithm, such as the Quadratic Assignment Procedure (QAP; Dekker, Krackhardt, & Snijders, 2003), can be used for the derivation of standard errors. Software, such as R, Stata, or UCInet, facilitates the use of the above techniques natively or through extensions (e.g., by using the package R "sna"; Butts, 2016). Associated research questions typically relate to the identification of systematic differences between actors' networks, to the explanation of actor-level variables through network characteristics, and to the explanation of network characteristics through actor-level variables.

20.4.3 Specifying Dependencies

Researchers can acknowledge that the dependence of network ties is foundational to social network thinking and that it distinguishes social network data from other types of data. Following this perspective, dependencies become a focus of theoretical interest, rather than a methodological issue.

When researchers specify dependencies in their models, it means they assume that the existence of a tie is dependent on the existence of other, neighboring ties (Pattison & Robins, 2002). Such dependencies typically operate at the dyadic level (e.g., reciprocity; see 20.3.2) and at the triadic levels (e.g., balance and transitivity; see 20.3.3) but can also involve more than three actors. For instance, the social circuit dependence assumption posits that a tie between two actors, A and B, who are connected to actors C and D, respectively (i.e., A–C and B–D exist), may be conditionally dependent on the existence of a tie between C and D (Lusher, Koskinen, & Robins, 2013). A number of statistical models have been developed that enable researchers to explicitly specify dependence assumptions. We review them based on the type of data (cross-sectional, longitudinal, continuous) they are most suitable for.

Models for Cross-Sectional Data. Exponential Random Graph Models are autologistic regression models that simultaneously treat the social network as dependent and independent variables (Robins, Pattison, Kalish, & Lusher, 2007). By enabling researchers to specify multiple forms of dependencies, ERGMs permit distinguishing between complementary mechanisms of network formation and quantifying the relative contribution of each mechanism. For instance, the observation of clusters in a network might be driven by tendencies of actors to become friends with friends of their friends (transitivity as a structural effect) or by a preference to become friends with individuals that are demographically similar to them (attribute homophily). An ERGM can give an indication of which mechanism was actually responsible for the observed network pattern (e.g., Wimmer & Lewis, 2010).

However, ERGMs do not provide one objectively true answer about the factors at work in organizing a particular network. The answer depends on how the question is asked (i.e., how the model is specified). Based on this specification, the algorithm simulates a large number of random networks and compares the simulated networks to the empirically observed network to identify whether the specified patterns occurred empirically more or less frequently than would have been expected by chance. Fit statistics help researchers assess how well a model is able to reproduce the empirically observed network structure. ERGMs can be estimated in the software PNet (Wang, Robins, & Pattison, 2009) and in R, using the package "ergm" (Hunter, Handcock, Butts, Goodreau, & Morris, 2008). They are typically used for single networks (directed or undirected), but extensions for bipartite networks and multi-level networks (MERGM; Wang, Robins, Pattison, & Lazega, 2013) are available.

Models for Longitudinal Data. Contemporary statistical models also allow researchers to analyze networks longitudinally and make inferences about changes that occur in the network structure. A prerequisite is usually a panel-like data collection, during which networks between the same actors are collected at multiple points in time (minor deviations are acceptable). The most prominent modeling approaches here are derivatives of ERGMs and Stochastic Actor Oriented Models (SAOMs).

ERGM Derivatives. Temporal ERGMs (or TERGMs; Robins & Pattison, 2001) estimate network change by starting from a fixed network, when simulating random networks, instead of starting from an empty graph. A stochastic algorithm attempts to model the network structure that was empirically observed at time 2, based on the network that was empirically observed at time 1 (again, using the specified parameters). TERGM parameter estimates can be difficult to interpret because they do not distinguish well between the creation of a tie and its persistence. Separable Temporal ERGMs (STERGMs; Krivitsky & Handcock, 2014) solve this problem by predicting, separately, tie formation and dissolution. Finally, Longitudinal ERGMs (LERGMs; Snijders & Koskinen, 2013) assume that ties change one at a time between adjacent network observations, making the model similar to the SAOM.

Stochastic Actor Oriented Models. Stochastic Actor Oriented Models (Snijders, Van de Bunt, & Steglich, 2010) are similar to temporal ERGMs in that they stochastically simulate networks to allow inference about the mechanisms that govern network evolution. However, the key difference to ERGMs is that SAOMs focus on simulating actors' decisions to change and adapt the social structure they are embedded in instead of simulating the structural change per se. In SAOMs actors are assumed to actively change their outgoing ties according to characteristics of other actors and their current network position. As a consequence, the method is primarily used for directed network data, yet an extension for non-directed network data exists. Additionally, SAOMs offer the possibility to model actors' networking decisions in conjunction with behavioral dynamics, which can be useful to separate influence processes from selection processes (e.g., to understand whether substance abuse is a consequence of friendship choices or vice versa; see Steglich, Snijders, & Pearson, 2010). The analysis of multi-level networks is also possible. The well-documented R package ("RSiena") is recommended for model estimation (Ripley, Snijders, Boda, Vörös, & Preciado, 2016).

Models for Continuous Data. The availability of continuous records of interactions (e.g., emails, chats, social media) allows researchers to apply statistical models that do not require temporal aggregation (i.e., instead of using waves of networks, the models rely on sequences of interactions). Temporal aggregation may be problematic because it removes a wealth of information about the timing and sequence of network changes (Quintane, Conaldi, Tonellato, & Lomi, 2014). Inspired by event history analysis, the relational event framework was developed as a way to capture the temporal dependencies in interactions and predict the occurrence of the next interaction (while respecting the social interdependencies between actors, such as reciprocity, transitivity, etc.; Butts, 2008). Consequently, this methodology is particularly useful in contexts where past behavior strongly influences future behavior.

20.5 Conclusion

SNA is a method that focuses on the relationships between actors. It enables researchers to show that relations or specific positions in the social structure can have a disproportionate effect on outcomes. For example, individuals are more likely to obtain a new job through distant (weak) relations (Granovetter, 1973), and individuals in brokerage positions tend to do better (Burt, 1992).

Researchers interested in using the method should carefully consider the tradeoffs during data collection and, especially, whether they want to use an ego network or a complete network design. There are a multitude of statistical measures and methods that can be used to either describe or explain network data. While analyzing network data, researchers should be aware of the dependence of observations, which has implications for the choice of statistical models.

KEY TAKEAWAYS

- Social network analysis focuses on relations between social actors. It is used to understand actor-, group-, or system-level outcomes.
- Patterns of social relations can be described and explained using a variety of quantitative measures and models. Network data can also be visualized graphically.
- Social relations are not independent of each other. Researchers need to decide how to handle the dependence of observations when choosing a statistical inference technique.

IDEAS FOR FUTURE RESEARCH

- By taking a longitudinal approach to social networks, what can we learn about the evolution of networks and actors' choices in adapting relational structures?
- What processes might differ between and across levels that would be best examined using a multi-level approach to SNA?
- What social and behavioral processes might differ between and across levels that would be best examined using a multi-level network analysis?

SUGGESTED READINGS

Borgatti, S. P., Everett, M. G., & Johnson, J. C. (2013). *Analyzing Social Networks*. London: Sage.

Borgatti, S. P., Mehra, A., Brass, D. J., & Labianca, G. (2009). Network analysis in the social sciences. *Science, 323*(5916), 892–895.

Freeman, L. C. (2004). *The Development of Social Network Analysis. A Study in the Sociology of Science*. Vancouver, BC: Empirical Press.

Robins, G. (2015). *Doing Social Network Research: Network-Based Research Design for Social Scientists*. London: Sage.

Wasserman, S., & Faust, K. (1994). *Social Network Analysis: Methods and Applications*. Cambridge: Cambridge University Press.

REFERENCES

Barrera, M. (1980). A method for the assessment of social support networks in community survey research. *Connections, 3*(3), 8–15.

Bastian, M., Heymann, S., & Jacomy, M. (2009). *Gephi: An Open Source Software for Exploring and Manipulating Networks*. International AAAI Conference on Weblogs and Social Media.

Bonacich, P. (1987). Power and centrality: A family of measures. *American Journal of Sociology, 92*(5), 1170–1182.

Borgatti, S. P., Everett, M. G., & Freeman, L. C. (2002). *UCInet for Windows: Software for Social*

Network Analysis. Harvard, MA: Analytic Technologies.

Borgatti, S. P., Everett, M. G., & Johnson, J. C. (2013). *Analyzing Social Networks*. London: Sage.

Borgatti, S. P., Mehra, A., Brass, D. J., & Labianca, G. (2009). Network analysis in the social sciences. *Science, 323*(5916), 892–895.

Breiger, R. L. (1974). The duality of persons and groups. *Social Forces, 53*(2), 181–190.

Burt, R. S. (1981). Studying status/role-sets as ersatz network positions in mass surveys. *Sociological Methods & Research, 9*(3), 313–337.

Burt, R. S. (1992). *Structural Holes: The Social Structure of Competition*. Cambridge, MA: Harvard University Press.

Burt, R. S. (2004). Structural holes and good ideas. *American Journal of Sociology, 110*(2), 349–399.

Butts, C. T. (2008). A relational event framework for social action. *Sociological Methodology, 38*(1), 155–200.

Butts, C. T. (2016). sna: Tools for Social Network Analysis. R package version 2.4. Retrieved from https://CRAN.R-project.org/package = sna

Cameron, A. C., Gelbach, J. B., & Miller, D. L. (2011). Robust inference with multi-way clustering. *Journal of Business & Economic Statistics, 29*(2), 238–249.

Campbell, K. E., & Lee, B. A. (1991). Name generators in surveys of personal networks. *Social Networks, 13*(3), 203–221.

Cartwright, D., & Harary, F. (1956). Structural balance: A generalization of Heider's theory. *Psychological Review, 63*(5), 277–293.

Cranmer, S. J., Heinrich, T., & Desmarais, B. A. (2014). Reciprocity and the structural determinants of the international sanctions network. *Social Networks, 36*(1), 5–22.

Csárdi, G., & Nepusz, T. (2006). The igraph software package for complex network research. *InterJournal Complex Systems, 1695*(5), 1–9.

Davis, A., Gardner, B. B., & Gardner, M. R. (1941). *Deep South*. Chicago, IL: University of Chicago Press.

Dekker, D., Krackhardt, D., & Snijders, T. (2003). Multicollinearity robust QAP for multiple

regression. In *1st Annual Conference of the North American Association for Computational Social and Organizational Science* (pp. 22–25). NAACSOS.

Fleming, L., Mingo, S., & Chen, D. (2007). Collaborative brokerage, generative creativity, and creative success. *Administrative Science Quarterly, 52*(3), 443–475.

Freeman, L. C. (2000). Visualizing social networks. *Journal of Social Structure, 1*(1), 1–15.

Freeman, L. C., Romney, A. K., & Freeman, S. C. (1987). Cognitive structure and informant accuracy. *American Anthropologist, 89*(2), 310–325.

Fu, Y. (2008). Position generator and actual networks in everyday life: An evaluation with contact diaries. In N. Lin & B. H. Erickson (Eds.), *Social Capital: An International Research Program* (pp. 49–64). New York: Oxford University Press.

Goetze, C. (2010). An empirical inquiry into co-patent networks and their stars: The case of cardiac pacemaker technology. *Technovation, 30*, 436–446.

Goldberg, A., Srivastava, S. B., Manian, V. G., Monroe, W., & Potts, C. (2016). Fitting in or standing out? The tradeoffs of structural and cultural embeddedness. *American Sociological Review, 81*(6), 1190–1222.

Golder, S. A., Wilkinson, D. M., & Huberman, B. A. (2007). Rhythms of social interaction: Messaging within a massive online network. *Communities and Technologies, 2007*, 41–66.

Gould, R. V., & Fernandez, R. M. (1989). Structures of mediation: A formal approach to brokerage in transaction networks. *Sociological Methodology, 19*, 89–126.

Granovetter, M. (1973). The strength of weak ties. *American Journal of Sociology, 78*(6), 1360–1380.

Heckathorn, D. D., & Cameron, C. J. (2017). Network sampling: From snowball and multiplicity to respondent-driven sampling. *Annual Review of Sociology, 43*, 101–119.

Heider, F. (1946). Attitudes and cognitive organization. *Journal of Psychology, 21*, 107–112.

Hunter, D. R., Handcock, M. S., Butts, C. T., Goodreau, S. M., & Morris, M. (2008). ergm: A package to fit, simulate and diagnose exponential-family models for networks. *Journal of Statistical Software, 24*(3), 1–29.

Krivitsky, P. N., & Handcock, M. S. (2014). A separable model for dynamic networks. *Journal of the Royal Statistical Society: Series B (Statistical Methodology), 76*(1), 29–46.

Lauman, E. O., Marsden, P. V., & Prensky, D. (1983). The boundary specification problem in network analysis. In R. S. Burt & M. J. Minor (Eds.), *Applied Network Analysis: A Methodological Introduction* (pp. 18–34). Beverly Hills, CA: Sage.

Lazega, E. (2001). *The Collegial Phenomenon: The Social Mechanisms of Cooperation among Peers in a Corporate Law Partnership*. New York: Oxford University Press.

Lazega, E., Jourda, M.-T., Mounier, L., & Stofer, R. (2008). Catching up with big fish in the big pond? Multi-level network analysis through linked design. *Social Networks, 30*(2), 159–176.

Lazer, D., & Friedman, A. (2007). The network structure of exploration and exploitation. *Administrative Science Quarterly, 52*, 667–694.

Leavitt, H. J. (1951). Some effects of certain communication patterns on group performance. *Journal of Abnormal and Social Psychology, 46*(1), 38–50.

Lin, N., Fu, Y., and Hsung, R. (2001). The position generator: Measurement techniques for investigations of social capital. In N. Lin, K. S. Cook, & R. S. Burt (Eds.), *Social Capital: Theory and Research* (pp. 57–81). New Brunswick, NJ: Transaction Publishers.

Lusher, D., Koskinen, J., & Robins, G. (2013). *Exponential Random Graph Models for Social Networks. Theory, Methods, and Applications*. New York: Cambridge University Press.

Marsden, P. V. (1990). Network data and measurement. *Annual Review of Sociology, 16*, 435–463.

Mayo, E. (1949). *The Social Problems of an Industrial Civilization*. London: Routledge.

McEvily, B. (2014). Do you know my friend? Attending to the accuracy of egocentered network data. In D. J. Brass, G. (Joe) Labianca, A. Mehra, D. S. Halgin, & S. P. Borgatti (Eds.), *Contemporary Perspectives on Organizational Social Networks* (pp. 295–313). Bradford, UK: Emerald Group Publishing.

McPherson, M., Smith-Lovin, L., & Cook, J. M. (2001). Birds of a feather: Homophily in social networks. *Annual Review of Sociology, 27*, 415–444.

Milgram, S. (1967). The small-world problem. *Psychology Today, 1*(1), 61–67.

Moreno, J. L. (1934). *Who Shall Survive: A New Approach to the Problem of Human Interrelations*. Washington, DC: Nervous and Mental Disease Publishing Company.

Padgett, J. F., & Ansell, C. K. (1993). Robust action and the rise of the Medici, 1400–1434. *American Journal of Sociology, 98*(6), 1259–1319.

Pattison, P., & Robins, G. (2002). 9. Neighborhood-based models for social networks. *Sociological Methodology, 32*(1), 301–337. https://doi.org/10.1111/1467-9531.00119

Portes, A. (1998). Social capital: Its origins and applications in modern sociology. *Annual Review of Sociology, 24*, 1–24.

Powell, W. W., Koput, K. W., & Smith-Doerr, L. (1996). Interorganizational collaboration and the locus of innovation: Networks of learning in biotechnology. *Administrative Science Quarterly, 41*(1), 116–145.

Pustejovsky, J. E., & Spillane, J. P. (2009). Question-order effects in social network name generators. *Social Networks, 31*, 221–229.

Quintane, E., Conaldi, G., Tonellato, M., & Lomi, A. (2014). Modeling relational events: A case study on an open source software project. *Organizational Research Methods, 17*, 23–50.

Reagans, R., & McEvily, B. (2003). Network structure and knowledge transfer: The effects of cohesion and range. *Administrative Science Quarterly, 48*(2), 240–267.

Ripley, R. M., Snijders, T. A. B., Boda, Z., Vörös, A., & Preciado, P. (2016). *Manual for RSiena*. Oxford: University of Oxford, Department of Statistics, Nuffield College.

Rivera, M. T., Soderstrom, S. B., & Uzzi, B. (2010). Dynamics of dyads in social networks: Assortative, relational, and proximity mechanisms. *Annual Review of Sociology, 36*, 91–115.

Robins, G. (2015). *Doing Social Network Research: Network-Based Research Design for Social Scientists*. London: Sage.

Robins, G., & Pattison, P. (2001). Random graph models for temporal processes in social networks. *Journal of Mathematical Sociology, 25*(1), 5–41.

Robins, G., Pattison, P., Kalish, Y., & Lusher, D. (2007). An introduction to exponential random graph (p*) models for social networks. *Social Networks, 29*(2), 173–191.

Sauder, M., Lynn, F., & Podolny, J. M. (2012). Status: Insights from organizational sociology. *Annual Review of Sociology, 38*(1), 267–283.

Schoch, D. (2017). Periodic table of network centrality. Retrieved from http://schochastics .net/sna/periodic.html.

Smith, M., Milic-Frayling, N., Shneiderman, B., Mendes Rodrigues, E., Leskovec, J., & Dunne, C. (2010). NodeXL: A free and open network overview, discovery and exploration add-in for Excel 2007/2010. Social Media Research Foundation. Retrieved from www .smrfoundation.org

Snijders, T. A. B., & Koskinen, J. (2013). Longitudinal models. In D. Lusher & G. Robins (Eds.), *Exponential Random Graph Models for Social Networks: Theory, Methods and Applications* (pp. 130–139). New York: Cambridge University Press.

Snijders, T. A. B., van de Bunt, G. G., & Steglich, C. E. G. (2010). Introduction to stochastic actor-based models for network dynamics. *Social Networks, 32*(1), 44–60.

Steglich, C., Snijders, T. A. B., & Pearson, M. (2010). Dynamic networks and behavior: Separating selection from influence. *Sociological Methodology, 40*(1), 329–393.

Tortoriello, M., & Krackhardt, D. (2010). Activating cross-boundary knowledge: The role of Simmelian ties in the generation of innovations. *Academy of Management Journal, 53*(1), 167–181.

Van Der Gaag, M., & Snijders, T. A. (2005). The Resource Generator: Social capital quantification with concrete items. *Social Networks, 27*(1), 1–29.

Wang, P., Robins, G., & Pattison, P. E. (2009). *PNet: Program for the Simulation and Estimation of Exponential Random Graph Models*. Melbourne School of Psychological Sciences, University of Melbourne.

Wang, P., Robins, G., Pattison, P., & Lazega, E. (2013). Exponential random graph models for multilevel networks. *Social Networks, 35*(1), 96–115.

Wasserman, S., & Faust, K. (1994). *Social Network Analysis: Methods and Applications*. Cambridge: Cambridge University Press.

Wimmer, A., & Lewis, K., 2010. Beyond and below racial homophily: ERG models of a friendship network documented on Facebook. *American Journal of Sociology, 116*, 583–642.

Wyatt, D., Choudhury, T., Bilmes, J., & Kitts, J. A. (2011). Inferring colocation and conversation networks from privacy-sensitive audio with implications for computational social science. *ACM Transactions on Intelligent Systems and Technology, 2*(1), 1–41.

21 Meta-Analysis: An Introduction

Gregory D. Webster

Meta-analysis describes a family of processes for synthesizing quantitative data across multiple studies. Thus, meta-analyses are studies of studies. Historically, meta-analysis is a relatively new technique, emerging in the mid-1970s as a response to some of the inherent problems with purely narrative reviews (Glass, 1976), with the first popular books on meta-analysis appearing in the early 1980s (Glass, McGraw, & Smith, 1981; Hunter, Schmidt, & Jackson, 1982). Although narrative and quantitative literature reviewers are necessary and important, meta-analyses have grown substantially in popularity over time as most social and behavioral sciences have become more quantitatively focused. For example, a Google Scholar title search in January 2018 of *Psychological Bulletin* – the leading review journal in psychology – using the terms "meta analysis" OR "meta analyses" OR "meta analytic," revealed substantial growth in the proportion of articles with these terms in their titles (1980–2017). Figure 21.1 (top) shows the results of this search; the gray line shows the data, the black line shows a five-year moving average. For example, looking at the last thirty years, the five-year average proportion in 1988 was 0.056 articles; by 2017, it had reached 0.176 articles – a 3.14-fold increase in meta-analysis articles. Meta-analyses are also

on the rise in other social and behavioral sciences as well as the medical and biological sciences. For example, in *The Journal of the American Medical Association (JAMA)*, an identical search showed a similar trend on a different scale. As shown in Figure 21.1 (bottom), *JAMA*'s five-year average proportion of meta-analysis articles in 1988 was 0.0010; by 2017, it was 0.0108 – a 10.8-fold increase. Along with this growth, meta-analytic techniques have become more varied and sophisticated, including different weighting systems, corrections for biases, and more optimal estimation procedures.

21.1 Search, Selection, and Extraction

21.1.1 Published Works

Meta-analyses start with a focused hypothesis or research question that guides the literature search. Although some meta-analyses focus on changes in means over time (e.g., Twenge, Konrath, Foster, Campbell, & Bushman, 2008), most focus on an effect size, which reflects the (bivariate) relationship between two variables. Effect sizes can include standardized relationships between two continuous variables (correlations – r's), standardized differences between two (or more) groups

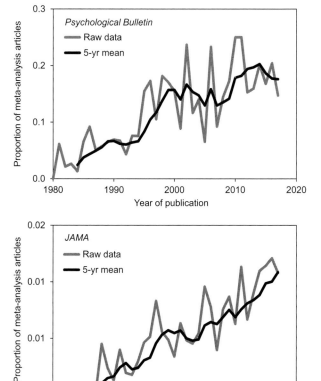

FIGURE 21.1 Proportion of meta-analysis articles published in *Psychological Bulletin* (top) and *JAMA* (bottom), 1980–2017

(Cohen's *d*'s), or effects on a binary outcome (odds ratios – *OR*'s). Thus, most researchers begin their meta-analysis by wanting to know the average effect size across multiple studies.

After establishing their research aims, meta-analysts scour multiple search engines (e.g., PsycINFO, Google Scholar, Web of Science), using relevant search terms. For example, if one were interested in finding articles on the relationships between self-esteem and aggression, then researchers would likely use various combinations of the following search terms: `self-esteem`, `esteem`, and `aggress*`. In some search engines, an asterisk (*) will return all permutations of the stem that precedes it. For example, `aggress*` might return articles containing "aggress," "aggressive," "aggressively," "aggression," "aggressing," and "aggressed."

Meta-analysts also attempt to find other relevant studies not identified by search engines. This involves selecting books, chapters, or review articles pertaining to the research question and scanning titles in their reference lists to locate additional candidate studies. Published works that feature such narrative reviews can offer meta-analysts a goldmine of primary sources, containing the desired effect sizes, which online search engines might miss.

21.1.2 Unpublished Works

Because larger effects and significant results are more likely to be submitted and published, locating only published effect sizes can bias meta-analytic results (see Publication Bias section below). To this end, meta-analysts seek both published and unpublished works. Luckily, there are multiple ways to gather unpublished effect sizes. Two of the more popular ways are group and personal email solicitation. First, meta-analysts often solicit unpublished effect sizes from colleagues by making a request using professional society email listservs. Such solicitations are often made to multiple listservs and frequently are successful. However, this method depends on the willingness of other researchers to send their unpublished research, data, or effect sizes. Second, meta-analysts can identify the "top," or most frequently published, researchers in the area related to the meta-analysis and solicit unpublished effect sizes directly from them. If researchers are cooperative, then this method can be effective and may even result in a growing network of "top" researchers, as those initially contacted bring their own students or coauthors into the fold. Using these procedures, meta-analysts can locate many – but likely not all – unpublished effect sizes.

21.1.3 Coding

Once this literature search is underway, one or more researchers read at least the title and abstract to decide which articles are relevant to the research aims. Ideally, the total number of works identified at each stage should be recorded and reported in the meta-analysis.

If articles are deemed relevant during this first review, then researchers should locate and extract relevant effect size information from the article (e.g., r's, d's, OR's). This often requires skimming – or sometimes carefully reading – the Method and Results sections of each study. Because this is time-consuming, meta-analysts often train a research team to identify and extract relevant effect sizes from articles. During this process, multiple researchers should assess a subset of common articles (e.g., 10–30%) to assure agreement and accuracy in effect size extraction. If agreement – assessed via an interclass correlation coefficient or similar metric – is sufficient, then the researchers can continue to extract effect size with confidence; if not, then they should reassess procedures.

Meta-analysts and their trained research assistants carefully code articles both for vital information (effect size, sample size) and other relevant information that may be used as moderator variables to explain variability in effect sizes. The study-level variables of interest will often depend on the research question, but many meta-analysts often code for publication year, ratings of study quality (internal and external validity; see Card, 2012, pp. 68–73; Wagner & Skowronski, Chapter 2), geographic location, lab affiliation, participant demographic characteristics (expressed as a percentage of sample), theoretical and methodological approaches (correlational, experimental, longitudinal), and the gender, affiliation, and status of lead author(s) (student vs. professor). Relevant effect size variables include scale reliabilities and construct validity. Meta-analysts often examine study-level variables as potential moderators and effect size-level variables to correct for bias (Schmidt & Hunter, 2015). For example, a meta-analyst may test for a decline effect (a reduction in absolute effect sizes over time; Schooler, 2011) by examining moderation by publication year. Regarding effect sizes, measures of reliability can be used to correct correlations for measurement error (Schmidt, 2010; Schmidt & Hunter, 2015). Ideally, meta-analysts should create a standardized checklist or coding sheet, which all coders use, to help assure consistency.

21.2 Meta-Analytic Approaches

After collecting effect sizes, analysis begins. There are multiple ways to meta-analyze effect sizes, including vote-counting, fixed versus random effects, raw versus bias-adjusted effect sizes, univariate, bivariate, and multivariate approaches, and standard versus structural equation modeling (SEM) approaches.

21.2.1 Vote-Counting

Among the first, simplest, and most problematic approaches developed was the vote-counting method, whereby meta-analysts simply count the number of significant (vs. non-significant) effects. The vote-counting method is problematic because the binary nature of statistical decision-making ignores both the magnitudes of the effects and their relative robustness as weighted by (a function of) their sample size. Consequently, the vote-counting method should be avoided.

21.2.2 Effect Size Heterogeneity

One reason why the vote-counting method can produce misleading results is that it does not allow for effect size heterogeneity – the extent to which effect sizes vary after controlling for such factors as the mean effect size, various artifacts, or moderators. Ideally, heterogeneity should be modest or statistically non-significant. If substantial residual

variance (heterogeneity) still exists in effect sizes, then, theoretically speaking, there is likely an additional moderator or artifact that is responsible or the sampled effect sizes come from two or more different populations.

Multiple heterogeneity measures exist (e.g., Q, τ^2, I^2), and these should be reported alongside the weighted mean effect size in meta-analyses. The Q statistic (or Hedge's Q test) is simply the sum of the squared differences between each study's effect size and the overall mean effect size, weighted (i.e., multiplied) by each study's inverse variance (or squared standard error). Q is distributed and evaluated as χ^2 with $k - 1$ degrees of freedom, where k is the number of studies or effect sizes. From Q, one can calculate τ^2, which estimates the population variability in effect sizes in random-effects meta-analysis (see next section). The τ^2 statistic provides meta-analysts with an estimate of population variability of effect sizes by accounting for the extent to which sampling error-based heterogeneity contributes to the total observed heterogeneity; the remainder reflects population variability. The I^2 statistic is simply a re-expression of either Q or τ^2 on a scale from 0 to 100%. For example, I^2 is $(Q - k - 1)/Q$ and then multiplied by 100 when $Q > (k - 1)$; when $Q \leq (k - 1)$, then I^2 is zero. I^2 is also the proportion of the estimate of population variance (τ^2) to total variability ($\tau^2 + \sigma^2$), re-expressed as a percentage, where σ^2 is the within-study variability. I^2s of 25, 50, and 75% reflect small, medium, and large amounts of effect size heterogeneity, respectively; an I^2 of 0% reflects homogeneity (Card, 2012; Huedo-Median, Sánchez-Meca, Marín-Martínez, & Botella, 2006). Despite its seemingly intuitive nature, I^2 is not an absolute measure of heterogeneity and should not be interpreted as such (Borenstein, Higgins, Hedges, & Rothstein, 2017). Assumptions about the heterogeneity of effect size have implications for whether a fixed- or random-effects meta-analysis is used.

21.2.3 Fixed- Versus Random-Effects Meta-Analysis

Meta-analysts must choose between fixed- and random-effects meta-analyses (Hedges & Vevea, 1998). Fixed-effects meta-analysis makes the assumption that all the effect sizes come from the same population, and that the meta-analysts have effectively sampled the entire population of effect sizes. In contrast, random-effects meta-analysis relaxes these assumptions, acknowledging that the theoretical mean effect size itself has a sampling distribution and that sampling all possible effect sizes is impractical, if not impossible. Because random- (vs. fixed-) effects meta-analysis makes fewer assumptions regarding effect size homogeneity, it is more conservative and, consequently, produces wider confidence intervals (CIs) around mean effect sizes. Most meta-analysts have reached consensus that random-effects meta-analysis should be the default choice. Indeed, fixed-effects meta-analysis should be implemented only after researchers have met its stricter assumptions (effect size homogeneity), which is difficult to attain in practice. For these reasons, random-effects procedures are discussed below.

21.2.4 Correcting for Bias and Artifacts: Raw Versus Adjusted Effect Sizes

Among the more controversial – and ongoing – debates in meta-analysis is whether to correct for bias or artifacts in effect sizes prior to meta-analysis. Most meta-analyses are conducted without controlling bias or artifacts. Nevertheless, Schmitt and Hunter (2015) have argued that much of the observed variance (heterogeneity) in effect sizes is a function of both sampling variance and artifacts related to each particular study. These artifacts include unreliability (measurement error), lack of validity, arbitrary dichotomizations, and range restriction (Card, 2012; Schmitt & Hunter, 2015).

Of these artifacts, the most frequently corrected for is unreliability. Correcting for unreliability in meta-analysis is similar to correcting correlations for attenuation – the correlation is corrected or adjusted by a factor of its reliability (Cronbach's alpha, test-retest reliability). Specifically, a correlation between two multi-item scales, r_{xy}, is divided by the square root of the product of the reliabilities of each scale (r_{xx}, r_{yy}) to yield a corrected correlation (r_c):

$$r_c = r_{xy} / \left(r_{xx} r_{yy} \right)^{1/2}. \tag{1}$$

In meta-analyses, however, one must also correct a correlation's standard error (*SE*) using the same correction:

$$SE_c = SE / \left(r_{xx} r_{yy} \right)^{1/2}. \tag{2}$$

Correcting effect sizes for unreliability or other artifacts has two consequences. First, the corrected effect sizes will be larger than uncorrected ones. Second, the resulting SE_cs will also be larger, resulting in wider, more-conservative CIs. This is because the CIs also take into account measurement error (or other artifacts). Thus, regarding measurement error, corrected effect sizes reflect those that would have been observed had unreliability been accounted for, just as latent variable or structural equation modeling (SEM) accounts for unreliability, which involves correcting variables for measurement error by differentially weighting the extent to which items (or parcels of items) contribute to the latent construct, using factor-analytic procedures (Kline, 2016). See Edlund and Nichols, Chapter 1 for a very brief explanation of factor analysis.

Many meta-analysts are reluctant to correct effect sizes for unreliability and other artifacts for both practical and theoretical reasons. First, many meta-analysts remain unaware that effect sizes can be corrected for artifacts, and such corrections add extra steps to the process. Second, on a theoretical level, meta-analyses of *raw* effect sizes generalize to what happens in the average study, whereas meta-analyses of *corrected* effect sizes generalize to what happens in a world free of artifacts, which exists only in theory, not in practice. For example, imagine a meta-analysis of the correlations between people's attitudes about gun ownership and capital punishment; each attitude is assessed with multi-item scales having imperfect reliability. A meta-analysis of *corrected* effect sizes would generalize to the association linking these attitudes as theoretical constructs; if *raw* effect sizes were analyzed, then the findings would generalize to the association between these attitude measures for the average study.

21.2.5 Univariate, Bivariate, and Multivariate Meta-Analysis

The number of variables that researchers focus on also guides their analyses. Researchers interested in examining changes in means over time perform univariate meta-analyses (Card, 2012, pp. 148–150). For example, researchers have used temporal meta-analyses to examine changes over time in self-esteem (Twenge & Campbell, 2001) and narcissism (Twenge et al., 2008). Although useful, some of this work has been criticized for at least two reasons (Wetzel, Brown, Hill, Chung, Robins, & Roberts, 2017). First, any changes in scale means over time could be due to multiple artifacts (history, cohort effects, cultural or geographic differences). Second, lack of measurement invariance (differential item functioning) in multi-item scales (across time, people, or both) calls into question the validity or generalizability of some temporal meta-analyses.

Meta-analyses of bivariate data are common because most researchers seek to address hypotheses involving two variables (e.g., drug vs. placebo effects on health, the correlation between two traits). Consequently, most meta-analytic techniques have been developed to address bivariate questions involving correlations, slopes, effects, or differences between two variables.

Multivariate meta-analysis involves relations among three or more variables (Becker, 2009; Card, 2012; Cheung, 2015b). For example, a psychologist investigates relationships among the three Dark Triad traits – narcissism, psychopathy, and Machiavellianism – to test if they are indeed three distinct constructs (Jonason & Webster, 2010; Paulhus & Williams, 2002; Webster et al., 2016) or whether psychopathy and Machiavellianism are part of the same superordinate construct (Vize, Lynam, Collison, & Miller, 2018). An example from the education literature examines the links between classroom management self-efficacy and three correlated facets of burnout: Depersonalization, emotional exhaustion, and decreased personal accomplishment (Aloe, Amo, & Shanahan, 2014). To answer such multivariate questions, meta-analysts seek complete sets of correlations – often in matrices – among all variables of interest (e.g., Webster & Duffy, 2016).

Although multivariate (Cheung, 2015b), or model-based (Becker, 2009), meta-analysis is a relatively new technique, it will likely continue to grow over time as meta-analysts' questions evolve from simpler, bivariate ones about simple associations, to more complete ones, involving multiple predictors and outcomes. One advantage of multivariate meta-analyses is that, once a meta-analytically derived matrix of weighted mean effect sizes (correlations) is obtained, researchers can use it to test a variety of models, including path and mediation models. One drawback of multivariate meta-analysis is that it is computationally complex. For example, analysts often adjust for covariances that exist among correlations (Card, 2012). Fortunately, recent R packages, including metafor (Viechtbauer, 2010) and metaSEM (Cheung, 2015a), provide platforms for running multivariate meta-analysis.

21.2.6 Structural Equation Modeling Approach

The past decade has also brought advances in structural equation modeling (SEM) to meta-analysis (Card, 2012; Cheung, 2015b). A SEM approach to meta-analysis often involves estimating a latent intercept that reflects the weighted mean effect size. SEM-based meta-analytic procedures often rely on maximum likelihood (ML) or restricted maximum likelihood (REML) estimation methods instead of the typical weighted least squares (WLS) approach. ML (vs. WLS) procedures typically produce estimates that more accurately reflect the true mean. SEM-based meta-analysis appears to be growing, especially with the recent publication of a book (Cheung, 2015b) and at least four programs or packages that support ML estimation procedures, including HLM (Raudenbush, Bryk, & Congdon, 2013), Mplus (Muthén & Muthén, 2012), metafor (Viechtbauer, 2010), and metaSEM (Cheung, 2015a).

21.3 Meta-Analysis Programs

Researchers have developed macros or syntax files specifically designed to conduct meta-analyses in all-purpose statistical packages such as SPSS (Wilson, 2001). Specialized modeling software, such as HLM (Raudenbush et al., 2013), can also perform ML-based meta-analysis, but the procedures can be complex. Both Card (2012) and Cheung (2015b) have provided excellent examples of how to perform SEM-based meta-analysis using Mplus software (Muthén & Muthén, 2012). Meta-analysts are also increasingly using free, open-source R packages specifically designed for meta-analyses. Two of the more popular R packages are metafor (Viechtbauer, 2010) and metaSEM (Cheung, 2015a). Metafor is more established, with a broader array of options and more documentation. In contrast, metaSEM is geared toward SEM-based meta-analyses. For more detailed reviews of meta-analytic programs and their features, see Bax, Yu, Ikeda, and Moons (2007) or Polanin, Hennessy, and Tanner-Smith (2017).

21.4 Outliers, Publication Bias, Decline Effects

Meta-analytic estimates can be biased by extreme or erroneous observations or by systematic problems beyond the researcher's control. These include both screening for outliers and assessing at least two forms of bias potentially unique to meta-analyses: Publication bias and decline effects.

21.4.1 Outliers

Screening for potential outliers is important because a single extreme or erroneous effect size can result in misleading weighted means. Outliers often take one of two forms: Erroneous or extreme values. Erroneous values are those that result from typos or coding errors and often reflect impossible values (e.g., a coder records a correlation as "1.09" instead of "0.09"); these should be removed or corrected.

Extreme-value outliers may involve some guesswork because outlying observations are often only considered as such relative to other values in the data. Thankfully, researchers can use many of the same formal procedures developed to detect outliers in typical bivariate data analyses, such as Cook's distance and studentized deleted residuals (McClelland, 2014). Most of these procedures recommend that analysts adjust their false positive error rates (e.g., α/k, where α is the false positive error rate and k is the number of studies or effect sizes) because each value is tested individually relative to other values.

Screening for outliers has been made easier by using the `influence` procedure, which is available in the metafor R package. The `influence` procedure presents analysts with the results of multiple outlier tests, and these can be plotted using the `plot` procedure, which marks cases that are potentially problematic outliers (see example in Section 21.5).

Although single outliers can – and often should – be removed, clusters of outliers may suggest that multiple populations of effect sizes are being studied, and, thus, separate meta-analyses may be required or relevant moderators may be added to the model. Excluding outliers from analyses is important because most researchers seek to present a model for most of the data, rather than a model based solely on one influential data point (McClelland, 2014). Nevertheless, meta-analysts should report which studies or effect sizes were removed and whether doing so substantially affected the results.

21.4.2 Publication Bias

Publication bias occurs when researchers refuse to submit "failed" (non-significant) studies for publication (the file-drawer effect; Rosenthal, 1979) or when editors and reviewers reject studies with small effects or non-significant findings (selective publication). Thus, publication bias can result in the systematic inflation of mean effect sizes because small or non-significant effects are omitted from the published literature. There are several ways to examine publication bias, ranging from statistical procedures (fail-safe N) to visual inspection (funnel plots) to combinations of both (trim and fill, Egger's test).

Fail-Safe N. The fail-safe N procedure (Orwin, 1983; Rosenthal, 1979) asks a simple question: How many studies would it take to make the mean effect size non-significant? Although this early test of the file-drawer effect was once popular, it has received multiple criticisms and should be abandoned (Becker, 2005).

Funnel Plot. Funnel plots provide a simple visual test of publication bias. Effect sizes and their respective inverse variances are typically plotted on the x and y axes, respectively; however, sometimes precision (i.e., the inverse standard errors) or (log) sample sizes are substituted on the y axis. The resulting plot should show a funnel shape – studies with higher precision (larger samples) should have effect sizes clustered close to the mean, whereas studies with lower precision (smaller samples) should have far more variable effect sizes. Extreme deviations from a (roughly) funnel shape often indicate the presence of severe publication bias. Specifically, if publication bias is present, then non-significant effect sizes – often those with smaller sample sizes – will be missing from the publication record, and, hence, the funnel plot will be asymmetric.

Trim and Fill. Trim and fill is also a funnel plot-based technique (Duval & Tweedie, 2000). Because asymmetric funnel plots reflect publication bias, trim and fill essentially asks what the funnel plot would look like without publication bias. To do this, one makes the funnel plot more symmetric by filling in the area of presumed missing effect sizes by similar effect sizes from the opposite side of the funnel plot, essentially mirroring them. Thus, one *trims* known effect sizes from one side of the funnel plot and *fills* them into the other side, where the effect sizes are believed to be missing. According to Duval and Tweedie (2000), this allows for a less-biased test of the overall effect size because the "missing" effect sizes have essentially been "filled in." Despite being a potentially innovative approach for correcting publication bias, trim and fill has been recently criticized on multiple grounds (see Bishop & Thompson, 2016; Carter, Schönbrodt, Gervais, & Hilgard, 2018). Instead, most meta-analysts currently stress regression-based approaches for empirically testing and adjusting for publication bias.

Egger's Test. Egger's regression test (Egger, Smith, Schneider, & Minder, 1997) provides a convenient empirical way to assess publication bias. Egger's test regresses the standard normal deviate (effect size/SE) on precision ($1/SE$). If the resulting intercept is significant, then publication bias may be present.

PET-PEESE. PET-PEESE stands for precision-effect test (PET) and precision-effect estimate with standard error (PEESE), respectively (Stanley, 2008; Stanley & Doucouliagos, 2014). It is basically a hybrid of Egger's test (Egger et al., 1997; the PET part) and a meta-regression predicting effect sizes from their respective variances. According to Stanley and Doucouliagos (2014), PET-PEESE's intercept should provide an unbiased estimate of the "true" effect size, corrected for publication bias. Despite some promising results using this new tool to detect publication bias, PET-PEESE has also recently received some criticism about its limitations (Stanley, 2017). It is possible that PET-PEESE may perform better in disciplines that do not often use standardized effect sizes (e.g., economics) than those that do (e.g., psychology). Research on PET-PEESE should continue to help meta-analysts understand the conditions under which it should be used or avoided (Stanley, 2017).

Cumulative Meta-Analysis. Cumulative meta-analysis provides another visual way to detect possible publication bias. It involves performing a series of sequential meta-analyses whereby the effect sizes are arranged from the largest to smallest samples (or smallest to largest standard errors) and the mean effect size recalculated after adding each additional effect size to the meta-analysis (Ioannidis & Trikalinos, 2005; Kepes, Banks, McDaniel, & Whetzel, 2012; McDaniel, 2009). Cumulative

meta-analysis can be performed using the `cumul` procedure in the metafor package for R (see example in Section 21.5).

For example, using the self-esteem instability correlations in Table 21.1, the Smith (2013) correlation would be first ($N = 124$, $r = .15$), and it would next be averaged with the Nezlek and Webster (2003) correlation because it has the second-largest sample size ($N = 115$, $r = .43$). Another meta-analytic mean correlation would be derived after adding the third-largest study, and yet another after adding the fourth, and so on. The effect size of the largest study, followed by the nine

Table 21.1 Self-esteem instability correlations, alphas, corrected correlations, and moderators

| Study | N | r | α | r_c | | Moderators | |
						Pub	Stamp
Kernis et al. (1989)	45	.15	.75	.17		Yes	No
Kernis et al. (1992)	112	.13	.78	.15		Yes	No
Webster et al. (2017)							
Study 1 [2002]	102	.45	.89	.48		Yes	Yes
Study 2 [2001]	110	.47	.88	.50		Yes	Yes
Study 3 [2000]	105	.41	.87	.44		Yes	Yes
Brunell & Webster (2010)	68	.28	.87	.30		No	Yes
Nezlek & Webster (2003)	115	.43	.91	.45		No	Yes
Smith et al. (2013)	124	.15	.85	.16		No	Yes
Webster (2000)	86	.31	.90	.33		No	Yes
Webster (2003)	20	.33	.88	.35		No	Yes
Sum or weighted mean	887	.32		.35			

Note. Table adapted from Webster, Smith, Brunell, Paddock, and Nezlek (2017).
N = Sample size
r = Correlation between two self-esteem instability measures (i.e., Rosenberg's 1965 Stability of Self Scale [RSSS] and temporal SD over repeated measure of state self-esteem)
α = Reliabilities for the RSSS
r_c = Correlation corrected for attenuation (i.e., unreliability or measurement error)
Pub = Publication status
Stamp = Whether or not study was time-stamped (nearly confounded with authorship)

meta-analytic mean correlations (based on the second-, third-, ... and tenth-largest studies, respectively) would then be plotted together (i.e., mean effect sizes as a function of number of studies contributing to those mean effect sizes). If strong publication bias is present, then smaller-sample studies will deviate more strongly from the established mean in a more extreme direction, thus moving the meta-analytic average away from zero as smaller and smaller samples are added to the cumulative meta-analysis. Publication year can be substituted for sample size to examine possible cumulative decline effects over time, which may also indicate publication bias. Although one can regress the resulting mean effect sizes onto number of studies (or years), this test will be biased because the data contributing to the means are inherently non-independent. Consequently, cumulative meta-analysis remains a largely visual tool in the publication bias toolbox.

Summary. The presence of publication bias suggests that non-significant effect sizes – frequently those based on studies with small sample sizes – are often omitted from the published literature, and thus, the resulting weighted mean effect size can be inflated (upwardly biased). Ideally, publication bias should be tested using multiple methods. However, this may be impractical in meta-analyses involving smaller numbers of effect sizes (due to lack of statistical power). Publication bias is more likely when the significance of the focal effect or association is related to decisions to submit or publish. In contrast, publication bias is less likely when the focal effect or association is not also the focus of the original study. Examples of

the latter might include examining a) changes in means over time (temporal meta-analysis) or b) correlations among scales that are assessed together (e.g., the Dark Triad, the Big Five).

21.4.3 Decline Effects

Decline effects occur when the absolute magnitudes of effect sizes decrease over time. Decline effects have been observed both within labs (Schooler, 2011) and across effect sizes over time in multiple scientific fields (Fischer et al., 2011; Jennions & Møller, 2001; Webster, Graber, Gesselman, Crosier, & Schember, 2014; Wongupparaj, Kumari, & Morris, 2015). There are multiple possible explanations for decline effects. Of course, one simple explanation is "true" change in effect sizes over time (e.g., declining negative attitudes toward same-sex marriage). Another possibility is regression to the mean, which can happen when a remarkable finding is published for the first time. To be remarkable, in the first place, an effect must often be large and novel. Other research may later attempt to replicate this effect and either fail to do so or find smaller effects. In this sense, the original finding may have been an outlier of sorts, whereas the replication attempts that follow it are closer to the "true" mean effect size. Because publication decisions may play a role in decline effects, decline effects are sometimes considered to be evidence of publication bias.

21.5 Example: Convergent Validity in Measures of Self-Esteem Instability

One use of meta-analysis is to assess the average convergent validity correlation between two related measures. For example, Webster et al. (2017) examined the correlations between people's self-reports on Rosenberg's Stability of Self Scale (RSSS, 1965; recoded to assess *in*stability) and their temporal self-esteem instability (*SD* of daily state self-esteem measures over time; see Kernis, Grannemann, & Barclay, 1989; 1992). Table 21.1 shows the correlations between these two self-esteem instability measures across ten studies. Table 21.1 also shows the reliabilities (Cronbach's alpha) for the RSSS. Below, meta-analyses are conducted on both the raw correlations and those corrected for attenuation (measurement error); see Appendix A for annotated R code, which is cited in the text by line number (e.g., "A.5–16" means Appendix A, lines 5–16).

21.5.1 Raw Correlations

After installing and calling the metafor package (A.1–4), we input the data from Table 21.1 into R (A.5–16), and use the `escalc` procedure (i.e., effect size calculation) to Fisher's *r*-to-*z* to transform the correlations for meta-analysis (A.17–18). We then conduct a random-effects meta-analysis and view the output (A.19–22). The output gives multiple heterogeneity statistics (e.g., $\tau^2 = 0.013$, $I^2 = 52.1\%$, $Q(9) = 18.50$, $p = .030$), which suggest some significant residual heterogeneity in effect sizes after estimating a weighted mean. The weighted mean effect size estimate and its 95% CIs are given in the *z* metric; we can reverse-transform them into the *r* metric using the hyperbolic tangent function – "tanh()" in most calculators and spreadsheets (e.g., tanh (.3336) ≈ .32). Thus, we obtain a weighted mean correlation of $r = .32$, 95% CI [.23, .41], which is significantly different from zero ($z = 6.56$, $p < .001$).

Detecting Outliers and Publication Bias. We can probe for outliers using multiple tests, the `influence` procedure, and plot the results for visual inspection (A.23–26). We can also examine a forest plot of the effect sizes and their 95% CIs (A.27–28). In this example, there are no obvious outliers. We can assess publication bias using multiple methods. In this example, a funnel plot (A.29–30) reveals little because the number of studies is small ($k = 10$). An Egger's test via random-effects meta-regression (A.31–33) reveals no presence of publication bias ($z = -0.39$, $p = .697$), but detecting it is limited by low power (small sample). Visual inspection of forest plots of cumulative meta-analyses by a) decreasing sample sizes (A.34–39) and b) year (A.40–44) shows no obvious signs of publication bias, but the sample is too small to detect meaningful trends.

Moderation. To interpret the weighted mean effect size, which is the intercept in meta-regression, it helps to first mean-center continuous moderators (Aiken, West, & Reno, 1991). After mean-centering year at 2001.3 (A.46–48), a test for moderation by year (A.49–52) shows no significant effect (slope $= 0.0024$, $z = 0.30$, $p = .768$). Moderation by publication status (A.53–55) also shows a null effect (difference $= 0.049$, $z = 0.47$, $p = .641$). In contrast, moderation by time stamp (A.56–58) shows a significant effect (difference $= 0.241$, $z = 2.15$, $p = .032$). Accounting for time stamp differences also reduces residual between-study heterogeneity to non-significance ($\tau^2 = 0.0064$, $I^2 = 34.9\%$, $Q(8) = 11.15$, $p = .194$). Because of how we coded time stamp (0 = no, 1 = yes), the intercept in this meta-regression is the simple effect of studies without time stamps. Reverse-transforming the intercept (0.1377) using the tanh() function yields $r \approx .14$ [$-.06$, .32]. We obtain the simple effect for studies with time stamps by recoding the moderator and rerunning the meta-regression (A.59–63) and again reverse-transform the intercept (0.3791), which yields $r \approx .36$ [.28, .44].

21.5.2 Corrected Correlations

To correct these correlations for attenuation, we can input the Cronbach's alphas for their respective RSSS scales, adjust the correlations, z-transform them, and also adjust their respective variances (A.64–72). We then rerun the random-effects meta-analysis using the new data (A.73–75). The output gives multiple heterogeneity statistics (τ^2 = 0.014, I^2 = 49.5%, $Q(9)$ = 17.32, p = .044), which suggest some significant residual heterogeneity but less than that of the raw correlations. The reverse-transformed weighted mean corrected correlation is r = .35 [.25, .44], which is significantly different from zero (z = 6.80, $p < .001$).

21.5.3 Additional Examples

Although the present example focuses on self-esteem, an important construct to psychologists, meta-analysis is a broad tool with applications to multiple disciplines. For example, a meta-analysis, led by a medical anthropologist, helped establish a definitive link between increased blood pressure and burning biomass fuels in homes, which is a common heat source for cooking in developing countries (Dillon, Webster, & Bisesi, 2018). Other meta-analyses help cross boundaries within the social and behavioral sciences. Bridging biological, developmental, and sociological

perspectives, meta-analysts have shown a small but significant link between biological father absence during childhood and earlier sexual maturation in girls and young women (i.e., younger age of menarche; Webster, Graber, Gesselman, Crosier, & Schember, 2014).

21.6 Meta-Analysis: Limitations and Alternatives

As with any modeling technique, meta-analysis has multiple limitations. There are also potential alternatives to meta-analysis that researchers should consider when they have access to raw data (vs. only effect sizes). Far from being a static technique, meta-analysis is constantly evolving.

21.6.1 Limitations

Many limitations of meta-analysis are tied to its assumptions. First, meta-analysis nearly always assumes linear bivariate associations. For example, studies producing Cohen's ds, which often reflect differences between control and experimental groups on a given outcome, implicitly assume linear relationships. This is true of both the original studies and the resulting meta-analysis, even if the levels of the predictor or outcome vary across studies. Consequently, most traditional meta-analytic techniques cannot easily assess possible nonlinear associations (e.g., quadratic or cubic relationships). If such nonlinear associations truly exist, then meta-analysis is likely to underestimate them or miss them entirely (i.e., they go undetected).

Second, meta-analysts are limited by the variables, conditions, and settings assessed by the authors of the original studies and by the statistics they choose to report. Omissions and substitutions of variables of interest across studies can easily frustrate even the most seasoned meta-analysts. In this sense, meta-analysts must be prepared to be scavengers – gleaning effect sizes from what has been handed to them as secondary-data analysts. For example, meta-analysts seeking to assess the link between sexual satisfaction and relationship satisfaction may find a study that includes only a measure of sexual frustration or dissatisfaction, which is not ideal, but could be salvaged after reversing the sign of the correlation. This lack of control or specification, however, is offset somewhat by the increase in overall sample size that meta-analyses typically achieve.

Third, when using the same scales or measures across different studies (or over time), meta-analysis assumes measurement invariance (i.e., that different participants from different times, places, and backgrounds are responding similarly to items or scales; see Hauenstein & Embretson, Chapter 19). Ideally, meta-analysts should attempt to test for measurement invariance across studies when using multi-item scales (Wetzel et al., 2017). However, doing so is often impractical because it requires raw, item-level data due to effect sizes alone providing insufficient data to assess measurement invariance. Thus, meta-analysts should at least acknowledge the limitation of being unable to establish measurement invariance and warn readers that different participants may respond to the same items or scales in different ways.

21.6.2 Overview of Alternatives

Two alternatives to meta-analysis include narrative reviews and integrative data analysis. Meta-analysis contrasts with narrative or qualitative literature reviews, which focus on a mixture of theoretical developments and summaries of selected empirical studies. Among the potentially serious limitations of narrative or qualitative reviews are selection bias and confirmation bias. Selection bias occurs when studies or empirical findings are cherry-picked from the literature; they may provide the readers with helpful anecdotes but may neither reflect nor generalize to the broader literature. Confirmation bias is a more specific form of selection bias whereby researchers a) select views, studies, or findings that support their position; b) ignore those that fail to support their position; or c) both. Because these biases conspire to skew literature reviews to favor one viewpoint or another, one goal of meta-analysis is to reduce bias by focusing on quantitative (vs. qualitative) information gleaned from all (vs. some) of the relevant literature (see also Wagner & Skowronski, Chapter 2).

When researchers have access to raw data from multiple studies (vs. solely effect sizes), then they may wish to use integrative data analysis (IDA; Curran & Hussong, 2009) instead of meta-analysis. Like meta-analysis, IDA requires that the independent and dependent variables be respectively similar across studies, preferably measuring the same underlying constructs. If different scales are used, but contain the same – or similar – items, then item response theory methods (Morizot, Ainsworth, & Reise, 2007; Hauenstein & Embretson, Chapter 19) can be used to assess differential item functioning across studies. After aggregating the data, analyses – typically multiple regressions – are done at the case- or person-level unit instead of the study level. This allows for greater flexibility than meta-analysis, including the ability to test non-linear and interactive effects (moderators). IDA is especially useful for aggregating data across one's own studies in multi-study papers because one often has the raw data and because IDA makes fewer implicit assumptions (e.g., linearity) than meta-analysis.

21.7 Conclusion

Despite some limitations, meta-analysis remains an indispensable tool for empirically deriving useful quantitative information from multiple studies, often with goals of estimating the "true" mean effect size, and assessing the extent to which study-level variables (e.g., publication year) moderate it. Although narrative and qualitative literature reviews remain useful, empirical and quantitative reviews continue to grow more common and influential. Of course, the best literature reviews incorporate elements of both, often using quantitative findings to make qualitative generalizations.

Although there are many variants of meta-analytic approaches, a simple, bivariate, random-effects approach is likely the appropriate and most accessible one for most research questions that social and behavioral scientists ask. Although SEM approaches may provide slightly more accurate estimates and correcting effect sizes for artifacts can be useful, these more sophisticated approaches often produce estimates that do not differ markedly from simpler, more traditional meta-analytic methods. In contrast, assessing and correcting for publication bias is becoming increasingly paramount, especially given that many effects in science are irreproducible (e.g., Open Science Collaboration, 2015; see also Soderberg & Errington, Chapter 14). Depending on their research questions, meta-analysts should also strive to test for potential study-level moderators or their effects. It is these study-level moderators that occasionally lead to breakthroughs in understanding why effect sizes vary. As more researchers and their institutions adopt open-source data-analysis packages, such as R, it is expected that R packages for meta-analyses (e.g., metafor, metaSEM) will continue to increase in ease of use and popularity. I hope that this chapter will encourage readers to consider using meta-analytic techniques to address the many creative research questions they pose.

KEY TAKEAWAYS

- Meta-analysis is a set of tools for quantitatively describing and testing effect sizes gathered from multiple studies.
- Preliminary steps involve 1) identifying a research question, 2) using search engines to scour the literature for relevant effect sizes in published and unpublished studies, 3) scanning articles for relevant effect sizes, and 4) coding for other relevant information (e.g., sample size, scale reliability, moderators).
- Data analysis includes 1) estimating the weighted mean effect size, 2) assessing heterogeneity, and 3) testing for outliers, publication bias, decline effects, and moderators.

- Most social and behavioral science questions will likely involve random-effects meta-analysis of bivariate correlations. Some questions, however, may require more advanced techniques – correcting effect sizes for bias (e.g., measurement error), multivariate meta-analysis, or a structural equations modeling (SEM) approach.

IDEAS FOR FUTURE RESEARCH

- Is there a way to test for publication bias that is well supported by most analysts?
- How can one assess and correct for non-independence in meta-analytic data (e.g., multiple effect sizes nested within studies, multiple studies nested within publications, and multiple effect sizes or studies nested within labs, authors, or groups of authors)?
- How might other statistical techniques, such as multi-level modeling and social network analysis, combine with meta-analysis to improve our knowledge of social behavior?
- How might spatial regression be used in meta-analysis to assess and adjust for spatial dependence among the effect sizes drawn from different cultures or geographic regions?

SUGGESTED READINGS

Card, N. A. (2012). *Applied Meta-Analysis for Social Science Research*. New York: Guilford.

Cheung, M. W.-L. (2015). *Meta-Analysis: A Structural Equation Modeling Approach*. Chichester, UK: Wiley.

Cooper, H. (2017). *Research Synthesis and Meta-Analysis: A Step-by-Step Approach* (5th ed.). Thousand Oaks, CA: Sage.

Schmidt, F. L., & Hunter, J. E. (2015). *Methods of Meta-Analysis: Correcting Error and Bias in Research Findings* (3rd ed.). Thousand Oaks, CA: Sage.

REFERENCES

Aiken, L. S., West, S. G., & Reno, R. R. (1991). *Multiple Regression: Testing and Interpreting Interactions*. Thousand Oaks, CA: Sage.

Aloe, A. M., Amo, L. C., & Shanahan, M. E. (2014). Classroom management self-efficacy and burnout: A multivariate meta-analysis. *Educational Psychology Review, 26*, 101–126. doi:10.1007/s10648-013-9244-0

Bax, L., Yu, L. M., Ikeda, N., & Moons, K. G. (2007). A systematic comparison of software dedicated to meta-analysis of causal studies. *BMC Medical Research Methodology, 7*, 40. doi:10.1186/1471-2288-7-40

Becker, B. J. (2005). Failsafe *N* or file-drawer number. In H. R. Rothstein, A. J. Sutton, & M. Borenstein (Eds.), *Publication Bias in Meta-Analysis: Prevention, Assessment, and Adjustments* (pp. 111–126). Hoboken, NJ: Wiley.

Becker, B. J. (2009). Model-based meta-analysis. In H. Cooper, L. V. Hedges, & J. C. Valentine (Eds.), *The Handbook of Research Synthesis and Meta-Analysis* (pp. 377–396). Thousand Oaks, CA: Sage.

Bishop, D. V., & Thompson, P. A. (2016). Problems in using *p*-curve analysis and text-mining to detect rate of *p*-hacking and evidential value. *PeerJ, 4*, e1715. doi:10.7717/peerj.1715

Borenstein, M., Higgins, J. P. T., Hedges, L. V., & Rothstein, H. R. (2017). Basics of meta-analysis: I^2 is not an absolute measure of heterogeneity. *Research Synthesis Methods, 8*, 5–18. doi:10.1002/jrsm.1230

Brunell, A. B., & Webster, G. D. (2010). [Couples' daily diary study]. Unpublished raw data.

Card, N. A. (2012). *Applied Meta-Analysis for Social Science Research*. New York: Guilford.

Carter, E. C., Schönbrodt, F. D., Gervais, W. M., & Hilgard, J. (2018, October 23). Correcting for bias in psychology: A comparison of meta-analytic methods. https://doi.org/10.31234/osf.io/9h3nu

Cheung, M. W.-L. (2015a). metaSEM: An R package for meta-analysis using structural equation modeling. *Frontiers in Psychology*, 5, 1521. doi:10.3389/fpsyg.2014.01521

Cheung, M. W.-L. (2015b). *Meta-Analysis: A Structural Equation Modeling Approach*. Chichester, UK: Wiley.

Curran, P. J., & Hussong, A. M. (2009). Integrative data analysis: The simultaneous analysis of multiple data sets. *Psychological Methods*, 14, 81–100. doi:10.1037/a0015914

Dillon, D. T., Webster, G. D., & Bisesi, J. H., Jr. (2018). *A Systematic Review and Meta-Analysis of the Contributions of Environmental Exposure to Biomass Fuel Burning to Blood Pressure Modification in Women*. Manuscript submitted for publication.

Duval, S., & Tweedie, R. (2000). Trim and fill: A simple funnel-plot-based method of testing and adjusting for publication bias in meta-analysis. *Biometrics*, 56, 455–463.

Egger, M., Smith, G. D., Schneider, M., & Minder, C. (1997). Bias in meta-analysis detected by a simple, graphical test. *British Medical Journal*, 315, 629–634.

Fischer, P., Krueger, J. I., Greitemeyer, T., Vogrincic, C., Kastenmüller, A., Frey, D., … Kainbacher, M. (2011). The bystander-effect: A meta-analytic review on bystander intervention in dangerous and non-dangerous emergencies. *Psychological Bulletin*, 137, 517–537. doi:10.1037/a0023304

Glass, G. V. (1976). Primary, secondary, and meta-analysis of research. *Educational Researcher*, 5, 3–8. doi:10.3102/0013189X005010003

Glass, G. V., McGraw, B., & Smith, M. L. (1981). *Meta-Analysis in Social Research*. Beverly Hills, CA: Sage.

Hedges, L. V., & Vevea, J. L. (1998). Fixed- and random-effect models in meta-analysis.

Psychological Methods, 3, 486–504. doi:10.1037/1082-989X.3.4.486

Huedo-Medina, T. B., Sánchez-Meca, J., Marín-Martínez, F., & Botella, J. (2006). Assessing heterogeneity in meta-analysis: Q or I^2 index? *Psychological Methods*, 11, 193–206. doi:10.1037/1082-989X.11.2.193

Hunter, J. E., Schmidt, F. L., & Jackson, J. E. (1982). *Meta-Analysis: Culminating Research Findings across Studies*. Beverly Hills, CA: Sage.

Ioannidis, J. P., & Trikalinos, T. A. (2005). Early extreme contradictory estimates may appear in published research: The Proteus phenomenon in molecular genetic research and randomized trials. *Journal of Clinical Epidemiology*, 58, 543–549. doi:10.1016/j.jclinepi.2004.10.019

Jennions, M. D., & Møller, A. P. (2001). Relationships fade with time: A meta-analysis of temporal trends in publication in ecology and evolution. *Proceedings of the Royal Society B: Biological Sciences*, 269, 43–48. doi:10.1098/rspb.2001.1832

Jonason, P. K., & Webster, G. D. (2010). The dirty dozen: A concise measure of the dark triad. *Psychological Assessment*, 22, 420–432. doi:10.1037/a0019265

Kepes, S., Banks, G. C., McDaniel, M., & Whetzel, D. L. (2012). Publication bias in the organizational sciences. *Organizational Research Methods*, 15, 624–662. doi:10.1177/1094428112452760

Kernis, M. H., Grannemann, B. D., & Barclay, L. C. (1989). Stability and level of self-esteem as predictors of anger arousal and hostility. *Journal of Personality and Social Psychology*, 56, 1013–1022. doi:10.1037/0022- 3514.56.6.1013

Kernis, M. H., Grannemann, B. D., & Barclay, L. C. (1992). Stability of self-esteem: Assessment, correlates, and excuse making. *Journal of Personality*, 60, 621–644. doi:10.1111/j.1467-6494.1992.tb00923.x

Kline, R. B. (2016). *Principles and Practice of Structural Equation Modeling* (4th ed.). New York: Guilford.

McClelland, G. H. (2014). Nasty data: Unruly, ill-mannered observations can ruin your analysis.

In **H. T. Reis** & **C. M. Judd** (Eds.), *Handbook of Research Methods in Social and Personality Psychology* (2nd ed., pp. 608–626). New York: Cambridge University Press.

McDaniel, M. A. (2009, April). Cumulative meta-analysis as a publication bias detection method. Paper presented at the 24th Annual Conference of the Society for Industrial and Organizational Psychology, New Orleans, LA.

Morizot, J., Ainsworth, A. T., & Reise, S. P. (2007). Toward modern psychometrics: Application of item response theory models in personality research. In R. W. Robins, R. C. Fraley, & R. F. Krueger (Eds.), *Handbook of Research Methods in Personality Psychology* (pp. 407–423). New York: Guilford.

Muthén, L. K., & Muthén, B. O. (2012). *Mplus User's Guide* (7th ed.). Los Angeles, CA: Muthén & Muthén.

Nezlek, J. B., & Webster, G. G. (2003). [A daily diary study of self-esteem.] Unpublished raw data.

Open Science Collaboration. (2015). Estimating the reproducibility of psychological science. *Science, 349*, aac4716. doi:10.1126/science.aac4716

Orwin, R. G. (1983). A fail-safe *N* for effect size in meta-analysis. *Journal of Educational Studies, 8*, 157–159. doi:10.2307/1164923

Paulhus, D. L., & Williams, K. M. (2002). The dark triad of personality: Narcissism, Machiavellianism, and psychopathy. *Journal of Research in Personality, 36*, 556–563.

Polanin, J. R., Hennessy, E. A., & Tanner-Smith, E. E. (2017). A review of meta-analysis packages in R. *Journal of Educational and Behavioral Statistics, 42*, 206–242.

Raudenbush, S. W., Bryk, A. S, & Congdon, R. (2013). HLM 7.01 for Windows [Computer software]. Skokie, IL: Scientific Software International, Inc.

Rosenberg, M. (1965). *Society and the Adolescent Self-Image.* Princeton, NJ: Princeton University Press.

Rosenthal, R. (1979). The file drawer problem and tolerance for null results. *Psychological Bulletin, 86*, 638–641. doi:10.1037/0033-2909.86.3.638

Schmidt, F. L. (2010). Detecting and correcting the lies that data tell. *Perspectives on Psychological Science, 5*, 233–242. doi:10.1177/1745691610369339

Schmidt, F. L., & Hunter, J. E. (2015). *Methods of Meta-Analysis: Correcting Error and Bias in Research Findings* (3rd ed.). Thousand Oaks, CA: Sage.

Schooler, J. (2011). Unpublished results hide the decline effect: Some effects diminish when tests are repeated. *Nature, 470*, 437–438. doi:10.1038/470437a

Smith, C. V., Hadden, B. W., & Webster, G. D. (2013). [Daily dairy study]. Unpublished raw data.

Stanley, T. D. (2008). Meta-regression methods for detecting and estimating empirical effects in the presence of publication selection. *Oxford Bulletin of Economics and Statistics, 70*, 103–127. doi:10.1111/j.1468-0084.2007.00487.x

Stanley, T. D. (2017). Limitations of PET-PEESE and other meta-analysis methods. *Social Psychological and Personality Science, 8*, 581–591. doi:10.1177/1948550617693062

Stanley, T. D., & Doucouliagos, H. (2014). Meta-regression approximations to reduce publication selection bias. *Research Synthesis Methods, 5*, 60–78.

Twenge, J. M., & Campbell, W. K. (2001). Age and birth cohort differences in self-esteem: A cross-temporal meta-analysis. *Personality and Social Psychology Review, 5*, 321–344.

Twenge, J. M., Konrath, S., Foster, J. D., Keith Campbell, W., & Bushman, B. J. (2008). Egos inflating over time: A cross-temporal meta-analysis of the Narcissistic Personality Inventory. *Journal of Personality, 76*, 875–902. doi:10.1111/j.1467-6494.2008.00507.x

Viechtbauer, W. (2010). Conducting meta-analyses in R with the metafor package. *Journal of Statistical Software, 36*, 1–48. Retrieved from www.jstatsoft.org/v36/i03/

Vize, C. E., Lynam, D. R., Collison, K. L., & Miller, J. D. (2018). Differences among dark triad components: A meta-analytic investigation. *Personality Disorders: Theory,*

Research, and Treatment, 9(2), 101–111. doi:10.1037/per0000222

Webster, G. D. (2000). [Daily diary study of domain-specific self-esteem]. Unpublished raw data. (2003). [A 14-week classroom diary study of self-esteem and aggression]. Unpublished raw data.

Webster, G. D., & Duffy, R. R. (2016). Losing faith in the intelligence–religiosity link: New evidence for a decline effect, spatial dependence, and mediation by education and life quality. Intelligence, 55, 15–27. doi:10.1016/j.intell.2016.01.001

Webster, G. D., Gesselman, A. N., Crysel, L. C., Brunell, A. B., Jonason, P. K., Hadden, B. W., & Smith, C. V. (2016). An actor–partner interdependence model of the Dark Triad and aggression in couples: Relationship duration moderates the link between psychopathy and argumentativeness. Personality and Individual Differences, 101, 196–207. doi:10.1016/j.paid.2016.05.065

Webster, G. D., Graber, J. A., Gesselman, A. N., Crosier, B. S., & Schember, T. O. (2014). A life history theory of father absence and menarche: A meta-analysis. Evolutionary Psychology, 12, 273–294. doi:10.1177/147470491401200202

Webster, G. D., Smith, C. V., Brunell, A. B., Paddock, E. L., & Nezlek, J. B. (2017). Can Rosenberg's (1965) Stability of Self Scale capture within-person self-esteem variability? Meta-analytic validity and test–retest reliability. Journal of Research in Personality, 69, 156–169. doi:10.1016/j.jrp.2016.06.005

Wetzel, E., Brown, A., Hill, P. L., Chung, J. M., Robins, R. W., & Roberts, B. W. (2017). The narcissism epidemic is dead; long live the narcissism epidemic. Psychological Science, 28(12), 1833–1847. doi:10.1177/0956797617724208

Wilson, D. B. (2001). Meta-analysis macros for SAS, SPSS, and Stata. Retrieved from http://mason.gmu.edu/~dwilsonb/ma.html

Wongupparaj, P., Kumari, V., & Morris, R. G. (2015). A cross-temporal meta-analysis of Raven's progressive matrices: Age groups and developing versus developed countries. Intelligence, 49, 1–9. doi:10.1016/j.intell.2014.11.008

Appendix A: Metafor Package R Code for Meta-Analysis Examples

```
1. # install metafor package
2. install.packages("metafor")
3. # call the metafor package
4. library(metafor)
5. # data are self-esteem instability convergent val-
   idity correlations see (Webster et al., 2017, JRP)
6. # input sample sizes
7. n <- c(45,112,102,110,105,68,115,124,86,20)
8. # input raw correlations
9. cor <- c(.15,.13,.45,.47,.41,.28,.43,.15,.31,.33)
10. # input study-level moderators
11. # input year of study
12. year <- c(1989,1992,2002,2001,2000,2010,2003,2013,
    2000,2003)
13. # input publication status; 1 = published, 0 =
    unpublished
14. pub <- c(1,1,1,1,1,0,0,0,0,0)
15. # input whether study was time-stamped; 1 = yes, 0 =
    no
16. stamp <-c(0,0,1,1,1,1,1,1,1,1)
17. # the escalc procedure calculates effect size
    information from the data (e.g., Fisher's r-to-z
    transform)
18. zcor <- escalc(measure = "ZCOR," ri = cor, ni = n,
    var.names = c("zr", "zrvar"))
19. # the rma procedure conducts a random-effects meta-
    analysis
20. rezma <- rma(yi = zr, vi = zrvar, data = zcor)
21. # view the output
22. rezma
```

```
23. # the influence procedure conducts multiple outlier
    analyses
24. infrezma <- influence(rezma)
25. # plot the influence procedure results for visual
    inspection
26. plot(infrezma)
27. # the forest procedure produces a forest plot of
    the effect sizes
28. forest(rezma)
29. # the funnel procedure produces a funnel plot of
    the effect sizes
30. funnel(rezma)
31. # the regtest procedure runs an Egger's (meta-)
    regression for testing publication bias; view
    output
32. eggersrt <- regtest(rezma, model = "rma", predictor
    = "se")
33. eggersrt
34. # the cumul procedure runs a cumulative meta-
    analysis for testing publication bias
35. # cumulative meta-analyis by decreasing sample size
    (n); view output
36. cumuln <- cumul(rezma, order=order(n, decreasing =
    TRUE))
37. cumuln
38. # visually inspect the forest plot
39. forest(cumuln)
40. # cumulative meta-analyis over time (year); view
    output
41. cumulyr <- cumul(rezma, order=order(year))
42. cumulyr
43. # visually inspect the forest plot
44. forest(cumulyr)
45. # find the mean of year of study or publication
46. mean(year)
47. # create new mean-centered year variable for moder-
    ation analysis; allows for a meaningful intercept
48. year2001 <- year - 2001.3
49. # moderation analyses
50. # moderation by year; view output
```

```
51. modyear <- rma(yi = zr, vi = zrvar, mods =
    ~ year2001, data = zcor)
52. modyear
53. # moderation by publication status; view output
54. modpub <- rma(yi = zr, vi = zrvar, mods = ~ pub,
    data = zcor)
55. modpub
56. # moderation by time stamp; view output; intercept:
    mean effect size for studies without time stamps
57. modstamp <- rma(yi = zr, vi = zrvar, mods = ~ stamp,
    data = zcor)
58. modstamp
59. # create simple effect variable for studies with
    time stamps
60. stampyes <- stamp - 1
61. # rerun moderation by time stamp; view output;
    intercept: mean effect size for studies with time
    stamps
62. modstampyes <- rma(yi = zr, vi = zrvar, mods =
    ~ stampyes, data = zcor)
63. modstampyes
64. # meta-analysis of correlations corrected for
    attenuation (i.e., unreliability of RSSS)
65. # input Cronbach's alphas for Rosenberg's (1965)
    Stability of Self Scale
66. alpha <- c(.75,.78,.89,.88,.87,.87,.91,.85,.90,.88)
67. # correct correlations for attenuation (i.e., unre-
    liability or measurement error)
68. rc <- cor/sqrt(alpha)
69. # Fisher's r-to-z transformation of corrected
    correlations
70. zrc <- atanh(rc)
71. # correct variance for attenuation (i.e., unreli-
    ability or measurement error)
72. zrcvar <- 1/(n-3)/alpha
73. # rerun random-effects meta-analysis for z-
    transformed corrected correlations; view output
74. rezcma <- rma(yi = zrc, vi = zrcvar)
75. rezcma
```

Index

Index

Index

Index